AF540656

BIOLOGY OF PROTOZOA

BIOLOGY
OF
PROTOZOA

By

Dr. D.R. Khanna
Reader
Reader in Zoology
Gurukul Kangri University
Haridwar (Uttaranchal)
&
Dr. P.R Yadav
Reader
Department of Zoology
D.A.V. College
Muzaffarnagar (U.P.)

DISCOVERY PUBLISHING HOUSE
NEW DELHI-110002

First Published-2004

ISBN 81-7141-906-2

Published by

DISCOVERY PUBLISHING HOUSE
4831/24, Ansari Road, Prahlad Street,
Darya Ganj, New Delhi-110002 (India)
Phone: 23279245 • Fax: 91-11-23253475
E-mail:dphtemp@indiatimes.com

Printed at:
Tarun Offset Printers,
Delhi–110 053

Preface

This fascinating title "Biology of Protozoa" has been carefully compiled and edited to meet the long felt needs of increasingly large number of those who have to deal with the different aspects of Protozoology in Colleges, Universities and research institutes. It provides a balanced and integrated treatment of the entire field of Protozoa. The title is intelligible to the educated layman but it deals with some complex ideas. It is an adequate text for all requirements in this area for most university students. Special efforts have been made to explain ideas in non-mathematical terms. The primary aim throughout has been clarity, simplicity and the high standard. It will definitely prove to be a boon to teachers, students and research workers in the field of Protozoology.

It has been the constant endeavour of the authors to furnish maximum substance, keeping in view the limitations of size of the volume. Efforts have been made to condense the matter as far as practicable. The book features both a text and a laboratory guide. It is hoped that this text will also significantly contribute to the efforts of instructors of microbiology to motivate students into increasing their knowledge concerning the activities and importance of protozoans as micro-organisms, and to establish a firm foundation that will be of value to students in related course work.

There can be no claim to originality except in the manner of treatment and much of the information has been obtained from the books and scientific available in different libraries.

Inspite of all care and caution, some errors might have crept in for which the authors are apologetic and to remove which the co-operation in the form of liberal criticism and valuable suggestion, from fellow teachers and students, is solicited.

The authors express their gratitute to Mr. Wasan and staff of M/s Discovery Publishing House for their whole hearted co-operation in the publication of this book.

Authors

CONTENTS

1

APPEARANCE OF PROTOZOA

The evolution of the protozoa is inevitably hypothetical, since many of them have no skeleton or other hard parts and thus only rarely provide a fossil record of their evolution. Hypotheses of the course of evolution which they have undergone, therefore, have to be based largely on deduction from the similarities and apparent relationship of the surviving groups with the *missing links* being filled in by judicious guesswork. It is generally accepted, because of the obvious similarities between the simpler members of both groups, that both animal and plant kingdoms had their origins in a single group of organisms; and it is widely, if not generally, believed that the present-day organisms which have deviated least from this primordial group are the Sarcomastigophora. This belief is based on the facts that (a) flegella (or cilia) are almost universally present among plants and animals, and (b) among the Mastigophora, species with chloroplast and species without them are inextricably mixed. From this primordial assemblage, evolution presumably proceeded along three main lines:

(i) The plants, organisms which retained chlorophyll and thus synthesized their own food;

(ii) The animals, which adopted the habit of eating the fruits of others labours of the others themselves; and

(iii) The fungi, with their saprophytic nutrition.

Thus in a sense all animals are parasitic on plants but not, of course, within the commonly used definition of this term. Within the animal group, evolution continued producing many flagellates and amoeboid organisms. The present-day survivors of this process are the Zoomastigophorea, Opalinata, Ciliphora, and the Sarcodina. From these, others branched off. The apicomplexa probably arose quite early in this process, since they possess both flagellate and amoeboid characteristics as well as specialization of their own.

The origins of the Microspora and Myxozoa are obscure: although at one time placed near the Apicomplexa they are in fact not particularly closely related to this group and are more likely to have arisen separately from the early sarcodine stock. This possibly occurred fairly close to the offshoot leading to the metazoan Coelenterata. Idea about evolution within the various protozoan groups are very spculative. Within the Ciliphora and Sarcodina, the symbiotic habit was probably adopted independently by organisms of several different groups.

Almost all symbiotic members of these groups are mainly or exclusively intestinal parasites, and it is easy to imagine how they could have entered the host's alimentary canal in food or water and multiplied there; eventually, mutation and selection would have given rise to organisms better suited to this relatively sheltered existence and, finally, to organisms unable to survive outside it unless protected by a cyst. Subsequent evolutionary experimentation by the Sarcodina would have led to forms able to invade tissues, as shown today by the faculatively parasitic genera *Naegleria* and *Acanthamoeba*, and by what are probably the most highly evolved parasitic members of the Sarcodina, *Entamoeba histolytica* and its close relatives.

Certain species of the ciliate *Tetrahymena* are probably today in the early stages of symbiosis since they appear to be able to live equally well inside or outside their hosts. Some ciliates have become highly adapted physiologically to a symbiotic life (e.g. those inhabiting the alimentary tracts of ruminants). Similarly, among the zoomastigophorea, symbiosis has probably arisen independently in several groups: examples are seen among the flagellates living in the intestines of vertebrates and invertebrates, some of which have by now become considerably specialized (e.g. the hypermastigid flagellates of termites and other insects). One group of flagellates has specialized *par excellence* in symbiosis the members of the suborder Trypanosomatina of the group Kinetoplastida.

These organism probably evolved from free-living Zoomastigophorea which became adapted to life within the alimentary canal of primitive invertebrate animals in the Pre-Cambrian period over 500 million years ago. From that time they evolved with their hosts, some becoming symbiotic in nematodes, some in molluscs, others in annelids or insects. The trypanosomatids underwent an evolutionary explosion in the insects diverging at first into two main groups: the *promastigote stock*, which retained the primitive anterior position of the flagellar basal body, and the *epimastigote stock*, which may well have arisen as an adaptation to life in the viscous contents of gut of insect. In this latter group the posterior displacement of the flagellar insertion made possible the development of an undulating membrane as an aid to locomotion.

The insect trypanosomatids are transmitted by the ingestion or resistant forms passed out in the feces of host. Some authorities have maintained that the genera *Trypanosoma* and *Leishmania* evolved from flagellates living in the intestines of vertebrates, but it seems more likely that they evolved from forms in the gut of invertebrates. When insects adopted the habit of sucking the blood of vertebrates, which they did at least as early as 40 million years ago in the Oligocene period, their trypanosomatid intestinal symbionts were given the possibility of entry into the vertebrate by contamination of the wound caused by the insect with the latter's feces which contained resistant forms of the trypanosomatid.

Some trypanosomatids were able to take advantage of this opportunity, and thus the genera *Leishmania* and *Trypanosoma* evolved—the former probably from the promastigote stock, and at least most species of the latter from the epimastigote stock. Some parasitologists believe that *T. cruzi* may have developed from promastigote stock; however, the forms of *T. cruzi* growing in the insect gut are epimastigotes, which suggests that *T. curzi* arose from epimastigote stock. The forms of *Leishmania* in insects are, on the other hand, promastigotes. The species of trypanosomes which inhabit aquatic vertebrates and leeches presumably evolved directly from the trypanosomatids of annelids.

When certain insects adopted the habit of feeding on plant juices, some of their promastigote symbionts became adapted to life in plants, thus giving rise to the genus *Phytomonas*. This development involved the adoption by the flagellates of transmission via the insect's proboscis, a route also adopted, presumably independently, by members of the genus *Leishamina* and by some of the species of trypanosomes which infect mammals. Alternatively, these (salvarian) trypanosomes may have evolved from leech transmitted trypanosomes of reptiles. Among the Apicomplexa, the other large group of protozoa which has specialized in symbiosis and of which some members have adopted the hematozoan or blood-dwelling habit, evolution has probably proceeded broadly as follows.

The Gregarinia, all of which live in invertebrates, presumably represents the basal stock of the Apicomplexa. From this stock evolution probably proceeded in a single main line to give rise to those organisms of which the Coccidia are the survivors. It was in this subclass that life within vertebrates was first embarked upon by Apicomplexa. The Coccidia are divided into three groups (ignoring the Protococcida)—Adeleina, Eimeriina and Haemosporina. Presumably the earliest representatives of the two forms groups were, as many still are, monoxenous symbionts (i.e. with one host, invertebrate or vertebrate, playing a part in their life-cycle).

At first the protozoa inhabited the cells of the host's alimentary canal; but later some of those that were living in vertebrates penetrated deeper into the body of the host, and some became adapted to spending part of their life-cycle in the circulating blood cells. This development was associated with the adoption of a heteroxenous life-cycle. This type of cycle, which added a phase in a blood-sucking invertebrate vector to the original phase within the vertebrate, increased the organism's chances of finding a new host and was thus of considerable survival value. This sequence of events appears to have occurred at least twice within the Apicomplexa. It occurred among the Adeleina, in those heteroxenous genera in which the sexual individuals (gamonts or gametocytes) and sometimes asexual stages also are found in blood cells, and also among the eimeriines.

In the latter group, most genera are still monoxenous, but in those few that have heteroxenous life cycles, such as *Lankesterella*, the transmissive organisms (sporozoites) are found in blood cells; all other stages occur within various fixed tissue cells of the vertebrate hosts. The heteroxenous adeleine type of development, in which sexual forms live in the blood cells and fertilization and post-zygotic multiplication (sporogony) occur in an invertebrate vector, is probably the type from which the Haemosporina (malaria parasites, etc.), all of which are heteroxenous, evolved.

In the heteroxenous Adeleina and Haemosporina, the part of the life-cycle which is undergone in the invertebrate host (sporogony) is that part which, in the monoxenous, more "primitive" Coccidia such as *Adelea* and *Eimeria*, occur outside the body of the host or in the lumen of its gut. This supports the view of many malariologists, that the monoxenous Apicomplexa which were directly ancestral to the Haemosporina were symbionts of vertebrates in contrast to the situation postulated above for the evolution of the Trypanosomatina. However, other authorities incline to the opposite view—that the Haemosporina evolved directly from monoxenous Coccidia with invertebrate hosts.

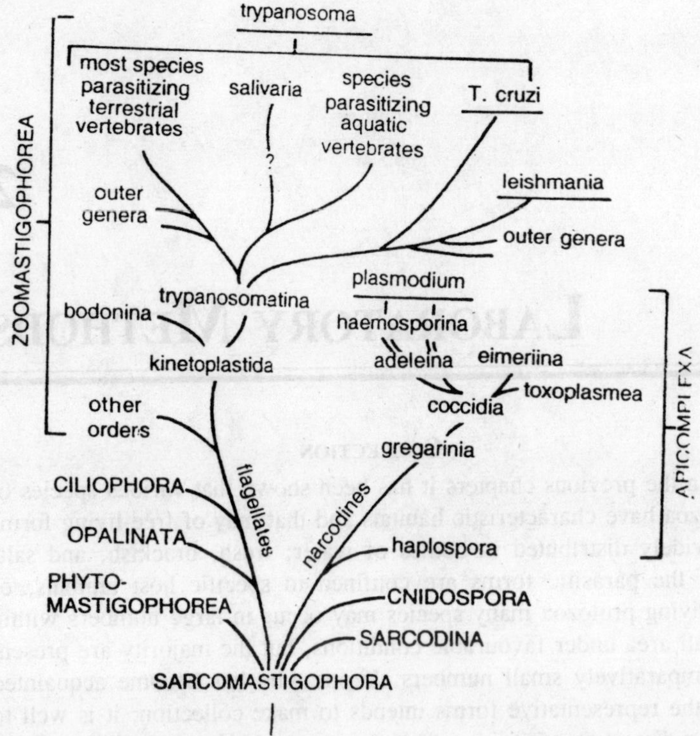

Fig. 1.1. Simplification phylogenetic tree of the protozoa, representing their presumed evolutionary development.

Among the Haemosporina, it is generally thought that *Plasmodium* is the most recently genus. This conclusion is based on the assumption that merogony in erythrocytes is a relatively recent acquisition. The malaria parasites of birds and reptiles probably diverged fairly early from those of mammals.

On the basis of the evolutionary sequence of their hosts, the species of *Plasmodium* infecting primates are thought to be the most recently evolved members of the genus; and among this group, probably the "quartan" species (those with a 72 hour cycle of erythrocytic merogony) are the representatives of the first to evolve, while, possibly, *P. falciparum* is the latest. However, the unity of the species infecting mammals is cast in doubt on the basis of DNA analysis, which indicates that *P. falciparum* may be more similar to rodent and avian malarias than to the other primate malarias. The views expressed above are summarized in the phylogenetic tree.

2

Laboratory Methods

Collection

In the previous chapters it has been shown that various species of protozoa have characteristic habitats and that may of free-living forms are widely distributed in bodies of water; fresh, brackish, and salt; while the parasitic forms are confined to specific host animals, of free-living protozoa many species may occur in large numbers within a small area under favourable conditions, but the majority are present in comparatively small numbers. If one who has become acquainted with the representative forms intends to make collection, it is well to carry a dissecting microscope in order to avoid bringing back numerous jars containing much water, but few organisms. Submerged plants, decaying leaves, surface scum, ooze, etc., should be examined under the microscope. When desired forms are found, they should be collected together with a quantity of water in which they occur. When the material is brought into the laboratory, it is often necessary to concentrate the organisms in a relatively small volume of water.

For this purpose, the water may partly be filtered rapidly through a fine milling cloth and residue quickly poured back into a suitable container before filtration is completed. The container should be placed in a cool moderately lighted room to allow the organisms to become established in the new environment. Stigma-bearing Phytomastigia will then be collected in a few hours on the side of the container, facing the strongest light, and the members of Sarcodina will be found among the debris on the bottom. Many forms will not only live long, but also multiple in such a container. For obtaining large freshwater amoebae, fill several finger bowls with the collected material and place one or two rice grains to each. After a few days, examine the bottom surface of the bowls under a binocular dissecting microscope.

If amoebae were included in the collection, they will be found particularly around the rice grains. Pipette them off begin separate cultures. In order to collect parasitic protozoa, one must, of course, find the host organisms that harbor them. Various species of tadpoles, frogs, cockroaches, termites, etc., which arc of common occurrence of easily obtained and which are hosts to numerous species of protozoa, are useful material for class work. Intestinal protozoa of man are usually studied in the faeces of an infected person. Natural movement should be collected. The receptacle must be thoroughly cleaned and dry, and provided with a cover. Urine or water must be excluded completely. The faeces must be examined as soon as possible, since the active tropozoites degenerate quickly once leaving the human intestine.

Cultivation

For extensive study or for class work, a large number of certain species of protozoa are frequently needed. Detection and diagnosis of human protozoa are often more satisfactorily made by culture method than by microscopical examination of the collected material. Success in culturing protozoa depends upon sevrral factors. As a rule, low temperatures seem to be much more favourable for culture than higher temperatures, although this is naturally not the case with those parasitic in homoiothermal animals. Furthermore, proper hydrogen ion concentration (pH) of the culture must be maintained. Mixed cultures of many free-living protozoa are easily maintained by adding from time to time a small amount of ripe hay-infusion or dried lettuce powder to the collected water. *Chilomonas*, *Peranema*, *Bodo*, *Arcella*, *Amoeba*, *Paramecium*, *Colpoda*, *Stylonychia*, *Euplotes*, etc. often multiply in such cultures.

To obtain a large number of a single species, individuals are taken out under a binocular dissecting microscope by means of a finely drawn-out pipette and transferred to a suitable culture medium. Such a culture is called a *mass* or *stock culture*. If a culture is started with a single individual, the resulting population makes up a *clone* or a *pure line*.

Free-living Protozoa

To deal with all the culture media employed by numerous workers for various free-living Protozoa is beyond the scope. Here only a few examples will be given.

Chromatophore-bearing flagellates

There are a number of culture fluids. Two examples:

(a)	Peptone or tryptone	2.0 gm.
	KH_2PO_4	0.25 gm.
	$MgSO_4$	0.25 gm.
	KCI	0.25 gm.
	$FeCI_3$	trace
	Sodium acetate	2.0 gm.
	Distilled water	1000 ml.
(b)	Peptone or tryptone	2.5 gm.
	KNO_3	0.5 gm.
	KH_2PO_4	0.5 gm.
	$MgSO_4$	0.1 gm.
	NaCI	0.1 gm.
	Sodium acetate	2.5 gm.
	Dextrose	2.0 gm.
	Distilled water	1000 ml.

Mast (1939) used the following media for *Chilomonas paramecium* and other colourless flagellates:

(a)	Glucose-peptone solution:	
	Peptone	8 gm.
	Glucose	2 gm.
	Water	1 litre
(b)	Acetate-ammonium solution:	
	Sodium acetate	1.5 gm.
	Ammonium chloride	0.46 gm.
	Ammonium sulphate	0.1 gm.
	Dipotassium hydrogen phosphate	0.2 gm.
	Magnesium chloride	0.01 gm.
	Calcium chloride	0.012 gm.
	Water	1000 ml.

Amoeba proteus and other freshwater amoebae

Fill a finger bowl with 200 ml. of glass distilled water, and place four rice grains in it. After a few days, seed with amoebae, add about 5 ml. of *Chilomonas* culture, and cover the bowl with a glass cover. In about two weeks, a ring of amoebae will be found around each rice grain, and if *Chilomonas* do not over- multiply, the amoebae will be found abundantly in another two weeks. If properly maintained, subcultures may be made every four to six weeks. *Chalkley* (1930) advocates substitution of the plain water with a salt solution which is composed of

NaCI	0.1 gm.
KCI	0.004 gm.
$CaCl_2$	0.006 gm.
Distilled water	1000 ml.

If the culture water becomes turbid, make subcultures or pour off the water and fill with fresh distilled water or the solution. Culture should be kept at 18-22°C.

Pelomyxa carolinensis

These amoebae grow well in a finger bowl with 150 ml. of redistilled water with 1-3 rice grains to which large numbers of *Paramecium* are added daily. *Pace* and *Belda* (1944) advocate the following solution instead of distilled water:

K_2HPO_4	0.08 gm.
KH_2PO_4	0.08 gm.
$CaCl_2$	0.104 gm.
$Mg_3\ (PO_4)_2\ 4H_2O$	0.002 gm.
Distilled water	1000 ml.

Small mono or di-phasic amoebae

Musgrave and Clegg's medium, modified by *Walker* is as follows:

Agar	2.5 gm.
NaCl	0.05 gm.
Liebig's beef-extract	0.05 gm.
Normal NaOH	2 ml.
Distilled water	100 ml.

Arcella and other Testacea

The testaceans commonly multiply in a mixed culture for several weeks after the collection was made. Hegner's method for *Arcella*: Pond water with weeds is shaken up violently and filtered through eight thicknesses of cheese cloth, which prevents the passage of coarse particles. The filtrate is distributed among Petri dishes, and when suspended particles have settled down to the bottom, specimens of *Arcella* are introduced. This will serve also for *Difflugia* and other testaceans. Hay or rice infusion is also a good culture medium for these organisms.

Actinophrys and Actinosphaerium

Belar cultivated these heliozonas successfully in Knop's solution:

Magnesium	0.25 gm.
Calcium nitrate	1 gm.
Potassium phosphate	0.25 gm.
Potassium chloride	0.12 gm.
Iron chloride	trace
Distilled water	1000 ml.

Freshwater ciliates

They are easily cultivated in a weak infusion of hay, bread, cracker, lettuce leaf, etc. The battery jars containing the infusions should be left standing uncovered for a few days to allow a rich bacterial growth in them. Seed them with material such as submerged leaves or surface scum containing the ciliates. If desired, culture may be started with a single individual in a watch glass.

Parasitic Protozoa

Intestinal flagellates of man

There are numerous media which have been used successfully by several investigators.

(a) *Ovo-mucoid medium* (*Hugue* 1921). White of two eggs are broken in a sterile flask with beads. Add 200 ml. of 7% NaCl solution and cook the whole for thirty minutes over a boiling water bath, shaking the mixture constantly. Filter through a coarse cheese cloth and through cotton-wool with the aid of a suction pump. Put 6 ml. of the filtrate in each test tube. Autoclave the tubes for twenty minutes under fifteen pounds pressure. After colling, a small amount of fresh faecal material containing the flagellates is introduced into the tubes. Incubate at 37°C.

(b) *Sodium chloride sheep serum water* (*Hogue*, 1922). Composed of 100 ml. sterile 0.95% NaCl and 10-15 ml. of sterile sheep serum water (dilution 1:3), 15 ml to each tube. *Trichomonas hominis*, *T. tenax*, and *Retortamonas intestinalis* grow well.

Trichomonas vaginalis

Johnson and *Trussell* (1943) reported the following mixture the most suitable medium:

Bacto-peptone	32 gm.
Bacto-agar	1.6 gm.
Cysteine HCl	2.4 gm.

Maltose	1.6 gm.
Difco liver infusion	320 ml.
Ringer's solution	960 ml.
NaOH(N/l)	11-13 ml.

Heat the mixture in a water bath to melt the agar; filter through a coarse paper; add 0.7 ml. of 0.5 per cent aqueous methylene blue; adjust pH to 5.8-6.0 with N/l HCl or NaOH; tube 8 ml autoclave. After colling, add aseptically 2 ml of sterile (filtered human serum. Incubate at least four days: store at room temperature for two to three weeks or as long as an amber "anaerobic" zone is apparent.

Trypanosoma and Leishmania

Novy, *MacNeal* and *Nicolle* (NNN) medium 14 gm. agar and 6 gm. of NaCl are dissolved by heating in 900 ml of distilled water. When the mixture cools to about 50°C, 50-100 ml of sterile defibrinated rabbit blood is gently added and carefully mixed so as to prevent the formation of bubbles. The blood agar is now distributed among sterile test tubes to the height of about 3 cm, and the tubes are left slanted until the medium becomes solid. The tubes are then incubated at 37°C for twenty-four hours to determine sterility and further to hasten the formation of condensation water (pH 7.6). Sterile blood or splenic puncture containing *Trypanosoma cruzi* or *Leishmania* is introduced by a sterile pipette to the condensation water in which organism multiply. Incubation at 37°C for trypanosomes and at 20-24°C for *Leishmania*.

E. histolytica and other amoebae of man

The first successful culture was made by *Boeck* and *Drbohlav* (1925) who used the following media:

(a) *Locke-egg-serum (LES) medium.* The contents of four eggs (washed and dipped in alcohol) are mixed with, and broken in 50 ml. of Locks's solution in a sterile flask with beads. The solution is made up as follows :

NaCl	9 gm.
$CaCl_2$	0.2 gm.
KCl	0.4 gm.
$NaHCO_3$	0.2 gm
Glucose	2.5 gm
Distilled water	1000 ml.

The emulsion is now tubed so that when coagulated by heat, there is 1-15 inches of slant. These tubes are now slanted and heated at

70°C until the medium becomes solidified. They are then autoclaved for twenty minutes at fifteen pounds pressure (temperature must be raised and lowered slowly). After colling the slant is covered with a mixture of eight parts of sterile Locke's solution and one part of sterile inactivated human blood serum. The tubes are next incubated to determine sterility. The culture tubes are incubated with a small amount of faecal matter containing active trophozoites. Incubation at 37°C *Yorke* and *Adams* (1926) obtained rich cultures by inoculating this medium with washed and concentrated cysts of *E. histolytica* in twenty-four hours.

(b) *Locke-egg-albumin (LEA) medium.* The serum in LES medium is replaced by 1 per cent solution of crystallized egg albumin in Locke's solution which has been sterilized by passage through a berkefeld filter.

(c) *Ringer-egg-serum (RES) or Ringer-egg-albumin (REA) medium.* Solid medium is the same as that of (a) or (b), but made up in Ringer's solution which is composed of

NaCl	9 gm.
KCl	0.2 gm.
$CaCl_2$	0.2 gm.
Distilled water	1000 ml.

The covering liquid is serum-Ringer or egg-albumin. The latter is prepared by breaking one egg white in 250 ml. of Ringer's solution which is passed through a Seitz filter. Before inoculating with amoebae, a small amount of sterile solid rice-starch (dry-heated at 180°C for one hour) is added to the culture tube. To inhibit bacterial growth in cultures of *Entamoeba*, various antibiotics have been tired. For example, *Spingarn* and *Edelman* (1947) found that when streptomycin was added in the amount of 1000-3000 units per ml. to culture of *E. histolytica*, the survival of the amoebae in culture was prolonged from an average of eight days to 33.7 days, which effect was apparently due to the inhibition of bacteria.

Plasmodium

Bass and John's (1912) culture is as follows: 10 ml. of defibrinated human blood containing *Plasmodium* and 0.1 ml. of 50 per cent sterile dextrose solution are mixed in test tubes and incubated at 37-39°C. In the culture, the organisms develop in the upper layer of erythrocytes. Since that time a number of investigators have undertaken cultivation of different species of *Plasmodium*.

Balantidium coli

Barret and *Yarbrough* (1921) first cultivated this ciliate in a medium consisting of sixteen parts of 0.5% NaCl and one part of inactivated human blood serum. The medium is tubed. Inoculation of a small amount of the faecal matter containing the trophozoites is made into the bottom of the tubes. Incubation at 37°C Maximum development is reached in forty-eight to seventy-two hours. Subcultures are made every second day.

Microscopical Examination

Protozoa should be studied as far as possible in life. Permanent preparations while indispensable in revealing many intracellular structures, cannot replace fresh preparations. The microscopic slides of standard size, 3" by 1", should be of white glass and preferably thin. The so-called No. 1 slides measure about 0.75 mm. in thickness. For darkfield illumination thin slides are essential. No. 1 coverglasses should be used for both fresh and permanent preparations. They are about 130-170μ thick. The most convenient size of the coverglass is about seven-eights square inch which many prefer to circular ones. The slides and coverglasses must be thoroughly cleaned before being used. Immerse them in concentrated mineral acids (nitric acid is best fitted) for ten minutes. Pour off the acid, wash the slides and coverglasses for about ten minutes in running water, rinse in distilled water, and keep them in 95% alcohol. When needed they are dried one by one with clean cheese cloth. Handle slides and covers with a pair of forceps. If thumb and fingers are used, hold them edgewise.

Fresh Preparation

In making fresh preparations with large protozoa care must be exercised to avoid pressure of the coverglass on the organisms as this will cause deformities. If small bits of detritus or debris are included in the preparation, the coverglass will be supported by them and the organisms will not be subjected to any pressure. The free-living ciliates swim about so actively as to make their observation difficult. However, an actively swimming ciliate will sooner or later come to stop upon coming in contact with various debris, air bubbles or margin of the coverglass to allow a study of its structure.

Various reagents recommended for retardation of swimming movements of ciliates, bring about deformities in the organisms and therefore, must not be used; but a drop of saturated or 10% solution of methyl cellulose may be added to a ciliate preparation to retard the active movement of the organism without causing any visible abnormality. A weak solution of nickel sulphate retards the movement of cilia and flagella and in *Paramecium* movement is said to be completely arrested in two hours and recovery is complete on returning to normal culture fluid. The phase microscope is highly useful for observation of cilia, flagella and other filamentous organelles as well as various internal structures in living protozoa. When treated with highly diluted solutions of certain dyes, living protozoa exhibit some of their organellae or inclusions stained without apparent injury to the organisms. These *vital stains* are usually prepared in absolute alcohol solutions.

A small amount is uniformly applied to the slide and allowed to day, before water containing protozoa is placed on it. Congo red (1:1,000) is used as an indicator, as its red colour of the salt changes blue in weak acids. Janus Green B (1:10,000-50,000) stains mitochondria. Methylene blue (1:10,000 or more) stains cytoplasmic granules, nucleus, cytoplasmic processes, etc. Neutral red (1:3,000-30,000) is an indicator: yellowish red (alkaline), cherry red (weak acid), and blue (strong acid). It also stains nucleus slightly. Golgi bodies are studied in it, through its specificity for this structure is not clear.

Physiological salt solution

Widely used concentrations of NaCl solutions are 0.5-0.7 per cent for cold-blooded animals and 0.8-0.9 per cent for warm-blooded animals.

Ringer's solution

Frequently used solution consists of

NaCl	0.8 gm.
KCl	0.02 gm.
$CaCl_2$	0.02 gm.
($NaHCO_3$	0.02 gm.
Glass distilled water	100 ml.

For demonstrating organellae, the following reagents which kill the protozoa upon application, may be used on living protozoa.

Ligol's solution

This is made up of potassium iodide 1.5 gm., water 25 ml. and iodine 1 gm. The solution deteriorates easily. Flagella and cilia stain clearly. Glycogen bodies stain ordinarily reddish brown. Cysts of intestinal protozoa are more easily studied in Lugol's solution. Sudan III and IV, 2% absolute alcohol solution diluted before use with the same amount of 45% alcohol. Natural fats are stained red. Methyl gree 1% solution in 1% acetic acid solution makes an excellent nuclear stain. In the case of faecal examination if the stool is dysenteric, a small portion is placed by a tooth-pick on a slide and covered with a cover glass.

Before placing the cover, all large particles must be removed quickly so that the smear will be uniformly thin. Smears of diarrhoeic stools can be made in a similar way. But if the faecal material is formed or semiformed, a small drop of warm (37°C) 0.85% NaCl solution is first placed on the slide, and a small portion of the faeces, particularly mucus, pus or blood, is emulsified in it. The whole is covered by a coverglass. The faecal smear should not be too thick or too thin for a satisfactory observation. If the smear is too thick, it will be impossible to distinguish objects clearly, and on the other hand, if it is too thin, there will be much time lost in observing widely scattered protozoa.

Permanent Preparations

Permanent preparations are employed, as was stated before, to supplement, and not to supplant, fresh preparations. Smear preparations are more frequently studied, while section preparations are indispensable in extensive studies of protozoa. Various fixatives and stains produce different results, care must be exercised in making and evaluating permanent preparations.

Smear preparations

Smears are made either on coverglasses or slides. However, coverglass-smears are more properly fixed and require a smaller amount of reagents than slide-smears. If a small drop of fresh egg-white emulsified in sterile distilled water is smeared on the coverglass very thinly with the tip of a clean finger, before mounting material for smear, more specimens will adhere to and remain on the coverglass upon the completion of the preparation. Let the smear lie horizontally for five to ten minutes or longer. Parasitic protozoa live in media rich in albuminous substances, and therefore, easily adhere to the coverglass in smear make uniformly thin smears on coverglasses. If the smears are made from dysenteric or fluid stools, they should be fixed almost immediately.

Smears made from diarhoeic or formed stools by emulsifying in warm salt solution, should be left for a few minutes. In any case, do not let the smear become dry except a narrow marginal zone. The smears are fixed next. The most commonly used fixative for protozoa is *Schaudinn's* fluid. This made up as follows:

Cold saturated mercuric bichloride (6-7%)	66 ml.
Absolute or 95% alcohol	33 ml.
Glacial acetic acid	1 ml.

The first two can be kept mixed without deterioration, but the acid must be added just before fixation. Fix at room temperature or warmed to 50°C. The fixative is placed in a square Petri dish and the smear is gently dropped on it with the smeared surface facing downward. With a little experience, air bubbles can be avoided and make the smear float on the surface of the fixative. After about one minute, turn it around and let it stay on the bottom of the dish for five to ten more minutes.

In case the smear is too thick, a thin coat of vaseline on the upper side of the coverglass will make it float. About six coverglass-smears may be fixed in the dish simultaneously. The coverglass-smears are now transferred to a staining jar for coverglass, containing 70 per cent alcohol for ten minutes, followed by two changes for similar length of time. Transfer the smears next to 50 per cent alcohol for five minutes, and then to a jar with water, which is now placed under gently running tap water for fifteen minutes. Rinse them in distilled water and stain.

Other fixatives frequently used protozoa are as follows:

Bouin's fluid

Picric acid (Saturated)	75 ml.
Formaldehydes	25 ml.
Glacial acetic acid	5 ml.

Fixation for five to thirty minutes; wash with 70% alcohol until picric acid is completely washed away from the smears.

Sublimate-acetic

Saturated sublimate solution	100 ml.
Glacial acetic acid	2 ml.

This is the original fixative for Feulgen's nuclear reaction. Fixation and after-treatment similar to Schaudinn's fluid.

Carnoy's fluid

Absolute alcohol	30 ml.
Glacial acetic acid	10 ml.

Fixation for five to thirty minutes; wash in 95 per cent alcohol.

Osmium tetroxide

The vapour from or the solution itself of 1 per cent Osmium tetroxide may be used. Fixation in two to five minutes; wash in running water.

Flemming's fluid

1% Chromic acid	30 ml.
2% Osmium tetroxide	8 ml.
Glacial acetic acid	2 ml.

Fixation for ten to fifty minutes; wash for one hour or longer in running water. The most commonly used stain is *Heidenhain's iron haematoxylin*, as it is dependable and gives a clear nuclear picture, although it is unsatisfactory for voluminous organisms or smears of uneven thickness. It requires a mordant, ammonio-ferric sulphate (iron alum) and a dye, *haematoxylin*. Crystals of iron alum become yellow and opaque very easily. Select clear violet crystals and prepare 2 per cent aqueous solution. Haematoxylin solution must be well "ripe". The most convenient way of preparing it is to make 10 per cent alcohol solution as it does not require ripening. By diluting this stock solution with distilled water, prepare 0.5 or 1 per cent slightly alcoholic solution which will be ready for immediate and repeated use.

Smears are left in the mordant in a jar for one to three hours or longer. Wash them with running water for five minutes and rinse in distilled water. Place the smears now in haematoxylin for one to three hours or longer. After brief washing in water, the smears are decolarized in Petridish in a diluted iron alum, 0.5% HCl in water or 50 per cent alcohol, or saturated aqueous solution of picric acid under the microscope. Upon completion, the smears are washed thoroughly in running water for about thirty minutes. Rinse them in distilled water. Transfer them through ascending series of alcohol (50 to 95 per cent). If counter staining with eosin in desired, dip the smears which were taken out from 70 per cent alcohol, in 1 per cent eosin in 95 per cent alcohol for a few seconds, and then in 95 per cent plain alcohol.

After two passages through absolute alcohol and through xylol, the smears are mounted one by one on a slide in a small drop of mixture of Canada balsam and xylol. The finished preparations are placed in a drying oven at about 60°C for a few days.

Delafield's haematoxylin

If the stock solution is diluted to 1:5-10, a slow, but progressive staining which requires no decolorization may be made; but if stock solution is used, stain for one to sixteen hours, and decolorize in 0.5% HCl water or alcohol. If mounted in a neutral mounting medium, the staining remains true for a long time.

Giemsa's stain

Shake the stock solution bottle well. By means of a stopper-pipette dilute the stock with neutral distilled water (5-10 drops to 10 ml.). Smears fixed in Schaudinn's fluid and washed in neutral distilled water are stained in this solution for ten minutes to six hours to overnight. Rinse them thoroughly in neutral distilled water and transfer them through the following jars in order (about five minutes in each): (a) acetone alone; (b) acetone: xylol, 8:2; (c) acetone: xylol, 5:5; (d) acetone: xylol, 2:8; (e) two changes of xylol. The smears are now mounted in cedar wood oil (which is used for immersion objectives) and the preparations should be allowed to dry for a longer time than the balsam-mounted preparations.

Blood film preparations

Thin film

The finger tip or ear lobe is cleaned with 70 per cent alcohol. Prick it with a sterilized needle. Wipe off the first drop with gauze and receive the second drop on a clean slide about half an inch form

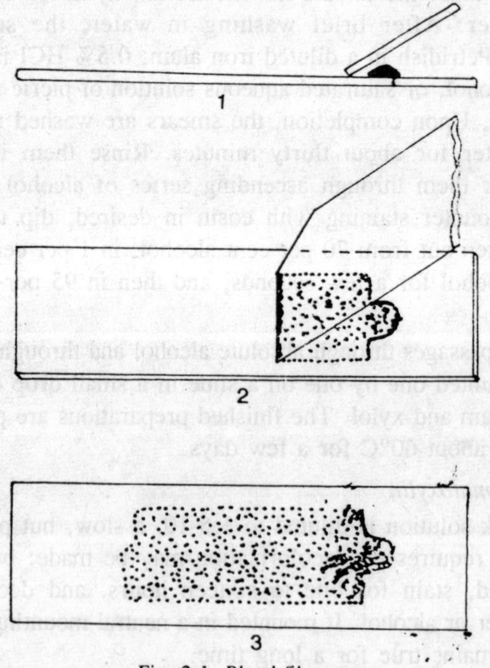

Fig 2.1. Blood film smear.

one end. Use care not to let the slide touch the finger. Quickly bring a second slide, one corner of which had been cut away, to the inner margin of the blood drop (1), and let the blood spread along the edge of the second slide. Next push the second slide over the surface of the first slide at an angle of about 45°C towards the other end (2). Thus a thin film of blood is spread over the slide (3). Let the slide lie horizontally and dry, under a cover to prevent dust particles falling on it and to keep away flies or other insects. If properly made, the film is made up of a single layer of blood cells.

Thick films

Often parasites are so few that to find them in a thin film involves a great deal of time. In such cases, a thick film is advocated. For this, two to four drops of blood are placed in the central half-inch square area, and spread them into an even layer with a needle or with a corner of a slide. Let the film dry. With a little practice, a satisfactory thick smear can be made. It will take two hours or more to dry. So not dry by heay, but placing it in an incubator at 37°C will hasten the drying. When thoroughly dry, immerse it in water and dehaemoglobinize it. Air dry again.

Thin and thick film

Often it is time-saving if thin and thick films are made on a single slide. Place a single drop of blood near the centre and make a thin film of it toward one end of the slide. Make a small thick smear in the centre of the other half of the slide. Dry when thoroughly dry, immerse the thick film part in distilled water and dehaemoglobinize it. Let the slide dry. Blood smears must be stained as soon as possible to insure a proper staining, as lapse of time or summer heat will often cause poor staining especially of think films.

Of several blood stains, Giemsa's and Wright's stains are commonly used. For staining with Giemsa's stain, the thin film is fixed in absolute methyl alcohol for five minutes. Rinse the slide well in neutral distilled water. After shaking the stock bottle well, dilute it with neutral distilled water in a ratio of one drop of stain to 1-2 ml. of water. Mix the solution and the blood film is placed in it for 0.5-2 hours or longer if needed. Rinse the slide thoroughly in neutral distilled water and wipe on water with a tissue paper from the underside and edges of the slide. Let the slide stand on end to dry. When thoroughly dry, place a drop of xylol and a drop of cedar wood oil (used for immersion objective) and cover with a coverglass. The mounting medium should be absolutely neutral Do not use Canada balsam for mounting, as acid in it promptly spoils the staining.

For *Wright's* stain, fixation is not necessary. With a dropper, cover the dried blood film with drops of undiluted Wright's stain, and let the film stand horizontally for three to five minutes; then the same number of drops of neutral distilled water is added to the stain and the whole is left for ten to thirty minutes. The stain is then poured off and the film is rinsed in neutral distilled water. Dry Mount in xylol and cedar wood oil. Use of coverglass on a stained blood film is advocated, since a cedar wood oil mounted slide allows the use of dry objectives which in the hand of an experienced worker would give enough magnification for species determination of *Plasmodium*, and which will very clearly reveal and trypanosomes present in the film.

Furthermore, the film is protected against scratches, and contamination by many objects which may bring about confusion in detecting looked-for organisms. Films made from spenic punctures for *Leishmania* or *Trypanosoma* are similarly treated and prepared.

Section preparations

Paraffin sections should be made according to usual histological technique. Fixatives and stains are the same as those mentioned for smear preparations.

3

CELL ORGANELLES

Nucleus, Its Somatic Function, and the Endoplasmic Reticulum

The importance of the nucleus in the living functions of the cell becomes apparent when it is removed from the cell. This can be achieved either by sucking out the nucleus with a micropipette in a micromanipulator or, since the Protozoa are generally able to regenerate, more simply through merotomy, cutting the cell into nucleated and enucleated pieces. After the cell has been divided up in this way the wound closes in all cases. Whilst motility, function of the contractile vacuole and, in some species, at first also food uptake are retained, the anucleate forms always lack the capacity to digest.

Comparative studies with the ciliate *Stentor* have shown that anucleate forms die for this reason in the same time as it takes nucleated forms to die of starvation when deprived of a food source. If amoebae which have accumulated glycogen and fats in the cytoplasm before merotomy are maintained after the operation in a medium lacking nutrients, the reserves in the nucleate half are largely broken down within a week, whilst the anucleate amoeba can no longer utilize the reserves. Only the amoeba with a nucleus remains functioning. In a ciliate, all framgments achieve complete regeneration if they contain parts of the segmented macronucleus; even single links of the chain-like nucleus are sufficient. In contrast, the segments lacking a nucleus are capable of surviving for only a short while.

In *Stentor coeruleus* the presence of the micronucleus in a merotomy fragment has no influence on regeneration, whilst in the ciliate *Euplotes patella* only pieces which also have a micronucleus are capable of life in the long term. Even so, fragments of ciliates

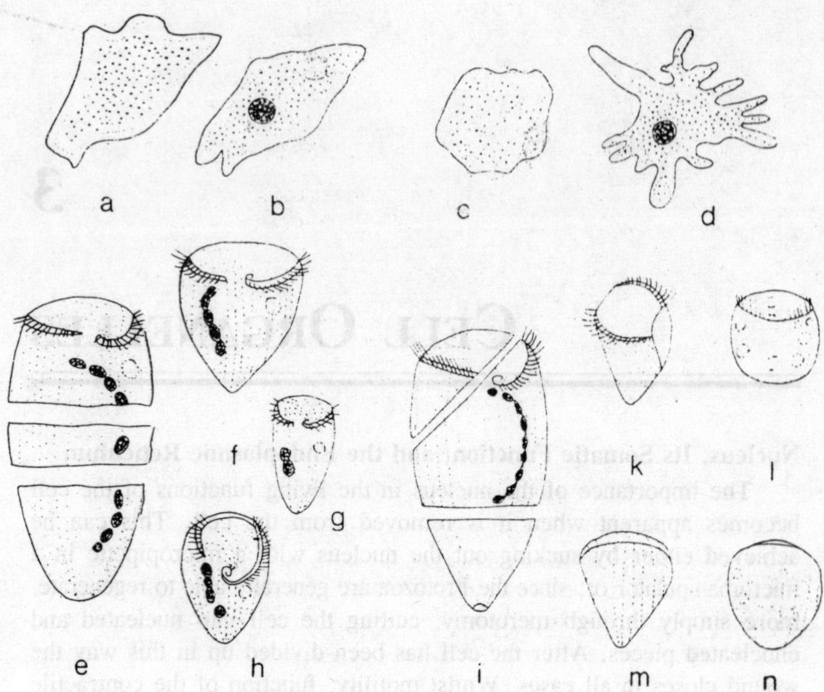

Fig. 3.1. Showing merotomy.

with only the micronuleus cannot regenerate. These studies show the significance of the cell nucleus in the somatic functions of the cell; these functions are dependent upon the macronucleus in the ciliates, which is therefore called the somatic nucleus. An analysis of nuclear structure can be made with the aid of light microscopy or electron microscopy. Because of its greater optical density, the nucleus of many Protozoa can be distinguished from the cytoplasm by light microscopy, even in unstained preparations. For example, the nucleus of many entamoeba species appears as a bright shining ring in transmitted light, whilst, under phase contrast, it appears darker than the cytoplasm. A clearer picture of the nuclear structure can be obtained by appropriate staining after fixing.

By this means, in suitable cases, four basic elements may be discerned in the nuclear structure, as can be seen, for example, in the gamont nucleus of the foraminiferan *Myxotheca arenilega*. The nuclear membrane separates the nucleus from the surrounding cytoplasm, and the nucleolar substance forms a substantial peripheral ring, here stained with iron haematoxylin. Within the nucleus lie the stained chromosomes

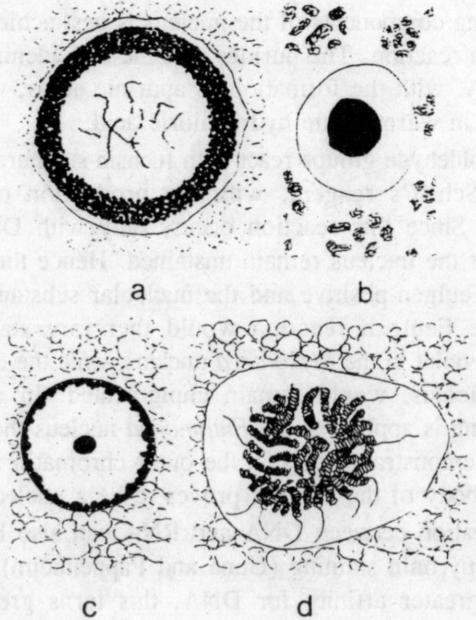

Fig. 3.2. Nuclear structure of Protozoa during interphase.

in various shapes and sizes, and all lighter spaces in between are filled with nucleoplasm. In the nuclei of most Protozoa the chromosomes can be discerned only when the nucleus enters mitotic division. In the intervening period, in the resting or interphase nucleus, the structural elements of the chromosomes are so finely distributed that they do not show clearly. The arrangement of the basic elements in the nucleus can differ widely. Figure 3.2 (b) shows a very simple nuclear structure, the vesicular nucleus of a limax amoeba, in which the "chromatin" component of the nucleus (which can be stained with haematoxylin or other basic stains) is a compact structure in the centre of the nucleus, while a broad zone of nuclear sap or nucleoplasm lies peripherally.

In the nucleus of *Entamoeba* species, on the other hand, the chromatin is divided between small central internal bodies and a peripheral chromatin ring under the nuclear membrane. The particular staining properties of the chromosomes and nucleolar substance within the nucleus derives from their nucleic acid content. Desoxyribonucleic acid (DNA) is the major component of the chromosomes, which contain only a small amount of ribonucleic acid (RNA), while the ability of the nucleolar substance to take up the customary nuclear stains is a function of the RNA content. Differentiation between the DNA- and

RNA-containing components of the nucleus is best achieved by the use of the Feulgen reaction. The purines guanine and adenine separate out from the DNA, with the formation of apurinic acids, when the DNA is hydrolyzed in warm dilute hydrochloric acid.

The free aldehyde groups react with fuchsin-sulphurous acid, which is added as Schiff's reagent, with the production of a red-violet pigmentation. Since this reaction occurs only with DNA, the RNA components of the nucleus remain unstained. Hence the chromosomal substance is Feulgen-positive and the nucleolar substance is Feulgen-negative. The Feulgen reaction would therefore stain the central chromosomes violet in the *Myxotheca* nucleus while the outer chromatin ring, the nucleolus, would remain unpigmented. In contrast, when Feulgen staining is applied to the *Entamoeba* nucleus the chromosomal substance is demonstrable only in the outer chromatin ring, while the central inner body of the nucleus proves to be a nucleolus.

Differentiation between DNA and RNA can also be achieved by methyl green-pyronin staining (Unna and Pappenheim). Since methyl green has a greater affinity for DNA, this turns green, whilst the RNA is stained red by the pyronin. Despite the ease with which chromosomal material can be demonstrated by the Feulgen reaction, it is by no means always possible in the interphase nucleus. In the *Entamoeba* nucleus, the chromosomes are first seen within the nucleus during mitosis. Only a few Protozoa show their chromosomal arrangement in the resting nucleus. Apart from *Myxotheca arenilega*, the chromosomes of the Euglenoidina, Dinoflagellata, and of some Polymastigina can be seen in the resting nucleus, sometimes even in unstained preparations. Chromosomes represent, through the organized arrangement of the various genes, the hereditary factors which work together to determine the morphological and physiological specificity of particular species. The whole set of chromosomes is known as the *genome*. A nucleus in which each hereditary factor is present only once, and which thus contains only one genome, is a *haploid* nucleus. Nuclei with 2 genomes are *diploid*; those with numerous genomes are *polyploid*.

The genes present in the genomes, which are made up of DNA, determine all that happens in the cell. They supply the information for the synthesis of proteins which, as structural elements, participate in the construction of the cell and, as enzymes, make possible the many different metabolic processes of the cell. mRNA (messenger RNA) carries the information from the DNA to the ribosomes, the site of

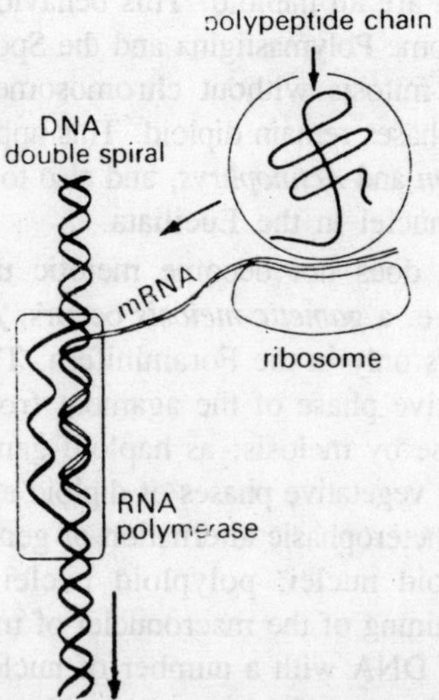

Fig. 3.3. Diagram illustrating transcription during protein synthesis.

protein synthesis. Fig 3.3. shows this relationship. The upper arrow shows the direction of movement of the ribosomes. The total number of chromosomes in a genome is very varied, so that great differences can occur within a single Order. For example, in the Polymastigina there are species known to have 2, 8, 10, 12, 14, 24, 26, 48 and 60 chromosomes in the genome.

Most nuclei in the Protozoa contain 1 or 2 genomes and are thus haploid or diploid. During sexual processes all gametic nuclei are haploid and, in the ensuring karyogamy, their fusion produces zygotic cells which are diploid (a). If division of the zygote proceeds as a reduction division, so that a *zygotic meiosis* occurs, the vegetative

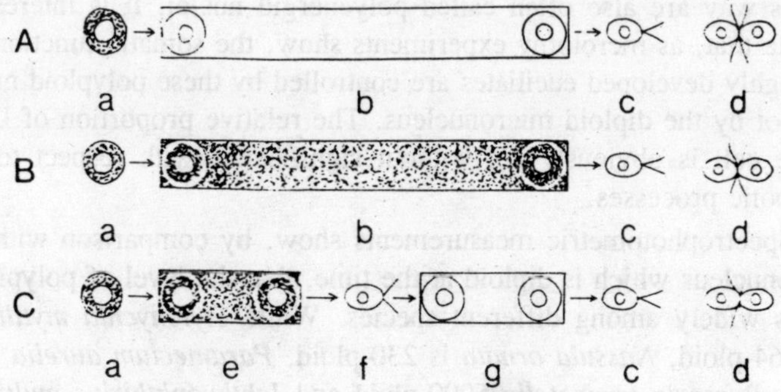

Fig. 3.4. Diagram of the alternation between haploid and diploid phases.

phases arising from it are all haploid. This behaviour has been seen in the Phytomonadina, some Polymastigina and the Sporozoa. If the zygote divides by a simple mitosis without chromosome reduction, all the following vegetative phases remain diploid. This applies to the heliozoan genera *Actinosphaerium* and *Actinophrys*, and also to some Polymastigina and to all the micronuclei in the Euciliata.

Nuclear division does not become meiotic until gametic nuclei are again produced, i.e. a *gametic meiosis* occurs. *Intermediate meiosis* in the Protozoa occurs only in the Foraminifera. The zygote gives rise to the diploid vegetative phase of the agamont from which the young haploid agametes arise by meiosis; as haploid gamonts these produce the gametic cells. The vegetative phases of diploid agamonts and haploid gamonts thus form a heterophasic alternation of generations. In addition to haploid and diploid nuclei, polyploid nuclei also occur in the Protozoa. Feulgen staining of the macronuclei of most euciliates shows a great abundance of DNA with a number of nucleolar zones lying in between. In diffuse macronuclei, e.g. as in *Nassula ornata*, division configuraiions of the chromosomal structures can be seen in the macronucleus too, at the time of division of the micronucleus. These configurations indicating chromosonal division can also be seen in the slowly growing macronucleus of the suctorian *Ephelota gemmipara* and during macronucleus formation in exconjugants of *Bursaria truncatella*, and in *Stentor polymorphus*.

The mass of chromosomal structures demonstrable here far exceeds the chromosomal content of the relevant diploid micronucleus from which the macronucleus develops anew after conjugation. In the production of the macronucleus there must therefore be repeated mitoses in the nuclear body, without nuclear division. This process is called *endomitosis* and it leads to endomitotic polyploidy. The nuclei produced in this way are also often called polyenergid nuclei. It is interesting to note that, as merotomy experiments show, the somatic functions of the highly developed euciliates are controlled by these polyploid nuclei and not by the diploid micronucleus. The relative proportion of DNA in the cell is obviously not without significance with respect to the metabolic processes.

Spectrophotometric measurements show, by comparison with the micronucleus which is diploid at the time, that the level of polyploidy varies widely among different species. While *Stylonychia mytilus* is only 64-ploid, *Nassula ornata* is 230-ploid, *Paramecium aurelia* 860-ploid, *Bursaria truncatella* 5000-ploid and *Ichthyophthirius multifiliis* even 12,600-ploid. It is not only the quantity of DNA but also the areas of surface contact between the nucleus and the cytoplasm that is raised in many cases by the form of the macronucleus. Few ciliates have a round macronucleus, which has the smallest surface area in relation to its volume.

In many cases the macronucleus has an oval shape, and further enlargement of the nuclear surface follows elongation of the nucleus, branching, chain formation, or the production of several individual macronuclei. Polyploid nuclei occur not only in the ciliates but also in the Tripylea of the Radiolaria. A highly magnified section of the nucleus from the central capsule of *Aulacantha scolymantha* shows the splitting of chromosomes during endomitotic polyploidy. If the central capsule divides during binary fission, an equatorial plate bearing a large number of ribbon-shaped chromosomes occurs at nuclear division; these chromosomes are distributed among the daughter nuclei with the formation of daughter plates. For swarmer production in the Radiolaria the membrane of the primary nucleus first disintegrates in the central capsule. After this, the numerous centrally-situated chromosomal genomes are distributed throughout the plasma of the central capsule and become secondary nuclei which undergo further mitoses to produce numerous swarmer nuclei.

In many cases an increase in the chromosomal complement is achieved in the homokaryotic Protozoa only by them becoming multinucleate. Examples of this are provided by the Distomatidae of the Protomonadina, by the Polymastigina, and by the amoeba *Pelomyxa palustris*, the heliozoan species *Actinosphaerium eichhorni* and the homokaryotic ciliates. Although most euciliates have polyploid macronuclei, some holotrichous genera have only diploid macronuclei like the heterokaryotic Foraminifera. This condition has already been mentioned in dealing with the reproduction and sexuality of the ciliates,

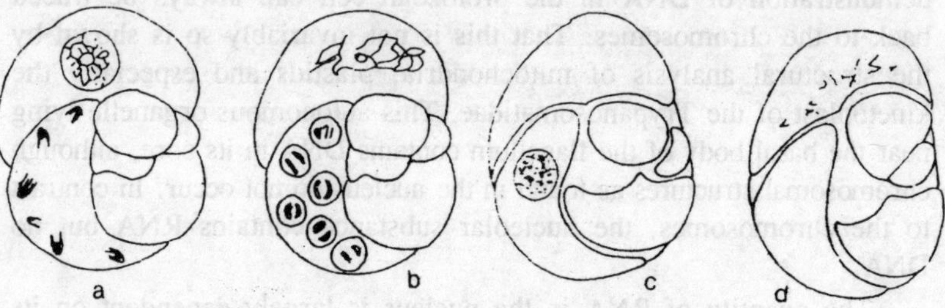

Fig. 3.5. Diploid macronuclei in Foraminifera.

and in these cases the macronuclei are not able to divide; they are always formed anew from diploid micronuclei. In these ciliates, too, an increase in DNA and in the surface area of the nucleus is achieved by the multinucleate condition.

In the heterokaryotic Foraminifera the chromosomes present in the macronucleus are clearly visible during meiosis of the micronuclei. When the micronuclei enter meiotic division, the macronucleus of the agamont in *Rotaliella roscoffensis* breaks up and disintegrates, thereby releasing the chromosomes. Aberrant agamonts which have a macronucleus but no micronucleus can occasionally arise. Like the ciliates, they are able to survive and grow without a micronucleus; but, when the agamont has reached the size at which meiosis normally begins, the macronucleus disintegrates just as it would in the course of meiosis, and the resultant cessation of somatic function leads to the death of the cell. Again as with ciliates, Foraminifera which have only micronuclei can never develop.

In contrast to these Foraminifera, ciliates with a polyploid macronucleus have an evidently unlimited capacity to live and reproduce. The capacity of the polyploid macronucleus for division has already been shown. Thus the ciliate races lacking a micronucleus, which are occasionally found in nature, can maintain themselves by asexual reproduction, as has been shown in cultures of *Tetrahymena pyriformis*, *Oxytricha hymenostoma*, *Oxytricha fallax*, *Urostyla grandis*. *Didinium nasutum*, *Colpoda steini* and *Paramecium bursaria*. These findings support the results of merotomy experiments, according to which the macronucleus is responsible for the somatic cell function.

As already described, the Feulgen reaction serves to demonstrate the DNA constituent in a light-microscopic study of the cell. Bearing in mind that the genes of the chromosomes also control the somatic functions of the cell, it is interesting to know whether the positive demonstration of DNA in the protozoan cell can always be traced back to the chromosomes. That this is not invariably so is shown by the structural analysis of mitochondria, plastids and especially the kinetoplast of the Trypanosomatidae. This autonomous organelle lying near the basal body of the flagellum contains DNA in its core, although chromosomal structures as found in the nucleus do not occur. In contrast to the chromosomes, the nucleolar substance contains RNA but no DNA.

The quantity of RNA in the nucleus is largely dependent on its physiological condition. Figure shows the nucleus of *Amoeba proteus*

(*Chaos difluens*) after staining with gallocyanin-chrome alum, which stains DNA and RNA. If the RNA is removed before staining, by means of the enzyme ribonuclease, the appearance of the nucleus of the amoeba after mitosis where the RNA fraction has been used up during nuclear division. *Gregarina cuneata*, a species found in the mealworm, develops to a considerable size during the growth period, and staining with the normal basic nuclear stains shows the presence of a very large nucleus.

When mitotic division begin after syzygy, the RNA fraction of the nucleus disappears. The DNA fraction, the "generative nucleus", which in this species was already visible, now enters into mitosis. Merotomy experiments suggest that the RNA fraction of the nucleus is transferred to the cytoplasm. The ability of the cytoplasm to take up basic stains is reduced within a matter of days in segments lacking a nucleus; the RNA content of the cytoplasm has already fallen to 60% of normal after 10 days. If amoebae are fed on foodstuffs containing radioactive ^{32}P. radioactivity is demonstrable in the nucleus. If such a nucleus is implanted into an untreated amoeba in a normal culture solution, the radioactivity is transferred to the cytoplasm.

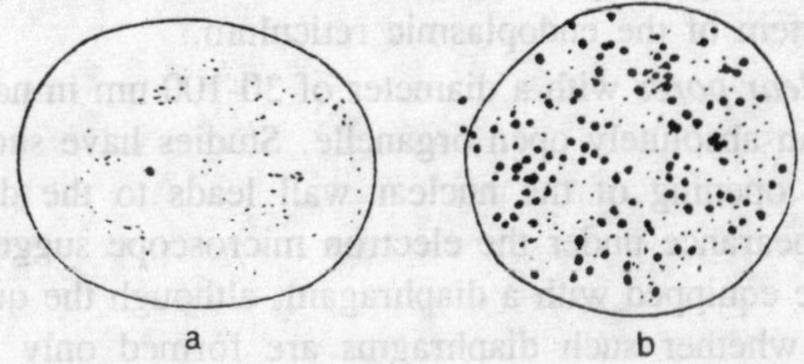

Fig. 3.6. Dependence of RNA production in the cytoplasm on the cell nucleus.

Further evidence for the importance of the nucleus in relation to the occurrence of RNA in the cytoplasm has been provided by experiments with ciliate *Tetrahymena pyriformis* (Hymenostomata). If, after merotomy, a fragment containing a nucleus and a fragment lacking a nucleus are maintained in a sterile nutritive solution with a component of RNA in a radioactive form, ^{3}H-cytidine, RNA formation, readily discerned by its radioactivity, occurs only in the cytoplasm of the fragment containing a nucleus. As studies with the giant chromosomes of Diptera have shown, the RNA occurs at specific sites on the chromosomes, the nuclear organizers. The nucleoli of the nucleus act as collecting centres for the temporary accumulation of RNA. The basic substance of the nucleoli is protein, and the RNA content usually amounts to only 5-10%. These relationships indicate how important the chromosomal fraction and hence also polyploidy is for the somatic function of the nucleus. We have up to now shown the importance of the nucleus with its genes to the somatic functions of the whole cell body, but the reverse is also true: an isolated nucleus cannot survive without it cytoplasm. Nucleus and cytoplasm together form a physiological entity.

The reciprocal relationship between nucleus and cytoplasm is demonstrated by the detailed findings of electron-microscopic studies, which show that the nucleus does not just lie in the cytoplasm as a discrete body enclosed in the nuclear wall. Each of these unit membranes has a thickness of about 7-8 nm. The space between the two membranes, the perinuclear space, can be of varying width with an average width of 10-15 nm. The two nuclear membranes and the perinuclear space differ from one another in their response to osmium. Whilst the membranes are osmiophilic, the perinuclear space is osmiophobic. There are three possible routes for the exchange of substances between the interior of the nucleus and the cytoplasm: an exchange via the pore system, by the formation of vesicles, and via the canal system of the endoplasmic reticulum.

The *nuclear pores* with a diameter of 30-100 nm in no way make the nucleus an absolutely open organelle. Studies have shown that an experimental opening of the nuclear wall leads to the death of the cell. The appearance under the electron microscope suggests that the cell pores are equipped with a diaphragam, although the question then arises as to whether such diaphragms are formed only temporarily when the need arises. It has nevertheless been shown that substances of high molecular weight such as serum proteins do not diffuse into the interior of the nucleus; this indicates that a selection process is operating. Just as with any other biological membrane, the nuclear pores are only semi-permeable; even substances with a high molecular weight can pass in the other direction, from the interior of the nucleus into the cytoplasm.

In some amoebae (*Amoeba proteus*, *Pelomyxa* sp.), cylindrical ring-shaped thickenings with a honeycomb-like or hexagonal prismatic structure (annuli) are deposited at the pores near the karyoplasm. These structures are also present in *Gregarina rigida*. Similar structures are deposited on the outer nuclear membrane of *Endomoeba blattae* which live in the cockroach. A further possible way for the nucleus to emit substances into the cytoplasm is through *production of vesicles* by the nuclear wall. This has been demonstrated in the amoeba *Pelomyxa carolinensis*. The nuclear wall protrudes into the cytoplasm, and this bulge is then cut off to form a vesicle, or utricle, which later dissolves in the cytoplasm, thereby releasing its contents. It is particularly worthy of note that the Protozoa also have a direct connection between the endoplasmic reticulum and the perinuclear space.

The *endoplasmic reticulum* is a fine branched system of canals which extends through the whole cytoplasm. Under the electron microscope it appears tubular in form, or with fine vesicles and flattened sacs, but this variable appearance may be caused by the tubes being cut through in different planes. The membrane of the endoplasmic reticulum contains proteins, phospholipids and abundant RNA. Like the internal nuclear membrane, it operates in the selection of substances which are transported via the lumen of the endoplasmic reticulum from the karyoplasm to the cytoplasm or vice versa.

The canal system of the endoplasmic reticulum, surrounded by a unit membrane, and embedded in the *ground cytoplasm* or *hyaloplasm*, is directly connected with the perinuclear space. Two outwardly different forms of the reticulum can be distinguished: smooth endoplasmic reticulum is on the right of the nucleus and rough on the left. There is no essential difference between these two forms. The appearance of the rough reticulum, which has also in the past been called the *ergastoplasm*, derives from the smooth form by the deposition of granules with a diameter of 10-15 nm. These granules, lying close to the reticulum in the ground cytoplasm, are rich in RNA and are called *ribosomes*. They can also lie in the hyaloplasm between the canals of the endoplasmic reticulum, particularly in cells with intensive protein synthesis, since protein synthesis occurs at the ribosomes. Figure shows once again, this time in electron micrographs of *Trypanosoma gambiense*, the open connection of the endoplasmic reticulum with the perinuclear space. The outer nuclear membrane passes over into the endoplasmic reticulum without interruption. However, this anchoring of the cell nucleus to the endoplasmic reticulum in no way results in an inflexible siting of the nucleus in the protozoan cell. These internal structures are not rigid elements.

During locomotion of some species of amoeba, the nucleus can be seen moving around in all parts of the cytoplasm. The siting of the nucleus in the cell body is fixed only in the Protozoa which have internal strengthening provided by fibrillar supporting elements, as in many ciliates. This occurs especially in the Entodiniomorpha, which are very much strengthened. The possession of a nuclear wall, which determines the classification of the Protozoa as eukaryonts, is not universal at all times. The nuclear wall, including the outer membrane which belongs to the endoplasmic reticulum, dissolves during mitosis in most eukaryotic cells, and does not form again until after telophase. This process is directly dependent on the condition of the nucleus and is thus, like all events in the cell, induced by the nucleus. Nevertheless, in some Protozoa, the dissolution process can follow a different course. For example in the amoeba *Amoeba proteus*, only the internal prisms (annuli) dissolve.

Golgi Apparatus and Parabasal Apparatus

In Protozoa and in other eukaryotic cells, the *Golgi apparatus* can be distinguished from the endoplasmic reticulum as a membrane system with its own ultrastructure. Figure shows the characteristic appearance of the cell organnele; its basic elements are small lamellar sacs filled with liquid, compressed at the centre, wider at the ends, bent into bowl shapes and stacked in a pile. This collection of sacs, which are called *cisterns*, usually comprises about 5-8 individual cisterns but in Euglenidae comprises 20-30; the whole constitutes the 0.5-2.0 μm dictyosome. From both ends of this, vesicles of various sizes become detached as transport organelles. Also, large vacuoles can fuse into a group, as is shown in figure on the concave side of the dictyosome. The quantity and size of the Golgi complex differ according to individual protozoan species.

The Golgi apparatus is either randomly arranged in the cytoplasm or it may have a specific position in relation to the nucleus. This applies particularly to the Polymastigina, in which osmium fixation reveals the so-called *parabasal apparatus* even under light microscopy. The normal Golgi apparatus can be detected under light microscopy only after treatment with silver nitrate or osmium tetroxide, because of the resultant precipitation of metal. In contrast, in parabasal bodies are seen more easily and, under electron microscopy, show both the actual membranous Golgi complex and a well-developed parabasal filament, a transversely striated thread of protein running the whole length of the parabasal body. The outward appearance of the parabasal apparatus varies; in *Tritrichomonas caviae* it is long and stretched out, in *Devescovina striata* and *Macrotrichomonas pulchra* the parabasal body, which lies anteriorly around the axostyle, ends in a parabasal spiral.

The parabasal apparatus of *Lophomonas blattarum* lies round the nucleus like a crown of rays, while the nucleus of *Trichonympha collaris* is surrounded by large thread-like structures of the parabasal apparatus. The normal Golgi apparatus has been described so far for very many flagellates, for Sporozoa and for various amoebae. In ciliates, the Golgi apparatus is apparently not a constant cell organelle. Thus it has been demonstrated in *Tetrahymena pyriformis* only during conjugation and not in the growth phase. The membranes of the Golgi apparatus correspond in magnitude to the membranes of the endoplasmic reticulum and like the smooth endoplasmic reticulum, they are free from ribosomes. Biochemical studies on centrifuged isolated Golgi substance show that practically no RNA is present. Phospholipids have been demonstrated in the membrane, and there is abundant acid phosphatase, an enzyme of the lysosome granules, which are significant in the assimilation of foodstuffs that have been taken up.

A continuity between the dictyosomes of the Golgi apparatus, the endoplasmic reticulum and the lysosomes can clearly be seen in some Protozoa, such as the flagellate *Chrysochromulina chiton*. The form and function of the dictyosomes of the Golgi apparatus are subject to the control of the cell nucleus. If the nucleus of the amoeba *Amoeba proteus* is removed, the size and number of cisterns in a dictyosome decrease. The significance of the Golgi apparatus for the cell is not yet completely clear. Studies so far indicate that the Golgi apparatus assists in the secretion of proteins and polysaccharides, and in glycoprotein synthesis.

In many Protozoa the Golgi apparatus lies in the region of the contractile vacuole, so it may possibly be involved also in the regulation of the water balance in the cell, particularly when the concentration of secretions in the cell is accompanied by an output of water. The light-refractive structures which are deposited as ejectisomes in the gullet wall of the Cryptomonadina arise in this way, e.g. in *Cryptomonas ovata* and *Chilomonas paramecium*. The scales deposited on the cell membrane of some flagellates, such as the Chrysomonadina, are also produced in the Golgi apparatus and transported by means of vesicles to the surface of the cell.

Mitochondria and Kinetoplast

Mitochondria also belong to the membranous cell organelles in the cytoplasm. These are the thread granules—granules with a tendency to form threads by association together. They are cell organelles with great plasticity (they measure 0.5-1.0 μm in diameter and up to 10 μm in length) which are free to move about in the ground cytoplasm and congregate particularly in regions of the cell which have great metabolic activity. They also tend to accumulate alongside the nucleus, the endoplasmic reticulum or the cell membrane. Vital staining with Janus green B is specific, and treatment with tetrazolium salts gives a very clear picture through the production of formazan, which gives a red-blue colour on account of the enzyme content of the mitochondria. After the mitochondria have been concentrated in the ultra-centrifuge, it is possible to identify them as the site of biological oxidations; they are of central significance in the whole oxidative metabolism of the cell, and especially in cell respiration.

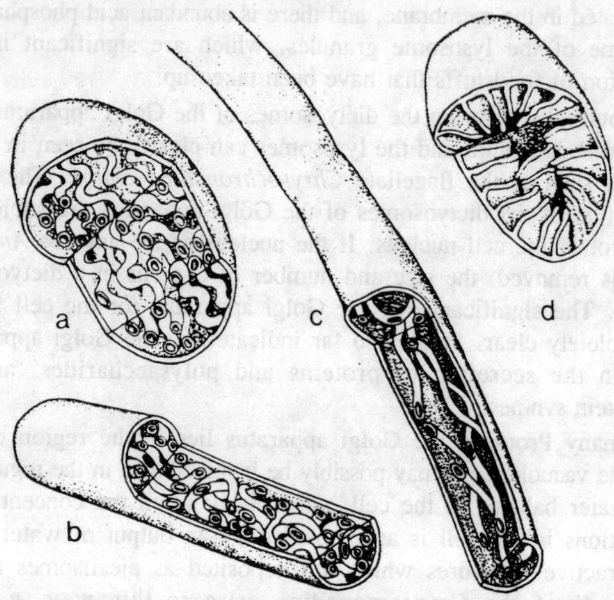

Fig. 3.7. Mitochondria from 4 protozoan species (diagrammatic): (a) from Pelomyxa carolinensis (amoeba); (b) Paramecium caudatum (ciliate); (c) Trypanosoma lewisi (flagellate); (d) Euglena spirogyra (flagellate).

The uniform membranous structural basis of all mitochondria was first shown by electron microscopy. Like the cell nucleus, the mitochondria are surrounded by two unit membranes which are separated by a space of about 7 nm. The smooth outer membrane constitutes the boundary with the cytoplasm, while the inner membrane produces a marked increase in surface area by means of numerous protrusions into the interior of the mitochondria. In the Protozoa, this increase in area leads in most cases to the formation of sinuous tubular processes which run through the internal mitochondrial cytoplasm, the matrix. This form of mitochondrion is known as *tubuli type*. In the Euglenoidina and some other flagellates the inner membranes of the mitochondria produce septa which project into the matrix and, depending on whether they are flat or sac-like, are called *cristae* or *sacculi*. The mitochondria also contain DNA, which differs, however, from the nuclear DNA in having a more homogeneous base composition and in lacking histones.

In this respect this mitochondrial DNA resembles bacterial DNA. The greater homogeneity of the mitochondrial DNA in comparison with the nuclear DNA derives from the smaller proportion of guanine and cytosine which, for example, in *Tetrahymena pyriformis* amounts to only 25%. In the ciliate *Tetrahymena* it can also be shown that multiplication of mitochondrial DNA does not proceed in synchrony with that of the nuclear DNA. The multiplication of the mitochondria themselves is effected by division, as light microscopy shows. However, quantitative considerations lead to the opinion that the DNA in the mitochondria cannot provide sufficient genetic information for the production of new daughter mitochondria. Dependence on the nucleus must thus be assumed for the mitochondria too, in respect of their division. It has already been said that the mitochondria are at the centre of cell respiration and oxidative processes.

The number and structure of the mitochondria give the first indication of the way in which the energy transformation reactions proceed in the Protozoa. *Tetrahymena* has 600-800 mitochondria. This number increases until it is doubled just before the next division. The erythrocytic trophozoite of mammalian plasmodia has only one body with a double membrane which can be regarded as the equivalent of a mitochondrion. The reactions of energy metabolism are determined by the type and amount of the substrate available to the Protozoa as fuel, as well as by the partial pressure of oxygen and carbon dioxide. It is therefore understandable in Protozoa which always live anaerobically (e.g. the intestinal amoeba *Entamoeba histolytica*; the amoeba *Pelomyxa palustris* which lives in stagnant marsh water; the flagellates of termites; and the ciliates in the rumen of ruminants) mitochondria are absent or, at least, the typical internal structure of the mitochondria cannot be detected.

In Protozoa which can alternate between the aerobic and the anaerobic way of life, the transition to anaerobic life, with the loss of the oxidative citric acid cycle, is accompanied by the disappearance of the mitochondrial tubuli. The *kinetoplast* of the Trypanosomatidae and Bodonidae, which both belong to the Protomonadina, can be regarded as a special form of mitochondrion. The kinetoplast of the Trypanosomatidae is not constant in size and shape, but has an average size of about 1 μm and always lies in the region of the basal body at the base of the flagellum. In a Giemsa preparation the kinetoplast is stained in the same way as the nucleus. With the Feulgen reaction the kinetoplast proves to be Feulgen-positive and therefore must contain a certain amount of DNA. Electron microscopy shows that the DNA-containing kinetoplast is only a special constituent of a sac-shaped mitochondrion. Only the section of the mitochondrion at the top of the figure has the appearance of a kinetoplast and contains the DNA which is easily detectable.

The DNA of the kinetoplast, like that of the normal mitochondria, contains no histone, in contrast to the nuclear DNA. It can be distinguished from that of the nucleus as the so-called satellite-DNA; it does not produce chromosomal structures, and in the purified state consists of small circular molecules with an individual size of around 0.45 μm. In this, the DNA of the kinetoplast is the same as the general mitochondrial DNA. The action of various substances on living Trypanosomatidae renders the kinetoplast Feulgen-negative, and the fibrillar DNA is no longer detectable. Such individuals with atypical kinetoplasts are called *dyskinetoplastic*.

In trypanosomes of the African *Trypanosoma brucei* group, which also includes *T. gambiense*, the causal agent of sleeping sickness, the development in the blood of the vertebrate host is not impaired by the absence of a normal kinetoplast, although such dyskinetoplastic trypanosomes are no longer viable in the insect vector; this is connected in particular with the effect on the synthesis of respiratory enzymes. The multiplication of the kinetoplast is achieved by autonomous binary fission, in which division of the kinetoplast begins before division of the nucleus. In the developmental cycle of the Trypanosomatidae, a stage is usually included in which the kinetoplast lies in close proximity to the cell nucleus. Even structural connections between these two organelles have been seen on more than one occasion. The functional significance of this relative position, and whether exchange reactions are occurring here, is not yet clear.

Plastids

The plastids (chromatophores) can be differentiated according to their colour under the light microscope as green chloroplasts, chromoplasts of other colours, and colourless leucoplasts. The appearance of the plastids in the electron microscope shows a basic correspondence with that of the mitochondria. A matrix, the stroma, is separated from the cytoplasm by a double membrane. The inner delimiting membrane undergoes an increase in active surface area by the formation of lamella-like processes extending into the stroma, the stromatic lamellae; each lamella comprises densely packed compressed membrane sacs, the *thylakoids*. Instead of producing continuous lamellae, the thylakoids can in certain places produce textured striped cylindrical grains, the *grana*, which can be seen even under the light microscope and which, however, occur only rarely in Protozoa.

On the other hand, zones with a densely granulate or fibrillar structure, the *pyrenoids*, are more frequently encountered in the plastids of the flagellates; starch or paramylum grains tend to be deposited on these rather more than they are elsewhere. Our ideas are still very hypothetical concerning the molecular structure and molecular function of the thylakoids, which are composed of structural proteins, lipids and chlorophyll, and which carry enzyme systems. However, it appears as though the primary reactions occur in the membranes, and the secondary reactions on their delimiting surfaces and in the stroma. When the active plastids change into the proplastids which form in the dark, the chlorophyll content disappears and the stromatic lamellae composed of thylakoids regress.

Under the influence of light, the lamellae are again produced and photosynthesis recommences. It can be deduced from this that the lamellae represent the actual structural site for photosynthesis. Exposure of *Euglena gracilis* to culture temperatures of 32-35°C, to ultra-violet irradiation, or a treatment with streptomycin and other antibiotics, has led to the production of strains which are not only temporarily colourless but which remain permanently without chloroplasts. They have lost the capacity to restore the chloroplasts. Such strains can be cultured in media containing suitable nutrients for an unlimited length of time.

The plastids and proplastids are capable of division. They are autonomous auto reproductive cell organelles. They are distributed between the daughter cells at cell division and, like the mitochondria and kinetoplasts, the plastids contain DNA. It has been demonstrated in *Euglena* that the DNA in the chloroplasts differs from the nuclear DNA in its base composition. Hence the two types of DNA can be separated from one another in a density gradient. About 4% of the total DNA fraction of the cell is accounted for here by the DNA of the chloroplasts. As to the function of this DNA, this is so far mainly conjecture, although it is known to play a part in the extrakaryotic inheritance in the flagellate *Chlamydomonas rheinhardi*.

Pellicle

The *pellicle* forms the outer boundary of the protozoan body, and vital life processes are linked with it. The pellicle participates in ion exchange and generally in selective exchange of substances between the cell and its environment; it is also involved in the detection and selection of nourishment and its uptake by molecular transport, pinocytosis and phagocytosis, and also in the aelimination of food wastes. It is concerned in sexual processes, particularly in the selection of a partner and the ensuing mutual orientation; in cell division; in locomotion; and particularly, in a protective capacity, in cases of mechanical damage and attack by antibodies. A very wide range of pellicle forms is seen, as determined by subpellicular structures. Even under light microscopy, clear differences between the pellicles of different protozoan species can often be detected. While the simple protozoan body, e.g. that of most amoebae, is evidently surrounded by only a thin cyoplasmic skin, other definite formations can be seen.

In the flagellate *Euglena ehrenbergi* the cell surface appears to be spirally striped. The pellicle of the ciliate *Vorticella monilata* has wart-like thickenings. The surface of *Paramecium caudatum* has a pattern of angular zones surrounded by prominent ridges, and the cilia originate in the middle of the sunken areas. In contrast, light microscopy clearly shows that the cilia of the protociliate *Opalina ranarum* arise from the rib-like thickenings of the pellicle. These complex cells coats of the ciliates are also called the cortex. An accurate analysis of the structure of the pellicle is possible only by electron microscopy.

Studies of the intestinal amoeba *Entamoeba histolytica* show the pellicle to be a closed system of membranes, the *plasmalemma*. Two layers of about 2.5 nm thickness, which show high contrast under the electron microscope, are separated by a lumen of the same width. This arrangement of membranes is so universal that the plasmalemma is also called a *unit membrane*. Primarily lipids and proteins, but also carbohydrates, are involved in its structure. There are as yet no uniform data concerning the molecular structure of the lipoprotein membranes. There consists either lipopration complexes or of two protein layers separated by a lipoid layer. In *Trypanosoma* and *Leishmania* a layer of fibrils lies beneath the plasmalemma. In an electron microscope these *fibrils* are seen as tube-shaped hardened cytoplasmic structures, the

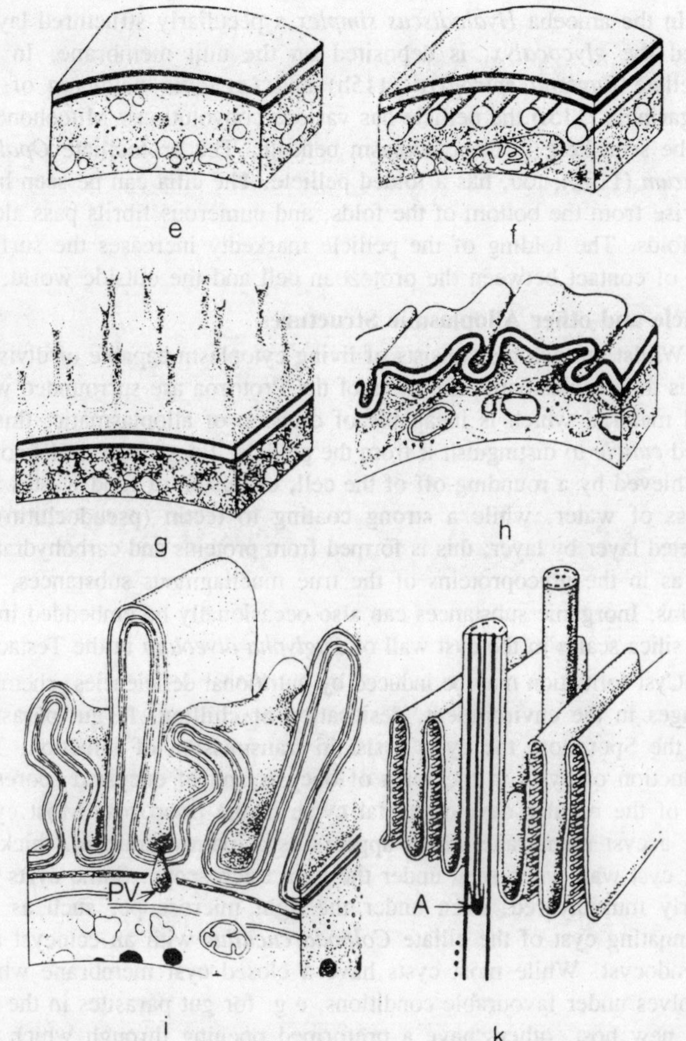

Fig. 3.8. Pellicular structure: (a) Entamoeba histolytica (intestinal amoeba); (b) Trypanosoma, Leishmania (flagellates); (c) Hyalodiscus simplex (amoeba); (d) Euglena spirogyra (flagellate); (e) Lecudina pellucida (gregarine) with pinocytosis vacuole PV; (f) Opalina ranarum (protociliate), A–axial grain of the cilium.

microtubules, with a diameter of about 24 μm. They assist in the general stabilization of the cell, linked with a varying ability to contract.

In the amoeba *Hyalodiscus simplex* a peculiarly structured layer, called the *glycocalyx*, is deposited on the unit membrane. In the flagellate *Euglena spirogyra* (115h) and *Lecudina pellucida* of the Gregarinida (115i) the pellicle has various convolutions. Mitochondria can be seen here in the cytoplasm beneath. The protociliate *Opalina ranarum* (115k), too, has a folded pellicle. The cilia can be seen here to arise from the bottom of the folds, and numerous fibrils pass along the folds. The folding of the pellicle markedly increases the surface area of contact between the protozoan cell and the outside world.

Cuticle and other Alloplasmic Structures

Whilst the pellicle consists of living cytoplasm capable of division and is thus euplasmatic, the cysts of the Protozoa are surrounded with dead material which is incapable of division of alloplasmatic; this is called *cuticle* to distinguish it from the pellicle. Encystment in Protozoa is achieved by a rounding-off of the cell, expulsion of food wastes and a loss of water, while a strong coating to tectin (pseudochitin) is secreted layer by layer; this is formed from proteins and carbohydrates, just as in the glycoproteins of the true mucilaginous substances, the mucins. Inorganic substances can also occasionally be embedded in it, e.g. silica scales in the cyst wall of *Euglypha alveolata* in the Testacea.

Cyst formation may be induced by nutritional deficiencies, chemical changes in the environment, desiccation or chilling. In gut parasites and the Sporozoa, the cysts assist in transmission of infection. The production of cysts in the form of oocysts and of encysted spores is part of the regular developmental cycle in the Sporozoa. Most cysts have a cyst membrane which appears as a mainly uniform, thick or thin, cyst wall when seen under the light microscope. Some cysts are clearly multilayered, even under the light microscope, such as the perennating cyst of the ciliate *Colpoda cucullus* with an ectocyst and an endocyst. While most cysts have a closed cyst membrane which dissolves under favourable conditions, e.g. for gut parasites in the gut of a new host, others have a preformed opening through which the vegetative stage can slip out. This opening can be closed by a plug in the resting condition. Differences in the formation of the cyst wall are also seen in the electron microscope. In the cyst of *Iodamoeba bütschlii* the plasmalemma is covered by a dense 0.05 μm-thick layer, with a 0.2 μm-thick outer layer of lower density.

The cyst of *Naegleria gruberi* is one of the cysts with a preformed pore, through which the vegetative form can emerge. The cyst wall consists here of an outer irregular coat, roughly 0.025 μm thick and

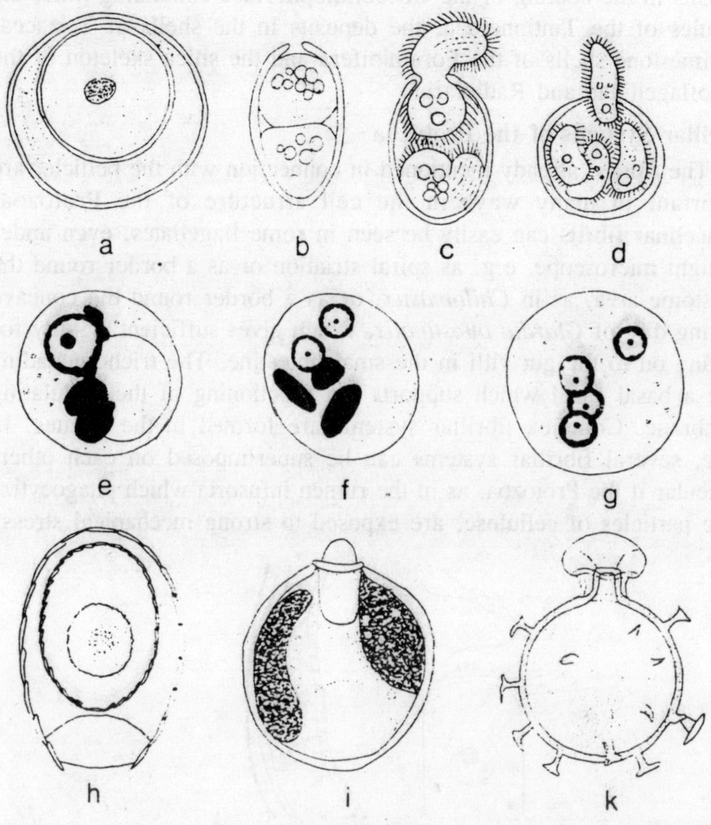

Fig. 3.9. Cysts.

an inner coat 0.2-0.4 μm thick, which are separated by a spongy layer. The pore, which is up to 0.6 μm across, is closed by a plug which has a membraneous plate. Division of the organism occurs in the cysts of some Protozoa. The ciliate *Colpoda cucullus* produces special division cysts for this. Division cysts also occur in the flagellate genus *Chlamydomonas*.

The perennating cysts of many intestinal amoebae are at the same time division cysts. In the amoeba *Entamoeba histolytica*, which causes dysentery in man, there is regularly a division into four and in the non-pathogenic *Entamoeba coli* even a division into eight. Many structures enveloping the vegetative stages belong among the alloplasmatic structures; these have already been referred to in the taxonomic section. They include the cellulose covering of most dinoflagellates (e.g. *Ceratum*) and many colony-forming flagellates, the deposits in the coating of the Coccolithophoridae containing lime, the capsules of the Tintinnoidea, the deposits in the shells of Testacea, the limestone shells of the Foraminifera and the silica skeleton of the Silicoflagellidae and Radiolaria.

Fibrillar Systems of the Protozoa

The fibrils, already mentioned in connection with the pellicle, are important in many ways in the cell structure of the Protozoa. Intracelluar fibrils can easily be seen in some flagellates, even under the light microscope, e.g. as spiral striation or as a border round the cytostome area, as in *Chilomastix*, or as a border round the concave sucking disc of *Giardia intestinalis*, which gives sufficient rigidity for sucking on to the gut villi in the small intestine. The trichomonadina have a basal fibril which supports the functioning of the undulating membrane. Complex fibrillar systems are formed in the ciliates. In these, several fibrillar systems can be superimposed on each other, particular if the Protozoa, as in the rumen infusoria which phagocytize large particles of cellulose, are exposed to strong mechanical stress.

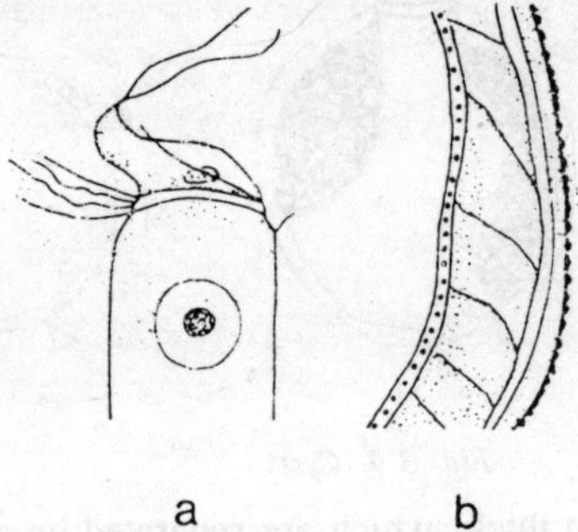

Fig. 3.10. Systems of fibrils. (a) Giardia intestinalis; (b) Eudiplodinium medium.

The strong trichites of the basket apparatus in the Gymnostomata are also examples of the fibrillar system of ciliates. The mechanical strength of the fibrils as they support the cell can be appreciably increased when they are arranged in bundles. Figure 3.11 shows an electron-microscopic view of a transverse section through the axopodium of a heliozoon with numerous fibrils associated together as the axis of the pseudopodium (*axoneme*). Bundles of fibrils are visible in transverse sections of the *axostyles* of the Polymastigina, the Trichomonadina and flagellates of termites when seen in the electron microscope. Figure shows the electron-microscopic view of *Tetrahymena pyriformis* as an example of the multidimensional arrangement of fibrils in ciliates.

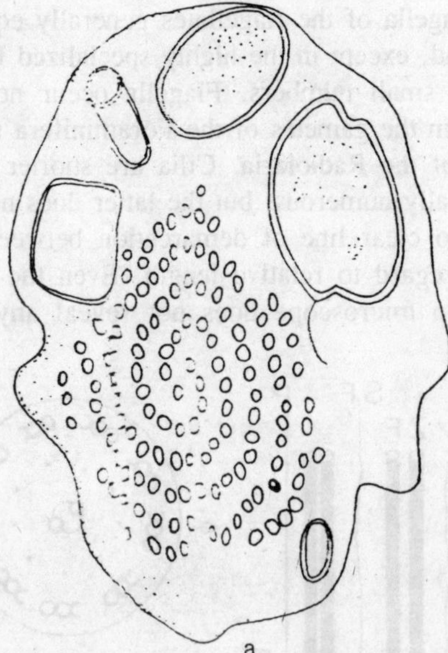

Fig. 3.11. Transverse section through an axopodium of the heliozoan Echinosphaerium nucleofilum; at the periphery is the cytoplasm of the pseudopodium; in the centre is the axis (axoneme) formed from numerous fibrils of the microtubuli type.

While the fibrils of the axostyle, axoneme and basket apparatus may have an exclusively static significance, other fibrils are characterized by a strong contractibility whose effect can also be enhanced by the bundling together of the fibrils. These fibrils are called *myonemes*. They are present, for example, at the place where the skeletal needles of the Acantharia emerge, but are particularly striking and widespread in various ciliates.

Among the heterotrichous ciliates, the genera *Spirostomum* and *Stentor* have strong contractibility. The peripheral fibrils are here easily visible as striations, even under the light microscope. The Folliculinidae, Tintinnidia and Loricate Peritrichiadae can draw themselves back into their capsules by means of the myonemes. But the aloricate peritrichous Vorticellidae are also largely contractile; here, the stalks suddenly contract spirally or in zig-zag form, and then slowly extend again. This reaction is controlled by the stalk muscles, i.e. the myonemes, which can be seen in the stalk cytoplasm even under the light microscope, but especially under phase contrast.

Locomotory Organelles

The locomotory organelles aid systematic classification; they include the flagella and cilia, which also belong to the fibrillar systems. Both appear as fine thread-like oscillating organelles under the light microscope. The flagella of the flagellates generally equal or exceed the cell in length and, except in the highly specialized Polymastigina, they occur only in small numbers. Flagella occur not only in the flagellates but also in the gametes of the Foraminifera and Sporozoa, and the swarmers of the Radiolaria. Cilia are shorter in relation to the cell and are usually numerous, but the latter does not hold for all ciliates. There is no clear line of demarcation between a flagellum and a cilium with regard to relative length. Even the fine structure seen in the electron microscope does not reveal any fundamental

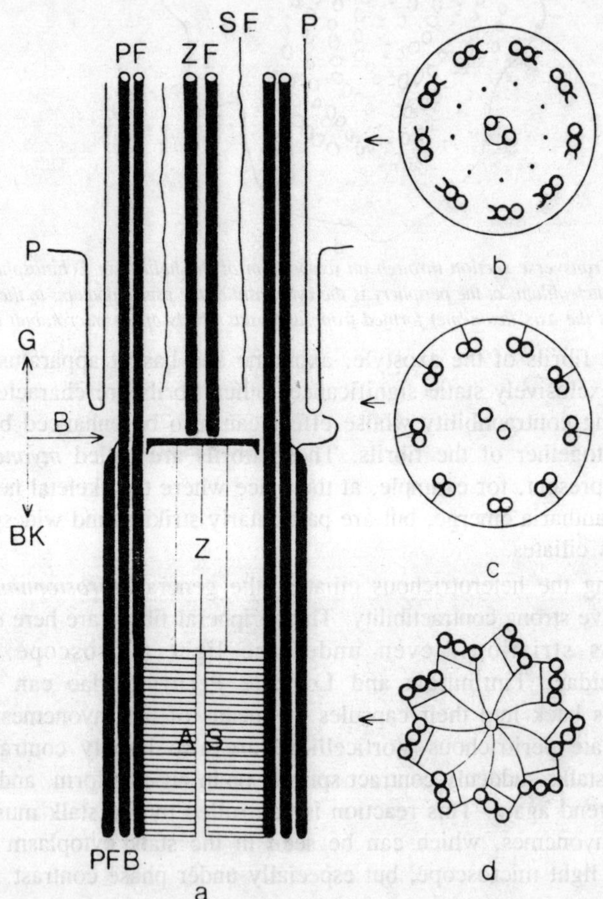

Fig. 3.12. Diagram of the structure of a flagellum a longitudinal section, b-d transverse section.

difference between flagella and cilia. Both have the same structural arrangement, which can be seen in figure 2.16 and which applies for all flagella and cilia of other eukaryotes.

The free portion of the flagellum, the shaft is surrounded, with its ground cytoplasm, by the pellicle. The effective fibrillar system lies embedded in the ground cytoplasm; there are two central longitudinal fibrils and nine pairs of peripheral double-fibrils in which one fibril of each pair is equipped with two lateral processes, the arms, as can be seen in the transverse section. Each of these 20 longitudinal fibrils of the microtubule type in the flagellum has a diameter of about 15 nm. Between the 2 central fibrils and the 9 pairs of peripheral fibrils are an additional 9, markedly thinner, simple secondary fibrils. The two central fibrils end basally at the basal plate, which can be developed as an axial grain (axosome) in ciliates. Below the basal plate, the flagellar shaft changes into the basal body, in which there are no central fibrils.

The basal body of ciliates is also often called a *kinetosome*. The nine peripheral pairs of fibrils are now usually strengthened by an additional fibril. These fibrils are arranged obliquely in the cross-section of the basal body, and the fibrillar arms are no longer present. In the centre of the basal body underneath the basal plate lies the cylinder. Sections from lower down can include spoke-like structures radiating from the middle. There can be variations in the structure of the flagellar shaft, and especially in the basal body in different Protozoa. For example, fibrillar anchorage granules can occur peripherally in the portion of sunken flagellum above the basal plate. The basal body is open to the cytoplasm. It is retained even in the transient amastigote stage of the flagellates, and the new flagellum is formed there.

Divisions of the basal body have never yet been observed. It is assumed that the site of a new basal body is determined by those already present, especially since it is always just near an existing basal body. Various surface structures can be seen in the flagella of the flagellates, even under the light microscope, after specific staining or in fresh preparations in a dark field. Whilst the thread-like flagella are uniformly smooth up to the end of the shaft, some flagellate species have flagella with hair-like processes, the *mastigonemes*. Flagella of this type are known as ciliary flagella or *Flimmergeissel*. The arrangement of the ciliary structures (flimmer) is typical for each systematic group. In the Euglenoidina the flagellum has a single line of ciliary structure wound spirally round it, and in the Chrysomonadina they occur all over the flagellum. If a member of the Chrysomonadina has another flagellum, this is a thread flagellum.

A further variation in flagellar structure is the lash flagellum, in which the peripheral layer of the flagellar shaft is shorter than the central core. In flagellates with several flagella, the size and arrangement of the flagella can also vary. When they differ in size, the longer is called the *main flagellum* and the smaller the *subsidiary flagellum*. Flagella which are directed backwards are called *trailing flagella* in contrast to the *swimming flagella*. For the dinoflagellates, we refer to a *longitudinal flagellum* and a *transverse* or *girdle flagellum*. The trailing flagellum becomes a *ribbon flagellum* when it is equipped with a wide margin of cytoplasm.

When a flagellum running the length of the cell body is joined to the cell surface it forms an *undulating membrane*. This is formed from a forward-pointing swimming flagellum in *Trypanosoma* and from a trailing flagellum in the Trichomonadina. The direction of movement of the flagellate is determined by the varying type of motion of the flagella. Figure shows this using the example of a forward-pointing swimming flagellum. A rapid down-stroke of the mainly outstretched flagellum, followed by a re-erection of the flagellum when bent, results in a movement of the cell forwards. Undulating wave-like beating to and fro results in a movement of the cell backwards and undulating to one side results in a movement of the cell sideways.

The wave-like motion of the flagellum is reinforced by the undulating membranes. Adenosine triphosphate (ATP) participates in the generation of energy necessary for the motion of the flagella or cilia. The arms of the nine outer peripheral fibrils are probably the site of adenosine triphosphatase activity. An adenosine triphosphatase known as *dynein* has been isolated from the ciliate *Tetrahymena pyriformis*. The outer membrane is apparently not necessary for motion; cilia lacking a membrane can be induced to resume beating by the application of ATP. On the other hand, in the flagellate *Chlamydomonas reinhardi*, the removal of the two central longitudinal fibrils by means of proflavin is linked with a loss of flagellar activity.

Despite these isolated findings, a thoroughly satisfactory elucidation of the motion of flagella and cilia on a molecular basis is not yet possible. The type of motion of cilia basically corresponds with that of flagella. In cilia, too, a forward thrust to the cell is achieved by a down-stroke in a mainly extended condition, followed by a re-erection of the flexed cilium. If the ciliature is over-all, as in the holotrichous

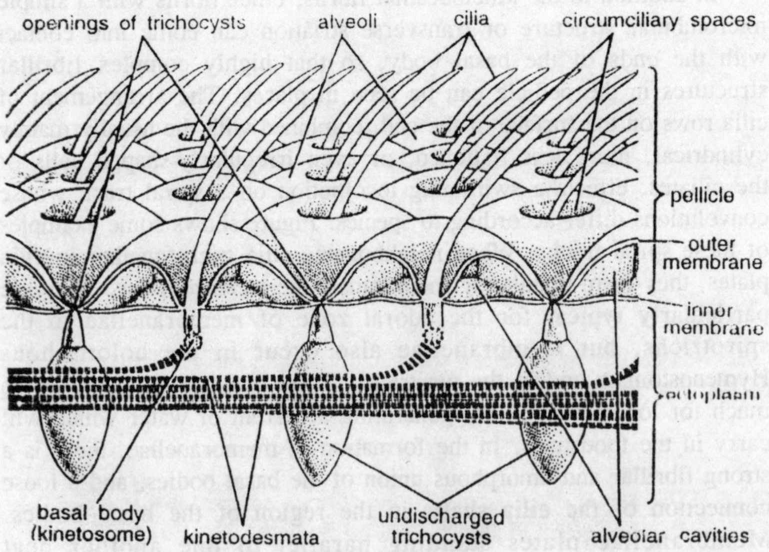

Fig. 3.13. Surface of Paramecium.

ciliates, cilia are arranged in parallel rows. Within each row, the continuity is broken only by the suture of the cilia rows.

The cilia which stand one behind another in a row beat metachronously, one after another, whilst each cilium beats synchronously with those at the same level in the neighbouring parallel rows. Thus the extension and flexion of the cilia during the course of locomotion leads to a rhythmical wave pattern like a cornfield in the wind. The arrangement of rows of cilia in the pellicle is determined by their anchorage in the surface structure. Figure has already shown, for *Paramecium caudatum*, the position of the cilia in the middle of the area between the ridges of the pellicle. A surface view of *P. caudatum* is shown, where rows of cilia are interrupted only by the ventral suture in which the entrance to the vestibulum is also situated. The cilia, anchored with their basal bodies in the centre of the cilia fields, are connected with the fibrillar systems of the pellicle. Fibrils come out laterally from the basal body; these are transversely striated when seen in the electron microscope and are called the *kinetodesmal fibrils*. All of these kinetodesmal fibrils run forwards in *Paramecium*. They end at the fourth or fifth cilium-field, but overlap the kinetodesmal fibrils of the neighbouring cilia, so that all of them together look like longitudinal fibrils. It is presumed that these fibrils can assist in the co-ordination of the beating of the cilia.

In addition to the kinetodesmal fibrils, other fibrils with a simple microtubular structure or transverse striation can come into contact with the ends of the basal body, so that highly complex fibrillar structures in the pellicle can be seen in places. The arrangement of cilia rows on the surface of the cell, combined with the usually mainly cylindrical, more less flattened, or even irregularly-shaped cells of the ciliates, effects a swimming locomotion on a spiral track whose convolutions differ according to species. Figure shows some examples of these spiral tracks. When neighbouring cilia stick together in cilia plates, they form undulating membranes, the *membranellae*. They are particularly typical for the adoral zone of membranellae in the spirotrichs, but membranellae also occur in the holotrichous Hymenostomata and in the peritrichs. These are usually used not so much for locomotion as for generating a current of water which will carry in the foodstuffs. In the formation of membranellae, there is a strong fibrillar and amorphous union of the basal bodies, and a loose connection of the cilia shafts in the region of the basal bodies. Membranellae plates standing parallel to one another beat metachronously like the cilia of one cilia row.

While fusion of the cilia to form membranellae increases the force of the beating motion, association of the cilia into styloid bundles increases their mechanical stability and carrying capacity; these are called *cirri* and function primarily as ambulatory organelles. They are characteristic of the hypotrichous ciliates. Their movement is by no means so co-ordinated as that of normal cilia or of membranellae; they tend to move more independently of one another. Cytoplasmic streaming, with the participation of the pellicle, can lead to locomotion. The resultant movements of the cell in free-swimming forms do not, however, invariably bring about a change of locality. In the Euglenodina, cytoplasmic bulges pass backwards over the cell body like waves. This process is called *metaboly*. A forwards movement of the whole cell results only when contact is made with a firm substratum. Flowing locomotion is specific particularly to the amoebae. This type of locomotion is best observed with an inverted microscope, i.e. viewed from below.

The active cytoplasmic streaming flows in the cell interior as an axial current from the back to the front. The forwards flow of the central cytoplasm causes the peripheral cytoplasm to flow passively to the posterior end. This is linked with a transition from endoplasm to ectoplasm at the front of the amoeba, and from ectoplasm to endoplasm

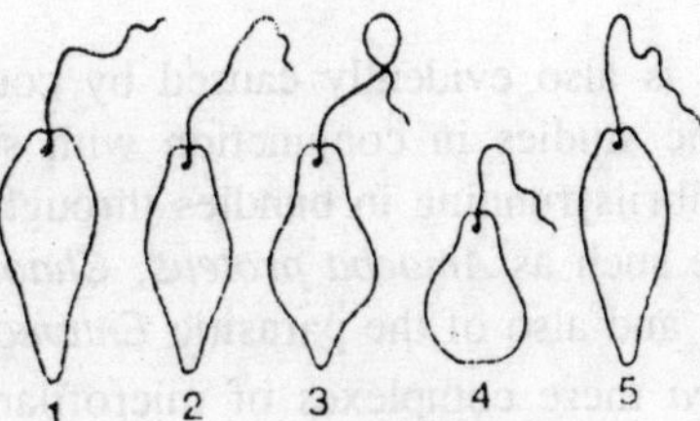

Fig. 3.14. Euglenoid movement.

at the back. This process is accompanied by morphological changes in the cytoplasmic structure. Under the light microscope the ectoplasm appears hyaline, glassy-transparent, whilst the endoplasm is granular and vacuolated. Figure shows the electron-microscope differences in structure between the two cytoplasmic zones in the free-living amoeba *Hyalodiscus simplex*; on the left is the denser fine-structure of the ectoplasm, and in the middle and on the right is the more diffuse endoplasm with numerous pinocytotic vesicles which allow the typical cell-wall structure to be seen at the places marked with arrows. Phagocytosis vacuoles and mitochondria can also be embedded here.

In most amoebae the ectoplasm produces branches of a particular formation, the *pseudopodia*. These mobile ectoplasmic processes can be more or less broad *lobopodia*, rounded at the front or slender *filopodia*, pointed at the front. While the filopodia consist of ectoplasm throughout, the lobopodia soon fill up inside with endoplasm which streams in. However, cytoplasmic streamings are possible not only into a pseudopodium but also in a backwards direction. These cause the pseudopodium to withdraw gradually back into the cell. Lobopodia occur not only in the naked amoebae but also in the Testacea and some flagellates; the same is true for the filopodia. The production of hernial-sac pseudopodia is a particular kind of lobopodium production. It occurs, for example, in *Entamoeba histolytica*. A local liquefaction causes the ectoplasm to swell out in scattered places to form sacs, until the endoplasm also starts to flows out here.

Close observation of *Amoeba proteus* has shown that the pseudopodia of some amoebae can occasionally be used as ambulatory organelles. Figure shows this type of stepping locomotion. Contractions are also evidently involved in the function of the pseudopodia as locomotory organelles; when the pseudopodia have found an attachment in front, the remainder of the cell body is pulled up after them. This is also true for the finely branched reticulate rhizopodia or reticulopodia of the Radiolaria and Foraminifera. However, the flowing motion of the amoebic cytoplasm is also evidently caused by contractile elements. Electron-microscopic studies in conjunction with suitable techniques have shown microfibrils running in bundles through the cytoplasm of free-living amoebae such as *Amoeba proteus*, *Chaos carolinensis* and *Acanthamoeba* spp. and also of the parasitic *Entamoeba histolytica*.

In *E. histolytica* these complexes of microfilaments are 3-4 μm long and 1-2 μm wide; the individual filaments themselves have a diameter of 2-6 nm. They are absent if there is no cytoplasmic streaming at the moment of fixing. It is assumed that the contraction first affects the forward-flowing cytoplasm, so that this temporarily assumes a solidified gel structure. In free-floating forms the rhizopodia of the Radiolaria and Foraminifera, like the axopodia of the Heliozoa, act as floating processes, corresponding to the floating spines of *Globigerina*.

The capacity of the Radiolaria to float is, however, also determined by their specific gravity. The carbon dioxide synthesized in the metabolism of the Radiolaria lowers the specific gravity of the cell at night by the production of vacuoles, so that the cell rises upwards. In daylight the carbon dioxide is used up by symbiotic zooxanthellae in assimilation processes; the Radiolaria, weighed down by their silica skeletons, sink down again. This vertical movement can amount to 200-350 m a day. Bubbles of gas has also been observed acting as organelles for buoyancy is some Testacea of the genera *Arcella* and *Diffugia*. The Gregarinida have a particular method of locomotion. A gliding motion can be seen under a normal light microscope, although actual organelles for locomotion cannot be discerned. Whereas it was formerly supposed that the Gregarinida secreted a slime which pushed them forwards as it gushed out, it now seems more probable that locomotion is achieved by means of very find contractions of the pellicle. Numerous fibrils are embedded in the pellicle and these could account for this type of locomotion. Undulating movements in the cell surface can occasionally be seen, even under the light microscope using time-lapse photography.

Food Uptake

The locomotory organelles are also employed in food uptake, by which the protozoan cell procures the raw material for structural processes and energy metabolism. Food uptake can proceed by *permeation* through the pellicle, by pinocytosis and by phagocytosis. The capacity for *permeation* has already been alluded to in the section dealing with the pellicle. As a unit membrane, the pellicle carries out a selective exchange of dissolved substances, in accordance with its chemical and physical constitution; in the course of this, smaller molecules and ions are selectively taken up. The simplest form of permeation is diffusion, a passive uptake of substances. In contrast to this, energy is needed for the uptake of matter by permeation in most cases.

The molecules which are to be transported are firstly bound to specific sites on the cell surface, then passed through the membrane, and finally freed on the inner side of the membrane. Substances known as *carriers* are involved in this type of transport. These are protein compounds called *permeases*. This function can apparently only be fulfilled with molecules up to a certain size which has not yet been determined. Larger particles of food are taken up in a dissolved state by pinocytosis and as solid food by phagocytosis. For *pinocytosis*, or "cell drinking", the cell produces invaginations of various forms into the cytoplasm. In *Entamoeba histolytica*, rounded vacuoles arise first and then migrate into the cell interior.

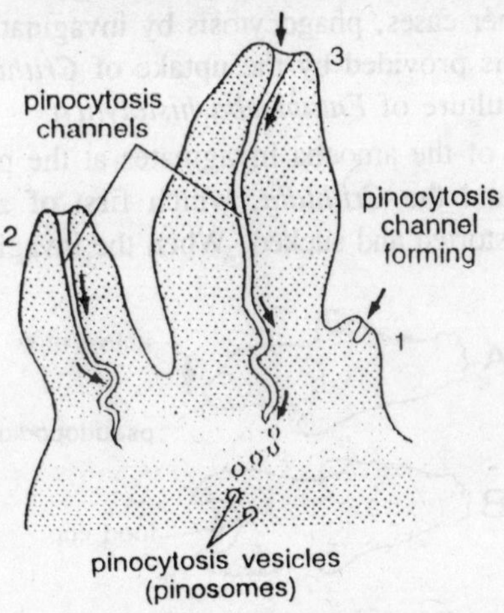

Fig. 3.15. Pinocytosis in Amoeba.

In the free-living amoeba *Amoeba proteus* small vesicles are cut off in the cell interior from the ends of thin funnel-shaped invaginations. These vesicles, enclosed in a membrane, which arise in pinocytosis can often be seen in large numbers in the endoplasm of amoebae; they can be recognized by the unit membranes which surround them. Pinocytosis can be stimulated by factors external to the amoebae, particularly by proteins and aminoacids; in the course of this, other dissolved substances which themselves do not induce pinocytosis, e.g. carbohydrates, are also taken up. The extent of pinocytotic activity is, however, also dependent upon the physiological condition of the amoeba, from which it can be inferred that pinocytosis is subject to certain control mechanisms. Although amoebae can produce pinocytotic vesicles anywhere on the cell surface, many Protozoa can obviously only do so at certain places on the surface.

When ferritin has been added to the culture medium of the frog trypanosome *Trypanosoma mega*, electron microscopy shows that the ferritin particles only enter the cell at certain places on the flagellar sac near the base of the flagellum. In *Opalina*, pinocytosis occurs at the base between the folds of the pellicle. This is also true for the gregarine *Lecudina pellucida*. The uptake of solid food by *phagocytosis* is easy to see, even under the light microscope, and has therefore been studied in detail for many years. In amoebae which produce vigorous extended lobopodia, two pseudopodia can flow round a food particle and then unite with one another distally, so that a food vacuole is formed. In other cases, phagocytosis by invagination may occur. An example of this is provided by the uptake of *Crithidia* (flagellates) in a bacteria-free culture of *Entamoeba histolytica*.

The pellicle of the amoeba invaginates at the point of contact, as in pinocytosis, and the *Crithidia*, which first of all usually become immobile, are distorted and sucked. When the invagination canal closes

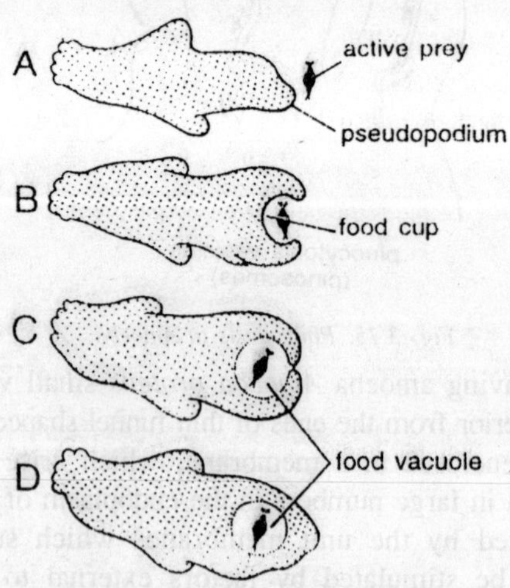

Fig. 3.16. Phagocyton amoeba.

distally, a food vacuole is formed within the cell of the amoeba. Electron-microscopic studies of this process show that a finely reticulate cytoplasm forms around the invagination canal; this is called the *phagocollar* and is evidently involved in the mechanism of invagination. If a food vacuole is formed, the phagocollar remains present only for a while. Digestion within the food vacuole begins after the phagocollar has dispersed. The vacuole, like the pinocytotic vesicles, is surrounded by the unit membrane of the pellicle. The process of phagocytosis, which has been studied particularly in the naked amoebae, also occurs in a corresponding form in the Testacea which have shells, and in many heterotrophic flagellates. Material well suited to the study of this is a culture of *Trichomonas vaginalis* after the addition of powdered rice starch, although the swimming form of this trichomonad does not produce pseudopodia.

In the Hypermastigidae, which are related to the Trichomonadidae, and which live in xylophagous termites, relatively large particles of wood are taken up at the posterior end of the cell. Free-living amoebae, too, can phagocytoze and enclose in vacuoles large food particles such as algal filaments which are considerably longer than the body of the amoeba. In the remaining Rhizopoda, some Testacea, the Heliozoa, Radiolaria and Foraminifera, food it trapped by rhizopodia and axopodia. After contact with the pseudopodia, the animal prey are usually rendered immobile by a toxic action, as are the *Crithidia* during uptake by *Entamoeba histolytica*. The trapped prey are subsequently transported to the cell interior by cytoplasmic streaming or contraction of the cytoplasm. Occasionally the prey begin to dissolve during transport along the pseudopodium. When larger objects are caught, the enveloping cytoplasmic mass can be augmented by the involvement of several pseudopodia together. In comparison with the axopodia and rhizopodia, the filopodia of the amoebae and Testacea function mainly in the capture of food and less as organelles of locomotion. Incorporation of food particles by cytoplasmic streaming also occurs in the Craspedomonadidae, of the Choanoflagellata. Here the flagellum is surrounded by a cytoplasmic collar, the *collare*, to which the food sticks to the carried away into the cell.

Gullet-shaped cytoplasmic differentiations, which occupy a specific site in the cell are called *cytostomes* or cell mouths. They are present in some flagellates such as in the genus *Chilomastix*, in which a shorter trailing flagellum at the anterior end washes the food into the cytostome. Electron microscopy has revealed permanent cytostome structures in

Sporozoa as well, so that permeation is not , as was originally supposed, the only process which occurs here. In the intra-erythrocytic phase of malaria parasites, the permanent cytostome structure has two concentric rings, so that a cytostome pore is present. Invagination of the cell membrane results in the uptake of erythrocytic cytoplasm at this point with the formation of food vacuoles. The cytostome structures of ciliates are particularly distinct, and they show a wide variation in morphology and function. In the "*whirlers*" the cytostome structure begins peripherally with the peristome cavity. If the peristome is narrowed below into a funnel or tube, this section is called the *vestibulum* or *cytopharynx*. The wall of the peristome and vestibulum is part of the cell surface and is ciliated. The cilia, often strongly developed, are so arranged in the region of the peristome that their beating produces a vortex of water which washes the food particles into the cytostome, where they are caught; secretion of slime here can contribute to the trapping of the prey. When the cilia adhere together to form membranellae or membranes, the effect of the peristome ciliature is enhanced. The peritrichous ciliates have a particularly highly developed peristome field; the whole expanse of the flattened front end is surrounded by a substantial spiral of cilia; this is an anticlockwise two-row spiral whose outer row extends as an undulating membrane into the vestibulum. In the chonotrichous ciliates, the sole cell cilia are those present within the spiral funnel. In these attached forms, the cilia function solely in producing the vortex for the uptake of food. When food which has been washed in collects at the bottom of the cytostome a food vacuole is cut off and travels into the cell, so that the food may be digested.

While the ciliates of the "whirler" type phagocytoze relatively small particles of food, the '*trappers*' are hunters which seize large prey and take them up through an elastic cell mouth. *Didinium nasutum* provides an example of this. The exposed cytostome structure has a protruding oral cone which is furnished at the front with *pexicysts*, spiky organelles for attachment. *Didinium nasutum* is specialized for preying on *Paramecium*. When it makes contact with *Paramecium* it holds it fast by means of the pexicysts. This stimulates the *Paramecium* to defend itself by ejecting its trichocysts. The *Didinium* is able to dodge this resistance, without releasing the prey, through the adoption by the oral cone, at this stage, of a tubular form. Meanwhile, a paralysing and lethal toxin is injected into the *Paramecium* by toxicysts which surround the pexicysts. The mouth opening of *Didinium* subsequently widens, grasps the prey and sucks it in. Like *Didinium nasutum*, other "trappers" are also usually specialized to prey on particular species. However, even dead cell matter can also serve as food, for example in *Coleps hirtus*. The deposition of solidified rods, the *trichites*, can lead to the production of a basket apparatus in the cytostome wall. This stiffening enables the ciliates to nip off algal filaments which serve as food. On the strength of this, Gymnostomata can be differentiated into *Cyrtophorina* with a basket apparatus, and *Rhabdophorina* usually with toxicysts.

The Suctoria can apparently be derived from the predatory Prostomata; they are still holotrichous in ciliature in their immature form as swarmers, but feed in a predatory manner by means of sucking tentacles after they have become attached and are aciliate. The tentacles of many Suctoria are used for both trapping and sucking. Haptocysts, short rod-shaped structures with a balloon-like thickening at the end, are situated in the distal enlargement of the tentacle, the tentacle head; when contact is made with the prey, its pellicle is pierced by these haptocysts which hold the ciliate fast so that it can be used for food. Here, too, a toxic substance can paralyse the ciliate victim. Fibrils are arranged peripherally within the tentacles, and the remaining tube-shaped lumen is filled with cytoplasm. The uptake of the cytoplasm from the prey during the act of sucking proceeds via this lumen. In some Suctoria two types of tentacle are present. For example, *Ephelota gemmipara* has long pointed trapping tentacles and shorter sucking tentacles. The trapping tentacles, which can move to the side, also have central fibrils of the microtubular type. After the prey has been captured, it is transferred to the shorter sucking tentacles.

Digestion

The nutrient taken up by the protozoan cell is enclosed by a vesicular constriction of the pellicle, in a food vacuole as soon as it gets in to the cytoplasm. The wall of the vacuole is a unit membrane, whether pinocytosis or phagocytosis has taken place. Enzymatic digestion begins in the food vacuole. After a certain amount of food has been washed in, a food vacuole forms at the end of the cytostome. In order to set digestion in motion, digestive enzymes must reach the food vacuole. These enzymes are hydrolases which split macromolecules up into fragments of low molecular weight. In ciliates these enzymes are produced in the region of the rough endoplasmic reticulum, and in other cases in association with the plastic apparatus. The enzymes first appear as 'granules' in the form of small vesicles enclosed in a membrane; these are called *lysosomes*.

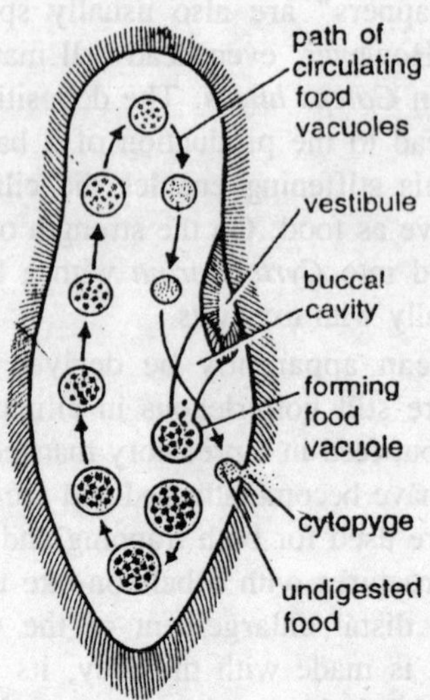

Fig. 3.17. Digestion in Paramecium.

The lysosomes are about 20-40nm. In the electron microscope they show no characteristic internal structures. They can be concentrated from a cytoplasmic suspension in an ultracentrifuge, according to their specific gravity, in a density gradient. They chiefly contain hydrolases such as α-glucosidase, acid phosphatase, desoxyribonuclease, ribonuclease, protease, α-amylase and β-N-acetylglucosaminidase. Acid phosphatase is the main enzyme of the lysosome. The production of a food vacuole induces rapid migration of lysosomes to the food vacuole, during which the lysosomes to some extent reabsorb the liquid taken in by pinocytosis. Only when the enzymes from the lysosome have invaded the food vacuole can the breakdown of the food begin. The products of breakdown which are valuable to the cell, the actual nutrients used for structural and energy metabolism, diffuse through the membrane of the digestive vacuole and form small cytoplasmic nutrient vacuoles in the surrounding area of cytoplasm; these then migrate to the site where they will be used.

The vacuole, previously a food vacuole and a digestive vacuole, thus becomes a defaecation vacuole, which delivers the indigestible remains of the food to the outside of the cell as balls of excrement. The path of the food vacuoles is not closely defined in the simply organized Protozoa such as the amoebae; it can merely be said that the food vacuoles always occur in the endoplasm. In the more highly organized ciliates, the food vacuole follows a quite definite path, the *cyclosis*. The food collected at the bottom of the vestibulum leads to the detachment of the food vacuole whose cyclosis runs according to the arrowed line. In the lower dotted section, the food vacuole temporarily shows an acid reaction with a pH value of around 4.5. The defaecation of the undigested refuse occurs anteriorly into the vestibulum.

The food vacuole of *Paramecium caudatum* has an oval circulation route. However, depending upon the type of food taken up, a large cyclosis which runs through the whole cell can annex this small cyclosis. The formation and path of the food vacuoles can easily be seen by feeding with carmine granules. Vital staining with neutral red shows a change in the pH value of the vacuoles in *Paramecium* too. Because of extraction of water, the vacuoles become smaller and show an increasingly acid reaction with a pH value of about 4.0; in this case the enzyme carriers, which are first detected on the periphery of the vacuoles, penetrate into the vacuole. The food vacuole then grows larger again by the uptake of water, and its contents change over to an alkaline reaction. The neutral red staining causes the acid vacuoles to appear red; the alkaline, yellow. This alternation of pH values corresponds to the conditions in the digestive tract of higher animals and is necessary, since the effective optimum of the various digestive enzymes requires different acid or alkaline conditions.

An electron-microscopic view show the amoeba *Pelomyxa carolinensis* which shows the transfer of breakdown products obtained from the food by enzyme reactions; these products represent the actual nutrients used for cell metabolism. The amoeba has been fed on *Paramecium* and *Tetrahymena*. The food vacuole, surrounded by the vacuolar membrane, shows a largely homogeneous zone, sited peripherally within the membrane, containing the nutrients which have already been split up and which must reach the cytoplasm of the amoeba for further metabolism. This proceeds at first by means of diffusion through the vacuolar membrane; after this small nutrient vacuoles of the cytoplasm form on the outer side of the food vacuole. These carry the nutrients to the other regions of the cell. These cytoplasmic nutrient vacuoles differ from pinocytotic vacuoles in their small size and in their lack of a unit membrane.

Like the digestive processes, all subsequent processes of structural and energy metabolism are also dependent on the participation of specific enzymes. These can be identified by biochemical methods in extracts from concentrated Protozoa, but some can also be demonstrated in the protozoan cell itself by means of cytochemical reactions. In these, a specific substrate added to the cell is chemically altered by the catalytic ability of the enzyme to such an extent that it either becomes a pigment itself or else it unites with an admixed pigment. The appearance of *Entamoeba histolytica* in a bacteria-free culture with *Crithidia* provided as food. The figure represents the reaction on acid phosphatase on non-specific esterases, and on lipoyldehydrogenase (NAD-dependent dehydrogenase), in which the acid phosphatase is here the most abundant.

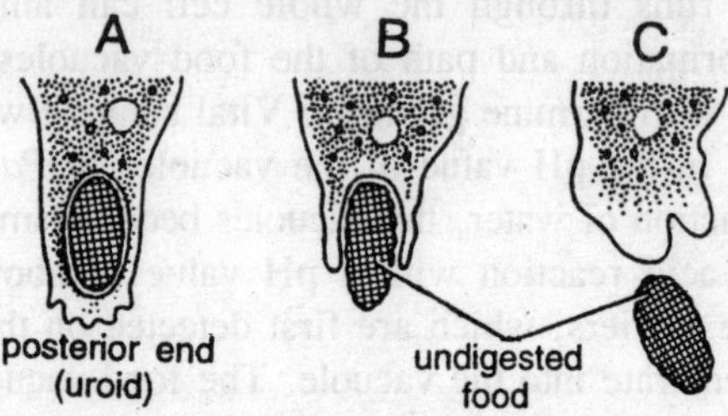

Fig. 3.18. Egestion in Amoeba.

The enzymes important for digestion are, as has already been described, prepared in the lysosomes already in the cytoplasm, which is protected from the action of the enzymes by the limiting membrane of the lysosomes. Experimental damage of this membrane by detergents, bile, salts, and alterations of the permeability of the membrane lead to the release of the enzymes and hence to plasmolysis. Under these unphysiological conditions, the lysosomal enzymes lead to the cell's autodestruction.

It has already been seen that the undigested remains in the food vacuole are discarded by evacuation to the outside. In the Rhizopoda, defaecation occurs at arbitrary sites on the cell surface. The vacuole, which has migrated to the pellicle, bursts and empties out its contents. In the ciliates, which have a fixed shape, there is a definite site for defaecation called the cell anus or *cytopyge*. The cytopyge either only appears during defaecation or else, when the pellicle is rigid, can be seen clearly in the interim as well, for example at the back of *Balantidium coli*, *Eudiplodinium maggii* and *Ostracodinium dentatum*. In some rumen infusoria the cytopyge can extend far into the interior of the organism as a cylindrical canal strengthened by fibrils; carbohydrate deposits make it particularly easy to see. However, defaecation does not invariably occur in the posterior part of the cell. It takes place in *Carchesium* into the vestibulum. In the Craspedomonadina, of the Choanoflagellata it occurs within the collar structure.

Reserve Substances

When food supplies are abundant, many Protozoa are able to store up reserve substances in their cells which can be used up when the food supply decreases. This is not important for the blood forms of the African trypanosomes which live in a medium with an excessive supply of glucose; it is therefore hardly surprising that they do not synthesize glycogen. If they are transferred to a glucose-free medium, they die within a few minutes. The ability to synthesize glycogen is vital for Protozoa which have to live under adverse conditions: hence the promastigote form of *Leishmania donovani* has glycogen sphericles, and the cysts of *Iodamoeba bütschlii* and *E. hystolytica* have glycogen vacuoles. The following energy-storing polysaccharides, which are all polymers of glucose, have been found in Protozoa: glycogen in Tetrahymena, Trichomonas, Prototheca and the above mentioned amoeba and Gregarinida; starch in Polytomella, Polytoma and Chilomonas; amylopectin in the rumen infusoria and *Eimeria*; cellulose in Acanthamoeba; and a glycogen polymer with β-1-3-linkages in Ochromonas, Astasia and Euglena (= paramylum).

In addition there is leucosin in Chrysomonadina and Heterochloridina, and oils and fats in Chrysomonadina, Cryptomonadina, dinoflagellates and Chloromonadina. Sufficient reserve substances are necessary, particularly for the metabolism of cysts which, after all, no longer take up food for themselves. Thus the cyst of *Entamoeba histolytica* produces not only a glycogen vacuole but also chromatoid bodies which stain intensively with nuclear strains and which are used up again during the maturation of the cyst which involves nuclear divisions. These chromatoid bodies are aggregates of ribonucleoprotein which can already be seen by electron microscopy, during the stage preceding cyst formation, as finely distributed crystalloid-stratified complexes in the cytoplasm of the amoeba. While the polysaccharide deposits are usually finely distributed in the cytoplasm in the vegetative stages, they can lead to the formation of granules in the Gregarinida. They also tend to occur more at particular places in the cell.

Special glycogen vacuoles are produced in the cysts of the intestinal amoebae; they stain intensively with the addition of iodine or other glycogen stains. If the cyst is expelled with the faeces, the glycogen vacuole, like the chromatoid bodies of *E. histolytica*, is slowly used up. In the rumen infusoria, abundant amylopectin is deposited at certain places on fibrillar structures. In some genera there are attenuated fine-meshed honeycomb-like nets of fibrils, the so-called *skeletal plates*, in addition to the universal peripheral fibrillar systems. All these fibrillar systems are places where storage tends to take place. The cylindrical anal tube above the cytopyge also has intensive deposits of carbohydrates. Above right are the two skeletal plates containing reserve substances; below is the macronucleus which has been cut through. After it has been fed with an abundance of starch, the whole cell gives an intense polysaccharide staining reaction. 50-60 hours after feeding, all the amylopectin contained in the cell has been used up.

Fats and oils in droplet form also frequently occur as reserve substances in the heterotrophic Protozoa; these are mainly produced by carbohydrate metabolism, and occur in particularly large amounts in Radiolaria, in the dinoflagellate *Noctiluca miliaris*, and in *Eimeria gadi*, a coccidian occurring in the air bladder of the cod. Excessive accumulation of fats can lead to pathological conditions in Protozoa, just as in Metazoa; fatty degeneration has been observed, for example, in *Polytoma uvella* in the Phytomonadina.

In some Protozoa, excluding the ciliates, staining reveals inclusion which have a strong affinity for basic stains; in this case, a change of colour, a metachromasis, can occur. Treatment with ribo-nuclease has shown that compounds containing RNA are involved here. The inclusions are rapidly used up in culture media lacking phosphorus, and the flagellates encyst or perish. These metachromatic deposits are called *volutin*. They are stores of phosphate, contain long-chain polyphosphates, and are important in cell division.

Osmoregulation and Excretion

The cytoplasm, with its organic substances and mineral salts, has higher osmotic pressure than the environment in the case of fresh-water Protozoa. Because of the permeability of the pellicle, water constantly diffuses from this hypotonic medium into the cytoplasm. In order to counteract this osmotic gradient, such Protozoa have a pump system in the form of the contractile vacuoles, through which the fluid which has been taken up is returned to the outside, so that physiological pressure is maintained. The most striking are the contractile vacuoles of some ciliates. *Paramecium caudatum* has two contractile vacuoles, which function in an alternate rhythm.

In the diastolic phase the central reservoir, the actual vacuole, is filled. A sudden contraction empties out the contents of the vacuole into the environment. This phase is called systole. The nephridial canals, now dilating, conduct more fluid to the central vacuolar vesicle, so that diastole of the central vacuole begins again. An analysis of this

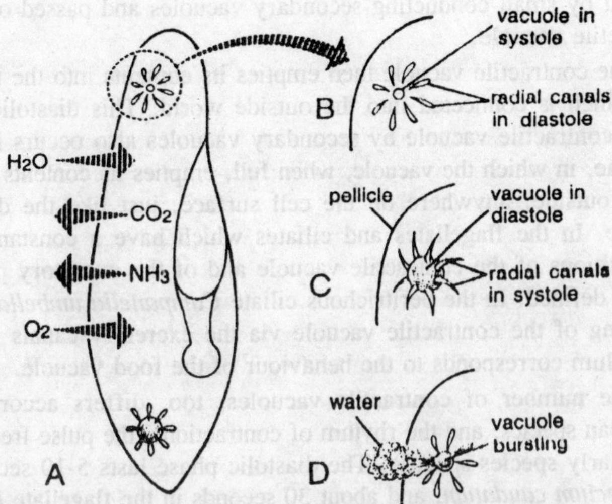

Fig. 3.19. Contractile vacuoles in Paramecium.

process is possible by electron microscopy. The nephridial canal is in diastole, and the central contractile vacuole in systole. The entry of the fluid from the cytoplasm into the nephridial canal proceeds by a system of reticularly-branched nephridial tubules which have an open connection with the nephridial canal on one side, and are in contact with the endoplasmic reticulum on the other side. The nephridial canal is linked with the contractile vacuole by the ampulla via the constricted injection canal. Contraction of the nephridial canal causes the fluid to pass on along this route to the central contractile vacuole; as this happens, contact between the nephridial canal and the network of nephridial tubules is broken to prevent the fluid flowing back into the cytoplasm.

In the subsequent systole of the vacuole, contractile fibrils of the vacuolar vesicle cause a contraction of the ampulla and its injection canal at the same time, so that the fluid which has collected is passed out through the pellicle into the environment via an excretory pore which can be closed by a membrane. The principle of filling the contractile vacuoles through nephridial canals by no means holds for all Protozoa. In the simplest case, simple vesicles, which differ from the normal fluid vacuoles only in that they empty their contents outside the cell, occur in the cytoplasm. In some Protozoa a process occurs which in the reverse for the distribution of dissolved nutrients emanating from the food vacuole. In *Euglena viridis*, for example, the fluid is brought by small conducting secondary vacuoles and passed on to the contractile vacuole.

The contractile vacuole then empties its contents into the flagellar sac, which is connected into the outside world. This diastolic filling of the contractile vacuole by secondary vacuoles also occurs in many amoebae, in which the vacuole, when full, empties its contents directly to the outside, anywhere on the cell surface, just like the digestive vacuole. In the flagellates and ciliates which have a constant form, the positions of the contractile vacuole and of the excretory pore are clearly defined. In the peritrichous ciliate *Campanella umbellaria*, the emptying of the contractile vacuole via the excretory canals into the vestibulum corresponds to the behaviour of the food vacuole.

The number of contractile vacuoles, too, differs according to protozoan species, and the rhythm of contractions, the pulse frequency, is similarly species-specific. The diastolic phase lasts 5-10 seconds in *Paramecium caudatum*, and about 30 seconds in the flagellate *Euglena ehrenbergi* and the ciliate *Coleps hirtus*. *Amoeba proteus* needs about 5 minutes for the production of its 50μm vacuole, and the marine suctorian *Acineria incurvata* requires 6-12 minutes. Basically, the pulse frequency in marine and parasitic Protozoa is slower than that of fresh-water forms, because of the lower osmotic difference. Species like *Actinophrys sol* and *Thecamoeba verrucosa*, which can become accustomed to sea water instead of fresh water, lose their contractile vacuole as they do so. The pulse frequency is influenced not only by the salt content of the habitat, but also by other environmental factors such as the composition of the medium and especially the temperature. In the hypotrichous ciliates *Stylonychia pustulata* and *Euplotes charon*, their respective pulsation rates change from 18 and 61 seconds at + 5°C, to 4 and 23 seconds at +30°C.

In addition to the fluids which are eliminated through the contractile vacuole, or directly through the pellicle, and the solid food remains which are emptied from the food vacuole by defaecation, secretions can also arise during metabolism which are deposited in solid form within the cell itself. These are called *secretions* if they are to use to the Protozoa for hardening the coats, shells, capsules, skeletons and stalk structures, but are called *excretions* if no such use can be seen. In *Paramecia* these excretory crystals consist mainly of calcium phosphate. They are distributed throughout the cytoplasm and shine brightly by birefringence when the unstained fresh preparation is looked at in polarized light. In the ciliate *Stentor coeruleus* the excretory inclusions in the form of a blue-green pigment result in pigmentation of the whole cell. In the malaria parasites a brown pigment, like haematin, forms in the cytoplasm of the parasite derived from the haemoglobin of the blood corpuscles. It shows it granular structure in a macrogametocyte of *Plasmodium berghei*, a parasite of small rodents, in a Giemsastained preparation. This pigment formation becomes clearer in an unstained smear seen by incident light using reflected-light microscopy.

In some Protozoa the excreted substance can be eliminated by the formation of excretory vacuoles and the discharge of these at the surface of the cell. In ciliates, the excretions perhaps also disintegrate in the region of the contractile vacuoles, whose significance in the excretory process can, however, not yet be satisfactorily evaluated. In the Foraminifera, the pigmented excretions known as *xanthosomes* are cemented with the faecal mass into spherical or ellipsoidal stercomes and eliminated at defaecation. Because of their durability, these stercomes form large deposits in places where there are large numbers of Foraminifera.

Extrusomes

Extrusome is the name given to organelles of various species which are shot out of the organism in response to a stimulus. They include the polar threads of the Cnidosporidia which have already been described; these lie in the polar capsule in the Myxosporidia and Actinomyxidia but only in a simple vacuole in the Microsporidia. They are organelles for attachment, which are released when stimulated by the digestive juices in the gut of a new host. The nematocysts, "sting capsules", function as defence organelles; these occur in the phylogenetically interesting dinoflagellates in the genera *Polykrikos* and *Nematodinium*. They resemble the sting capsules of the coelenterates. They arise autogenetically by a special form of self-differentiation. The trichocysts, which are produced in many holotrichous ciliates, are also defence organelles.

In the genera *Frontonia* and *Paramecium*, they lie assembled under the cell surface in large quantities. They form inside the cell in cytoplasmic vesicles and migrate to the pellicle as rod-shaped structures. In *Paramecia* they are anchored to the ridges of the pellicle, whereas the cilia arise from the centres of the fields formed by the ridges. The trichocysts shoot out through window-like openings in the ridge structure when strong mechanical or chemical stimuli influence the ciliates, e.g. when attacked by *Didinium nasutum* or experimentally by dilute acetic acid. While the trichocysts of some holotrichs (*Paramecium*) are completely expelled, those of other species (*Frontonia*,

Lionotus) remain with their ends lodged in the pellicle. It shows the electron-microscopic appearance of a subpellicular trichocyst in *Paramecium çaudatum*.

The ripe trichocyst consists of a large shaft upon which sits a spike which is itself surrounded by a cap. After ejection, the appearance of the spike remains unchanged, while the shaft (which has increased in length about ten-fold) is transversely striated with a periodicity of 55 nm. Salts are also deposited in this shaft which, however, consists chiefly of filamentous proteins; thus the trichocysts also have a certain osmoregulatory function. This is also suggested by the occasional apparently spontaneous ejection of trichocysts. Trichocysts occur not only in ciliates but also in flagellates, namely in many dinoflagellates. In the holotrichous predators, the Gymnostomata, extrusomes are used as weapons for attack. When they eject, a paralysing and lethal toxin penetrates into the prey. These extrusomes known as *toxicysts*, which are produced in cytoplasmic vesicles as are the trichocysts,, are deposited in the region of the cytostome. They take the form of tube-shaped, often slightly twisted capsules from which a thread shoot out as in the sting capsules in a telescopic fashion or by evagination like the finger of a glove; the toxin reaches the prey through this thread. In *Didinium nasutum* the toxicysts surround the pexicysts by which the prey is at first held fast.

The *Paramecia* which have been attacked react to this by exploding their trichocysts, which the *Didinia* evade by the adoption of a tubular structure by their oral cone. Phagocytosis of the prey by the widened mouth opening of *Didinium nasutum* proceeds only after the toxicysts have had their paralysing effect. The extrusomes deposited in the pellicle of the vestibulum in Cryptomonadina are called *ejectisomes*; these too expel a thread in an explosive manner. The structures are at first in two contiguous parts and are relatively-short externally-tapering cylinders surrounded by a membrane. The cylinders are evidently formed by the tight coiling of the thread which is later released. Finally, subpellicular extrusomes containing slime (mucocysts) occur in some ciliates. Here, too, evacuation represents a reaction to an outside stimulus. These mucocysts have been demonstrated in, for example, the genera *Colpidium* and *Tetrahymena* their; extruded slime allows the ciliate to produce a protective coat.

Sensory Organelles and Reactions to Stimuli

The cytoplasm of the Protozoa can react to environmental influences with behavioural changes; it is, in other words, susceptible to stimuli.

This is provided that the stimulatory effect reaches a sufficient threshold value. There are various types of reaction. They include, for example, food uptake, the onset of sexual processes, changes of shape such as contractions and, in many cases, movements in response to stimuli which are called *taxes*. If the induced movement is towards the stimulus, the taxis is positive and, if away from the stimulus, the taxis is negative. A direct alignment towards or away from the stimulus to *topotaxis*. An escape movement which is merely evasive is *phobotaxis*.

The responsive movements are named according to the type of stimulus operating: *thermotaxis* is response to temperature, *chemotaxis* to chemicals, *phototaxis* to light, *galvanotaxis* to electric current and voltage, and *barotaxis* to pressure stimuli in general, comprising *thigmotaxis* to contact, *rheotaxis* to water current, and *geotaxis* to the earth's gravitational pull. The amoebae, with few defined organelles, are able to react; this shows the capacity of the cytoplasm to respond to stimuli, even when no specific cytoplasmic structures are present. The stimulus may be detected by the whole protozoan cell, or certain parts of the cell may prove to be particularly sensitive to stimuli. When living paramecia are cut up, the individual pieces remain sensitive to contact; i.e. they continue to be thigmotactic. If the cut is just in front of the cell mouth, the posterior part continues to be capable of chemotaxis and thigmotaxis, but not of thermotaxis.

The sensitivity to temperature must therefore be localized in the anterior part of the cell. If the cut passes through the cell mouth, chemotaxis is also lost. If this sort of localization of response to stimuli is linked to specific cell differentiations, these are called *sense organelles*. In the flagellate amoeboid forms, the Rhizomastigina, the flagella function as tactile organelles, not as organelles for locomotion. The teactile function is also attributed to the flagella of other flagellates. The cilia of the ciliates are also sensitive to pressure and current, particularly certain stiffer individual cilia, e.g. on the dorsal surface of Hypotricha or in the caudal tuft of *Paramecium caudatum*. They help to stabilize locomotion. In this process, *Paramecium* is positively rheotactic in reaction, as are many other ciliates, and hence swims against the water current.

The trailing flagella of some flagellates also show rheotactic sensitivity. But some cilia can detect not only mechanical stimuli but also chemical stimuli. In studies of food uptake in Entodiniomorpha, many animals when feeding will at first wash granules of charcoal or carmine in by means of their membranellae, just as they do grains of

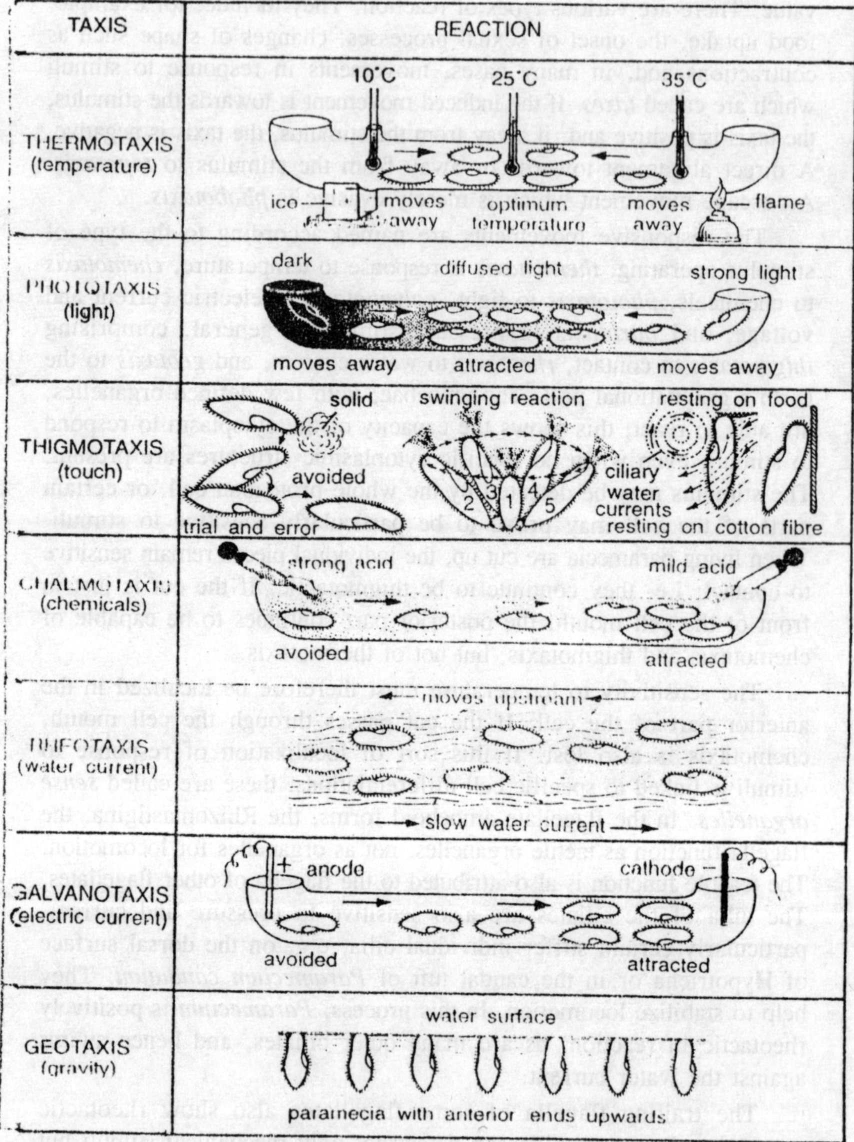

Fig. 3.20. Responses to different stimuli in Paramecium.

rice starch. Soon, however, only rice starch is taken up. Powdered chalk is never taken up; indeed, the cilia are often suddenly retracted.

Positive *phototaxis* in the Protozoa occurs particularly in the pigmented phytoflagellates, whose metabolism requires light. For this reason, some of them produce specific photo-sensory organelles, the *stigma* or *eyespots*. Some marine dino-flagellates form highly developed eyespots. Stratified hyaline bodies which appear homogeneous by light microscopy but which by electron microscopy show differentiation, form a lens system which effects an intensification of the light incident on a light-sensitive absorbing pigment. In this, light energy is converted to chemical energy. In simple cases, for example in *Euglena*, a pigmented eyespot, a stigma, serves merely to cast a shadow in a region of light-sensitive cytoplasm, so that the direction of the incident light is ascertained. The stigmata are often differentiations of the chloroplasts; however, in the Euglenoidina they lie laterally at the bottom of the flagellar sac. As in some Astasia, species of Euglenodidina which no longer have plastids may also lack a stigma. While the autotrophic flagellates are positively phototactile, most heterotrophic Rhizopoda and ciliates show negative phototaxis, provided they do not contain autotrophic symbionts, the zoochlorellae, as found in *Paramecium bursaria*.

Chemotaxis has great significance. Dilute fatty acids, e.g. butyric acid in the proportion 1:10^6, have a strongly positive effect on some flagellates (*Euglena*, *Astasia*). The pH value has been found to be important in the chemotaxis of paramecia, with an optimum for *P. caudatum* at pH 5.4-6.4. If a drop evoking a negative chemotactic response is added to a uniform suspension of living paramecia, the paramecia draw from this place straight away. Conversely, they collect together very quickly in a chemotactically positive drop. If the latter solution is present at too high a concentration, however, it has chemotactically negative effect, so that the paramecia accumulate in the region of optimal diffusion. A bubble of carbon dioxide has a chemotactically positive action. After increasing diffusion, the zone of optimal concentration is sought out in this case too. If the paramecia have swum into a chemotactically positive drop, this has a trap effect. A phobotactic evasion motion occurs now and then at the outer limit of the drop of fluid, with an aimless dodging movement and return, so that this process is repeated often.

Chemotaxis is very important in the selection of food, for location of a sexual partner and, in parasitology, in the form of organotropy. The malarial sporozoites can be taken as examples of this, after the oocysts have burst, they circulate with the haemolymph of the mosquitoes throughout the whole body and are finally gathered, by the trap effect, in the salivary glands. After transmission to the blood of the vertebrate host, the causal agents of malaria in man collect only in the parenchymatous cells of the liver. Only the later merozoites have an organotropy towards the erythrocytes. Chemotaxis thus varies between the individual developmental stage of the same parasite.

Thermotaxis, too, proceeds mainly as phobotaxis, as in the hypotrichous ciliate Oxytricha. In this case, too, there are species-specific optimal ranges which, for example, lie between 24 and 28°C for *Paramecium*. The reaction of the Protozoa to electro-stimuli is also species-specific. *Trachelomonas*, amoebae and many ciliates normally move to the cathode. *Spirostomum* adopts a transverse position across the current. *Polytoma*, *Chilomonas* and *Opalina* migrate to the anode. As is the case for many stimuli, intensity of stimulus also has an effect. Whilst *Paramecium* migrates to the cathode when the current is weak, a stronger current activates the cilia at the posterior end, which beat in opposition to the others, thereby causing a return to the anode.

Geotaxis. Some ciliates have special statocysts for detecting the earth's gravitational pull; these are vesicles in which enclosed granules exert pressure on whichever is the lower side at the time. These also include the conspicuous excretory vacuoles (*Muller corpuscles*) of the genera *Remanella* and *Loxodes*. In the rest of the Protozoa, the inclusions of the normal excretory and food vacuoles distributed in the cytoplasm assume the role of geotactic indicators. Most ciliates show a negatively geotactic response. Thus *Paramecium* rise to the surface. After phagocytosis of finely powdered iron, the pressure effect in the food vacuoles of paramecia can, however, be altered in a magnetic field, resulting in a motion along the lines of magnetic force.

This example of geotaxis in the ciliates also indicates a conduction of the stimulus from the sensory sites, the vacuoles, to the responsive organelles, the cilia, through the cytoplasm itself without special stimulus-conducting organelles. If paramecia are constricted in the middle with a glass thread, the beating of the cilia on the anterior end becomes independent of that of the cilia of the posterior end as soon as the internal cytoplasmic column is interrupted. On the other hand, if the surface of a ciliate is severed, the organized beat of the cilia is interrupted, from which it can be concluded that the beat is co-ordinated among the cilia by the subpellicular fibrillar system.

Functional Morphology

A consideration of the various cell organelles reveals that the functions of a protozoan cell are linked at times to certain structural elements and forms. These are, however, by no means always constant, but can change in a temporal sequence. In the Suctoria, only the swarmers are ciliated in order to seek out a new habitat, while the non-ciliated adult stages develop a stalk in order to attach themselves in this habitat, and they develop tentacles in order to trap and consume their prey. Each stage of development has a morphology and function suited to the needs of the moment. This changeability is clear even with light microscopy, particularly in Protozoa with a distinct life cycle, e.g. in the life cycle of the malarial parasite in the production of slender motile sporozoites, whose duty it is to penetrate into new host cell, in contrast to the subsequent much-larger rounded schizonts and gamonts. In this example, the morphological and functional differences between the motile microgametes and the nutrient-rich larger microgametes is appropriate. This change in form, which is dependent on function, can be recognized in the fine structure seen in the electron microscope, as well as in the structures which can be discerned under the light microscope. The numerous membrane systems which can be seen in a protozoan cell are not static structures. Rather, they are subject to constant variations. Their condition is dependent upon the physiological state of the organism and can be influenced by external factors.

In the amoeba *Pelomyxa illinoisensis*, profound changes in the mitochondrial tubules appear during mitosis. Differences in structure, position and number of the motochondria as a function of age are known in the ciliate *Tetrahymena pyriformis*. When *Plasmodium fallax* is exposed to the action of 8-aminoquinolines, the mitochondria become distended. Membranes are, however, not only capable of structural changes but can, as is typical for the Golgi apparatus, cut off vesicles, undergo complete reversion, and be rebuilt as part of the ground cytoplasm. The dynamics of these processes are therefore aptly called "membrane flow". In Protozoa, structures necessary for a particular function can arise very quickly. In the transition from the amoeboid to the flagellate stage, the flagella of *Naegleria gruberi* grow at a rate of 0.5 μm in length per minute. In the Protozoa with a cyclical development there is a regular change also in structures seen by electron microscopy.

The developmental cycle of the *Plasmodium* spp. is characterized by an alternation of hosts, involving insects on the one hand and various vertebrates on the other. The individual stages develop intracellularly in various tissues. Electron-microscopic studies of the sporozoites and the merozoites of the tissue phase and blood phase have shown that structure are produced for particular functions, and regress when they are no longer required. In this case, electron microscopy has made in important contribution to the understanding of functional morphology.

Sporozoites and merozoites are stages which must penetrate into cells in order to develop further. Before this penetration, special organelles and structures are added to the anterior region of the cell in these stage of development. These include: apical structures, comparable to the conoid of other Sporozoa, provcided with concentric rings; this paired organelle; an additional inner cell membrane and subpellicular fibrils—structures which belong to the apical complex. Since these structures disperse immediately after penetration into the cell, it is assumed that they are required to overcome the resistance of the host cell membrane. In this case, the apical structures would be responsible, in conjunction with the paired organelle, for the penetration. The inner cell membrane assists in strengthening the parasitic cell, and the fibrils effect motility.

The cyclostome, which has already been laid down at this stage, is small and probably not yet functional. Shortly after penetration into a new host cell, the apical structures, the paired organelle, the inner cell membrane and the fibrils all disperse. The cystostome expands and food uptake begins. The structure of the cell nucleus changes, and nuclear division is accompanied by enlargement of the mitochondria, multiplication ribosomes, and an increase in the endoplasmic reticulum. These changes are taken to be signs of increased protein synthesis. In the course of nuclear division, the dispersed structures are again laid down. The first indications of a fresh production of these organelles are the thickening of the cell membrane as the inner membrane become established, and the appearance of the precursor of the paired organelle. Before they leave the cell, all merozoites are again equipped with the organelles necessary for penetration. The ookinete is surrounded by an outer and an inner cell membrane, and subpellicular fibrils are present. The ookinete penetrates the stomach wall of the fly.

Later it lies intracellularly. The oocyst is bound to the basal lamina of the epithelial layer and surrounded by a three-layered membrane. The cytoplasm includes mitochondria, abundant ribosomes and pigment inclusions in vacuoles, in addition to the cell nucleus. The oocyst cytoplasm, which is at first multinucleate, is divided up by the formation of splits and vacuoles. The sporozoites arise in these sporoblasts. The production of sporozoites starts with a thickening of the sporoblast membrane, and the nuclei lie in the immediate vicinity. The apical structures, the paired organelle, the inner cell membrane and the subpellicular fibrils are formed, and each sporozoite also contains mitochondria. Thus equipped, the sporozoite is able to invade the cells of the reticuloendothelial system after transmission.

The formation of the organelles required for penetration proceeds in the same way in sporozoites, and in exoerythrocytic and intraerythrocytic merozoites. This process is always preceded by a thickening of the cell membrane at a narrowly defined site. Nothing is yet known about the mechanisms which trigger off the formation anew and retrogression of these structures, not about the means by which these processes are co-ordinated.

The different developmental stages of a protozoon, which have different functions, develop particular structures for these purposes at appropriate times. Comparative studies reveal, however, that there is by no mans always only one structure possible for any given function. This is evident, for example, in the gamonts of the Coccidia in the genus *Eimeria*. These parasitic stages which live intracellularly take their food from the host cell and release products of their metabolism into it. The composition of the boundary surface between the parasite and its host cell is significant for this exchange of material. Figure shows how varied the boundary structures for this function can be within a single genus, as they appear by electron microscope. Figure shows part of microgamont of *Eimeria bovis*, in which the double cell membrane of the parasite lies closely applied to the inner membrane of the host. The parasite protrudes at places where the host cell has no mitochondria, thereby increasing its surface area. In figure the gamont of *Eimeria stiedae* lies in a large vacuole in the host cell which has formed as a result of the parasitic infection. Micropores in the membrane of the parasite evidently encourage exchange of substances.

It is morphologically clear in the gamont of *Eimeria auburnensis* that substances from the host cell reach the parasitotrophic vacuole. The host cells form protrusion into the parasitotrophic vacuole, which are cut off as small vesicles and migrate into the vacuole which disintegrating. This process is reminiscent of the small transport vacuoles which carry substances into the cytoplasm from the food vacuoles. In a species of *Eimeria* of the desert jerboa, the host membrane is enlarged by numerous villous tubes, the *microvilli* while in *Elimeria perforans* tube-shaped strucures form a connection from the parasite through the parasitotrophic vacuole, direct to the host cell. These few examples give an insight into the multiplicity of the morphological solutions to a single functional problem in the realm of the ultrastructure.

Sexuality and Heredity

The basic property of all living organisms is their capacity for reproduction while retaining their species-specific constitution. This capacity derives from the ability of the DNA molecules to increase by replication; since they contain the code for all the properties of the cell, these are the carriers of heredity, the genes. Since the morphology and functions of the cell are induced by the nucleus, nuclear division has particular significance in the propagation of the cell. Whereas sexuality and reproduction are closely linked and directly dependent on one another in the higher Metazoa, it appears in the Protozoa that replicative multiplication as the basic property of life by no means has to be instigated by sexual processes.

Numerous Protozoa reproduce solely by asexual division, by agamogony, in which they themselves always live on in the developing daughter cells. In this sense, the protozoan cell is immortal. When sexuality occurs, this is always characterized by karyogamy, the fusion of two haploid nuclei, although the form it takes may vary. These nuclei are called *gametic nuclei*. At syngony, the haploid gametic nucleus together with its own cytoplasm normally constitutes a gamete. The fusion of two gametes produces the zygote. In just a few cases, actual gamete production does to occur. In *Notila proteus* of the Polymastigina and the amoeba *Sappinia diploidea*, the gametic nuclei fuse within a common cytoplasm which was already there.

There are no free gametes in the conjugation of ciliates either these only produce gametic nuclei, also called *prenuclei*. In this case, however, too gametic nuclei arise in each conjugant; one of each pair, the migratory nucleus, passes over into the other conjugant and forms a synkaryon with the stationary nucleus there. In all these cases the sexual process begins with two different cells, for which reason it is also called *amphimixis*. For this, it is necessary for two sexual partners to find one another. The affinity of the two partners for one another is determined by specific sexual hormones, the *gamones*. Figure shows the effect of these gamones in, for example, the flagellate *Chlamydomonas eugametos*; the male gametes agglutinate after the addition of a filtrate of female gametes. Carotinoids are known to be gamones. Since the ciliates are hermaphrodite, a female gynogamone cannot be differentiated from a male androgamone. On the other hand, conjugation types can be recognized in ciliates. Progeny of a single ciliate, i.e. ciliates of a clone, show no sexual reaction among themselves; such reaction is only possible with other suitable strains of the same species.

If two suitable conjugation types are mixed together, there is a rapid agglutination from which the conjugation pairs subsequently emerge. Whereas the gamones of flagellates can be secreted into the culture medium, the sexual hormones of the ciliate remain bond in the cell. They are induced in the cytoplasm by the macronucleus. The sexual hormones can first occur and be functional in the gametes or even in the gametic mother cells. In the former case only the gametes find each other and a gametogamy takes place. In gamontogamy, in which the gametic mother cells locate each other (syzygy), the reciprocal affinity of the gametes originating from them is, however, also retained from then on, as in gametogamy. The behaviour of the Foraminifera shows that gametogamy and gamontogamy can occur between closely related species. In the euciliates, only gamontogamy is possible, in which the outward appearance of a gametogamy is simulated secondarily by total conjugation; however, in true gametogamy each partner would only contain one gametic nucleus.

Syngamy in the form of oogamy corresponds to the sexual act in the higher Metazoa, but conjugation shows that a great multiplicity of sexual processes is possible in the Protozoa. This is also demonstrated by *autogamy*, which is carried out by a single protozoan cell without a partner and, as *automixis*, is contrasted with amphimixis. In the heliozoans *Actinosphaerium eichhorni* and *Actinophrys sol*, two daughter cells form in one outer coat and these fuse with one another as gametes. In the foraminiferan

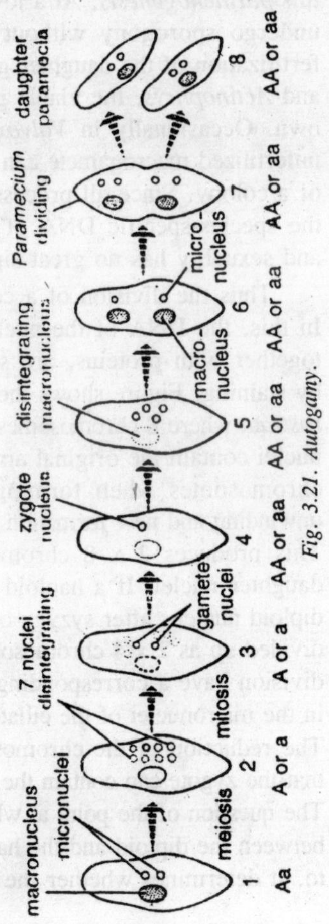

Fig. 3.21. Autogamy.

Rotalliella heterocaryotica, the gametes of a single individual fuse. Autogamy has also been demonstrated in some ciliates, e.g. *Paramecium bursari*, *P. aurelia*, *Euplotes patella*, *E. minuta*, *Tetrahymena rostrata*; here, the nuclear processes take their course as in conjugation. The haploid nuclei produced by meiosis are distributed between two daughter cells. Subsequent mitosis of these nuclei produces another 2 gametic nuclei in the daughter animals which fuse to form a diploid synkaryon from which the new nuclear apparatus develops, as in conjugation.

Autogamy is called cytogamy when the individual is in purely superficial contact with another ciliate or with a conjugating pair. In some cases, the further development of a gametic cell can proceed without fertilization. *Eucoccidium dinophili* provides an example of this *parthenogenesis*. At a low level of infection the macrogamete can undergo sporogony without fertilization. If there is no reciprocal fertilization of the daughter gametes in the heliozoan *Actinosphaerium* and *Actinophrys*, the viable gametes can continue to develop on their own. Occasionally in *Volvox aureus* of the Phytomonadina, too, an unfertilized macrogamete can lead parthenogenetically to the formation of a colony. Since all processes of life depend for their specificity on the species-specific DNA of the cell, its distribution in reproduction and sexuality has no great significance.

Thus the division of a cell is always limited to nuclear division. In this, the DNA of the nucleus collects in the chromosomes which, together with proteins, are structures which are easily demonstrated by staining. Figure shows these processes in the gregarine *Monocystis rostrata* where 8 chromosomes are formed. To ensure that both doughter nuclei contain the original amount of DNA, there is a doubling of the chromosomes when forming the chromatids, which is lined with unwinding and new formation of the double helix of the DNA molecule. This produces 2 x 8 chromosomes which are divided between the daughter nuclei. If a haploid gametic nucleus forms from the normal diploid nucleus after syzygy of two gregarines, the 8 chromosomes are divided up as 2 x 4 chromosomes in meiosis. The processes of nuclear division have a corresponding pattern in other diploid nuclei too, e.g. in the micronuclei of the ciliates *Kidderia mytili* and *Didinium nasutum*. The reduction of the chromosome number in meiosis is necessary so that the zygote can contain the normal diploid chromosome complement. The question of the point at which meiosis takes place in this alternation between the diploid and the haploid condition has already been referred to. It determines whether the normal vegetative phase of protozoon is

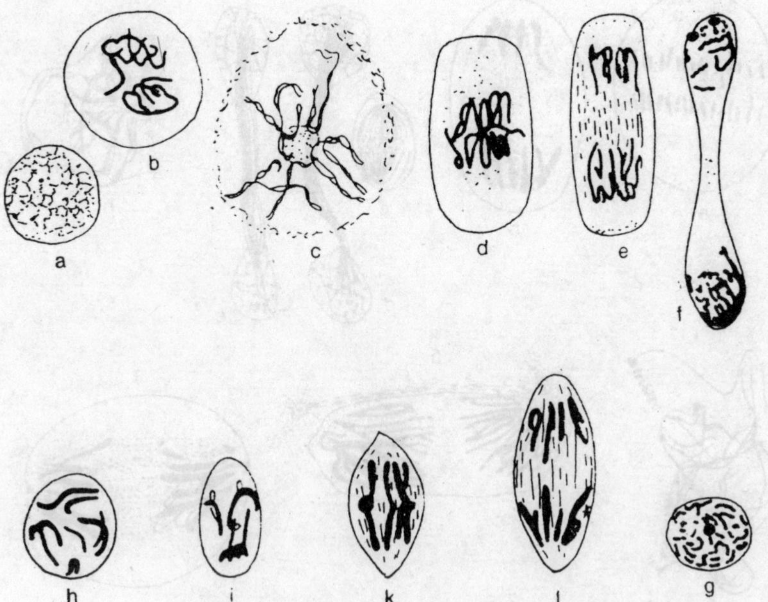

Fig. 3.22. Nuclear division in the gregarine Monocystis rostrata.

diploid or only haploid. The separation of the chromosome complement in mitosis and meiosis generally takes place in the field of force of a division spindle between two centrioles. They arise from the division of an existing granular centriole usually lying in the cytoplasm at the periphery of the nucleus, and eventually come to lie at opposite poles. In many cases the development of a special area around the centriole leads to the formation of a centrosome.

Centrioles and centrosomes function as a *kineto centrum* during the separation of the chromosomes. In many Protozoa the centrioles enter the nuclear region and form an intranuclear division spindle; in this case, the nuclear membrane breaks up sooner or later during division. In the Tetramitidae and Hypermastigidae, an extranuclear division spindle develops which participates in nuclear division only after this has begun. Since in this case the nuclear membrane is retained for a long time during mitotic chromosome division, the chromatids draw away from one another under the influence of special pulling-threads which emanate from a centromere lying by the nuclear membrane. The chromosomes generally become arranged in the equatorial plane at the beginning of nuclear division but this stage can be omitted in some cases. This peculiarity is also called *paramitosis*.

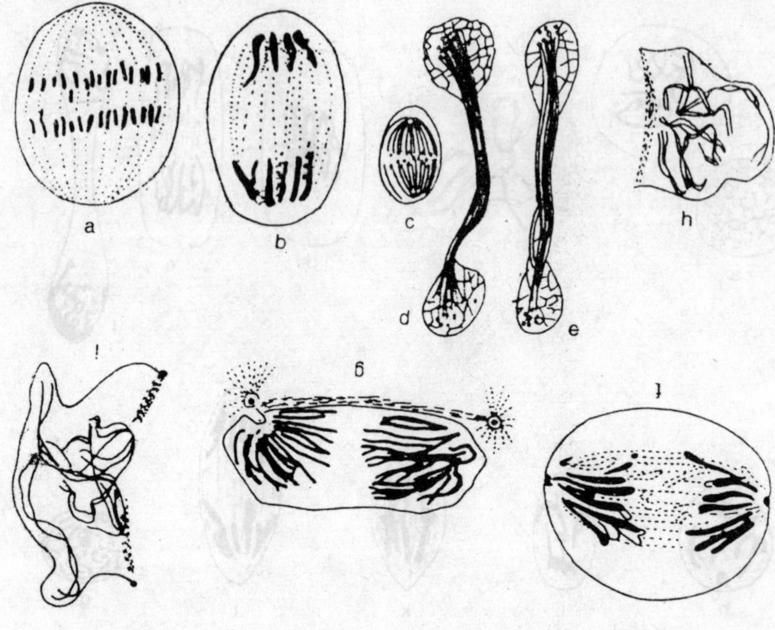

Fig. 3.23. Distribution of the chromosomes during nuclear division.

The spindle attachment, the *kinetochore*, can sometimes be seen under the light microscope in stained preparations with large chromosomes. A spiral structure is clear in the large chromosomes of some Polymastigina. With the aid of phase contrast microscopy, the mitotic processes can be seen in some of these flagellates, even while they are still alive. Electron-microscopic study of protozoan chromosomes has not contributed anything very significant; in section, only an internal tight-spiral structure is visible .

After two haploid gametic nuclei have united to produce a synkaryon, the zygotic nucleus contains maternal and paternal chromosomes so that characteristics of both parental cells are transmitted to the zygote. This *inheritance* in Protozoa is subject to the Mendelian laws operating in higher plants. If the genes of the chromosome of a haploid gametic nucleus match those of the corresponding chromosome of the other gametic nucleus, so that two similar alleles form a homologous chromosomal pair during synkaryon formation, the resultant diploid nucleus is homozygous with respect to this chromosomal pair. Meiosis in this zygotic nucleus then leads to 100% similar gametic nuclei or gones. If, however, the chromosomal

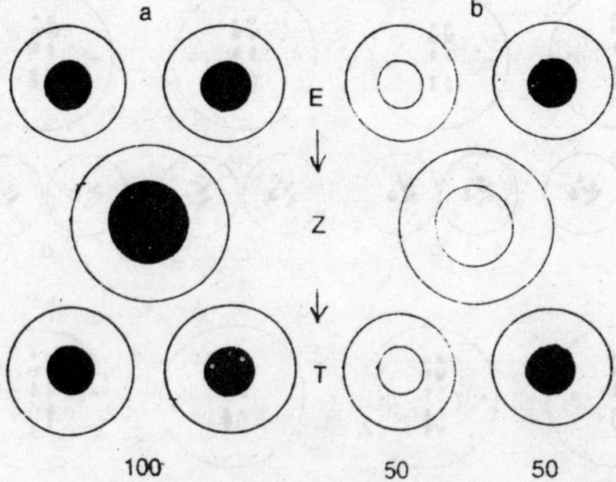

Fig. 3.24. Inheritance of a chromosomal pair in haplonts.

structures corresponding to each other embody different characteristics, the zygotic nucleus contains both attributes; it is heterozygous. Meiosis then leads to a 50% distribution of the characteristics of each parental cell in the gametic nuclei.

The path of inheritance is more complex in the diploid Protozoa, although the same rule applies. If both alleles of chromosome pair are similar to each other and to those of the sexual partner, i.e. if, for example, the two conjugants in the ciliates are similarly homozygous, conjugation produces 100% similar homozygous offspring. If both conjugants are still homozygous but dissimilarly homozygous (2 black alleles on one hand and 2 white alleles on the other), crossing, when the next generation of ciliates each inherit one allele from each

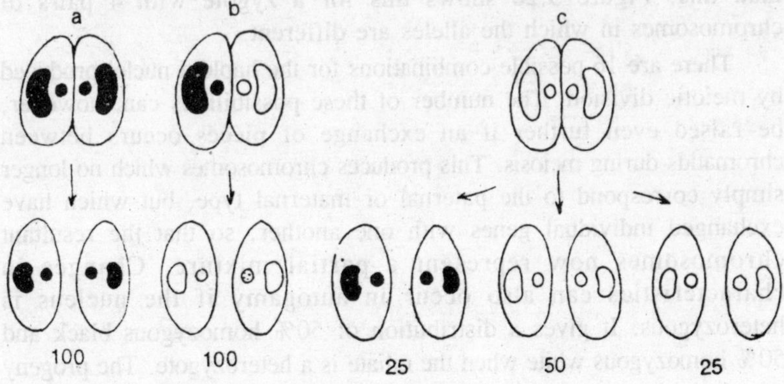

Fig. 3.25. Inheritance of a chromosomal pair in the conjugation of diploid ciliates.

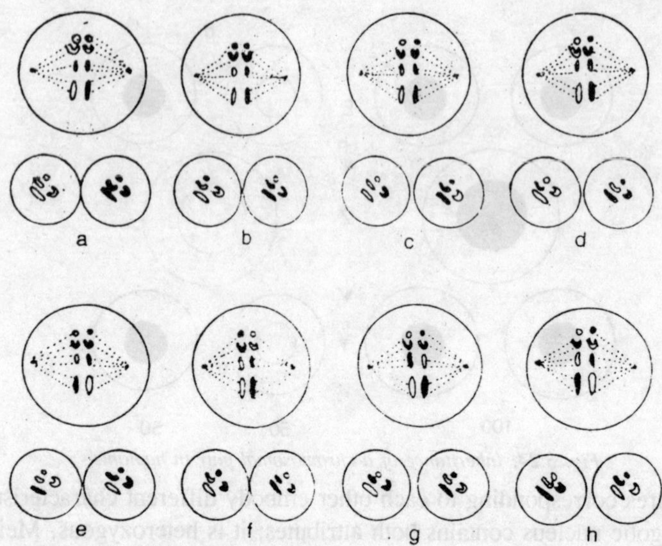

Fig. 3.26. The 4^2 possible distributions of 4 maternal and 4 paternal chromosomes.

conjugant, leads to 100% heterozygous ciliates. In this case a mixture of the characteristics can appear, unless one of the inherited characteristics is a recessive one with its effect masked by the other (dominant) one. If two heterozygous ciliates are crossed with one another, according to Mendelian rules 50% of the daughter cells will be homozygous, namely 25% homozygous black and 25% homozygous white, while the remaining 50% will be heterozygous. Since, however, one genome in the Protozoa always comprises more than one chromosome, inheritance of characteristics is much more complicated than this. Figure 3.26 shows this for a zygote with 4 pairs of chromosomes in which the alleles are different.

There are 16 possible combinations for the haploid nuclei produced by meiotic division. The number of these possibilities can, however, be raised even further if an exchange of pieces occurs between chromatids during meiosis. This produces chromosomes which no longer simply correspond to the paternal or maternal type, but which have exchanged individual genes with one another, so that the resultant chromosomes now represent a partial mixture. Changes in characteristics can also occur in autogamy if the nucleus is heterozygous. It gives a distribution of 50% homozygous black and 50% homozygous white when the ciliate is a heterozygote. The progeny of a homozygous ciliate are homozygous and similar to the parent.

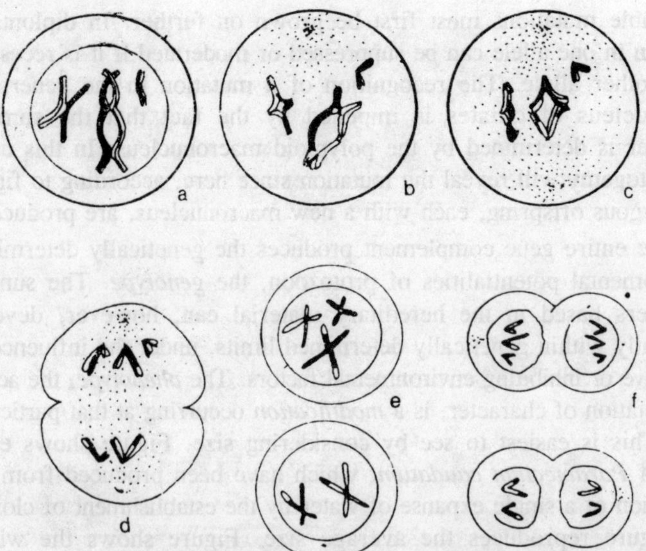

Fig. 3.27. Diagram of exchange of parts of chromatids in a 2-step meiosis.

The prerequisite for all new combinations is the occurrence of differences between the genes of the sexual partners. Such differences arise occasionally through spontaneous changes in the genes, through *mutations*. These changes in the genes can also be induced experimentally in Protozoa by the influence of mutagenic substances, such as urethane and antibiotics, but particularly by irradiation with X-rays or ultra-violet light. These changes are immediately effective in haplonts in which, of course, not all mutations are viable, and the

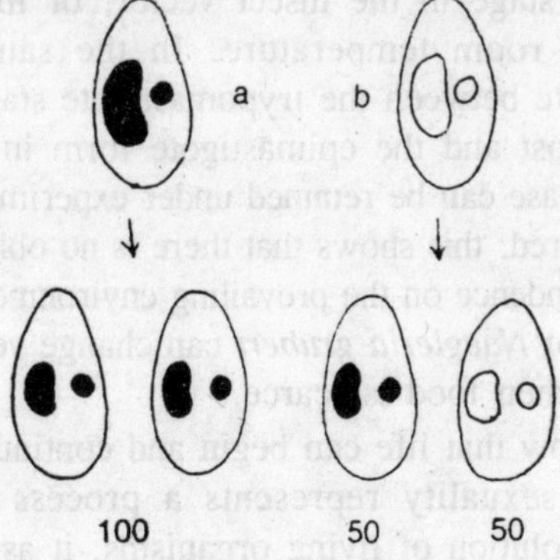

Fig. 3.28. Inheritance of a chromosomal pair in the autogamy of a ciliate.

few viable mutations must first be grown on further. In diplonts, a mutation in one allele can be suppressed or moderated if it is recessive to the other allele. The recognition of a mutation in the generative micronucleus of ciliates is impeded by the fact that the somatic character is determined by the polyploid macronucleus. In this case, only autogamy will reveal the mutation since here, according to figure homozygous offspring, each with a new macronucleus, are produced.

The entire gene complement produces the genetically determined developmental potentialities of protozoon, the *genotype*. The sum of characters based in the hereditary material can, however, develop differently within genetically determined limits, under the influence of conducive or inhibiting environmental factors. The *phenotype*, the actual manifestation of character, is a *modification* occurring at that particular time. This is easiest to see by considering size. Figure shows eight races of *Paramecium caudatum*, which have been produced from the population of a single expanse of water by the establishment of clones. The figure reproduces the average size. Figure shows the whole spectrum of modifications for the stippled races A, D and H, which show a normal (Gaussian) distribution, the average size (stippled) being the most frequent in the clone. The genetically determined range of modifications of the race shown here is retained, irrespective of the size of the individual within a pure race from which the new clones are produced. The modifications controlled by environment also include the respective phenotypes of many parasites.

Leishmania species of the Trypanosomatidae produce the aflagellate amastigote (leishmania-form) in the tissues of their vertebrate host or in tissue cultures at 37°C. The promastigote flagellate form arises from the amastigote stage in the insect vector, or in culture media containing blood at room temperature. In the same way, many trypanosomes alternate between the trypomastigote stage in the blood of their vertebrate host and the epimastigote form in the gut of the vector. Here, each phase can be retained under experimental conditions for as long as is required; this shows that there is no obligatory rhythm, but really only a dependence on the prevailing environmental conditions. The amoeboid form of *Naegleria gruberi* can change very quickly into the flagellate stage when food is scarce.

The Protozoa show that life can begin and continue even without sexuality, and that sexuality represents a process acquired only secondarily in the evolution of living organisms; it assumes an ever-increasing significance as they become more highly developed. However, even Protozoology reveals the significance which sexuality can assume. It facilitates a much greater potentiality for variation in all characteristics within a species; by this means, although many misdevelopments arise, some genotypes are produced which, according to *Darwin*, are better equipped for the struggle for survival. The examples of inheritance shown so far are all dependent upon the DNA of the nuclear chromosomes, upon the *genome*; i.e. they are karyotically determined. However, some cases of inheritance in the Protozoa are dependent upon extrakaryotic DNA. This extrakaryotic DNA occurs, as already mentioned, in the plastids of the phytoflagellates, in the mitochondria (including the kinetoplasts) and also in the centrioles.

In contrast to the genome, all the genes of the plastids comprise the *plastome* and the genes of the mitochondria the *chondriome*. The proportion of this DNA is usually small in relation to the nuclear DNA. In the flagellate *Euglena* the plastid DNA constitutes 1%, and in *Chlamydomonas* gametes over 6%. DNA can also be present in the cytoplasm of Protozoa because of symbiotic parasite. In the discussion of plastids and mitochondria it has already been established that their DNA is not quantitatively sufficient for an autogenous development of these organelles, so that these must be coded in addition by the nuclear DNA. This applies also to the killer property of paramecia which, as electron microscopy shows, can be attributed to bacterial cytoplasmic inclusions, the *kappa symbionts*. These symbionts, easily seen with Giemsa staining, are dependent upon the presence of a gene K in the

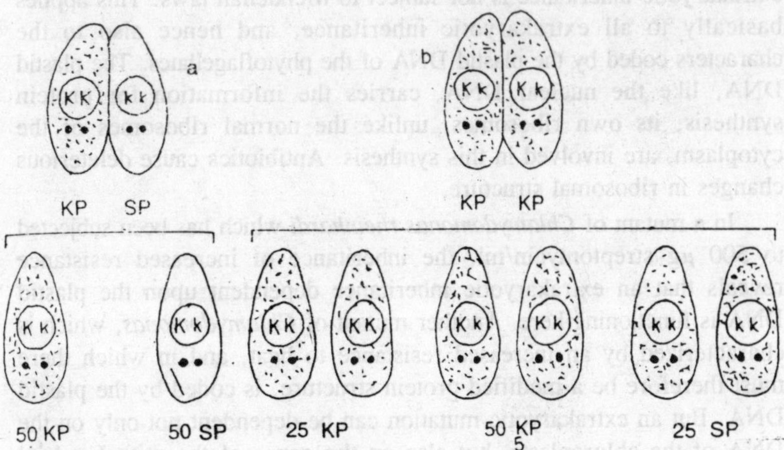

Fig. 3.29. Inheritance of the cytoplasmic killer-character, dependent on the gene K.

nucleus, whereas the allele k does not permit the development of the symbionts. Paramecia which contain the kappa symbiont in their cytoplasm are called killer paramecia because they kill other paramecia which lack the kappa symbionts by means of toxic secretions. The kappa-free paramecia are hence called *sensitive paramecia*.

Sensitive paramecia can cross with killer paramecia because they are resistant to the killer toxin during conjugation. The genes K and k are subject to the Mendelian rules, whereby the conjugation of homozygous K and a homozygous k leads to 100% heterozygous k. Nevertheless, crossing produces only 50% killer paramecia, and the other 50% are sensitive since during conjugation it is usually only the migratory nucleus, without cytoplasm and without kappa symbionts, which passes over into the sensitive partner. If two heterozygous killer paramecia conjugate according to Mendelian inheritance they produce 50% heterozygous and 25% homozygous killer paramecia. Although the remaining 25% homozygous paramecia with the genes kk at first contain kappa symbionts, they are sensitive paramecia since the kappa symbionts perish without the gene K.

The amoeba *Pelomyxa palustris* provides an example of the great significance the DNA of a symbiont can have for the protozoan cell, in contrast to the previous example. Since this amoeba has no mitochondria of its own, the symbiotic bacteria here take over the function of mitochondria. The possible results of crossing, cited with respect to the killer character, lead to the important finding that the extrakaryotic inheritance is not subject to Mendelian laws. This applies basically to all extrakaryotic inheritance, and hence also to the characters coded by the plastid DNA of the phytoflagellates. The plastid DNA, like the nuclear DNA, carries the information for protein synthesis; its own ribosomes, unlike the normal ribosomes of the cytoplasm, are involved in this synthesis. Antibiotics cause deleterious changes in ribosomal structure.

In a mutant of *Chlamydomonas rheinhardi* which has been subjected to 500 μg streptomycin/ml, the inheritance of increased resistance reveals that an extrakaryotic inheritance dependent upon the plastid DNA is functioning here. Another mutant of *Chlamydomonas*, which is characterized by an increased resistance to heat, and in which there must therefore be a modified protein structure, is coded by the plastid DNA. But an extrakaryotic mutation can be dependent not only on the DNA of the chloroplasts, but also on the genes of the mitochondrial DNA. There is a mutation in *Chlamydomonas rheinhardi* which is not only resistant to streptomycin but which is actually streptomycin-dependent; this can be traced to the presence of sd genes in the mitochondria. If these streptomycin-dependent strains are transferred to a medium lacking streptomycin, mutation in the mitochondria can lead to a reappearance of the streptomycin-sensitive ss gene. Selection leads to the formation of streptomycin-sensitive colonies in this medium. They are, however, at first only phenotypically sensitive to streptomycin, since to begin with there are always a few remaining sd genes. All the sd genes are finally lost only after a longer time in the streptomycin-free medium. Only then is a gradual transference back to a medium containing streptomycin no longer possible.

As long as some sd genes are retained, the strain can re-adapt to a medium containing streptomycin. This behaviour corresponds to the phenomenon of "*Dauer-modifikation*" with the protracted retention of an acquired characteristic. These long-lasting modifications are also known in the pathogenic trypanosomes, which can become resistant to drugs during therapy with arsenical preparations; experiments with animals show that this character can be retained, even without the influence of further medication, for months or years before it is reversed. Long-lasting modifications like this also arise against antibodies. In *Paramecium caudatum*, adaptations made towards arsenicals show that here the increased resistance is evidently linked to the alleles of the somatic macronucleus. When a new macronucleus is formed from the micronucleus after a conjugation, the resistance to arsenic is immediately lost. On the other hand, resistance to calcium in paramecia is not influenced by conjugation of autogamy.

Cells Associations and Cellular Differentiation

It was said at the outset that the word "protozoon" does not necessarily imply a unicellular condition. If the Protozoa are nevertheless generally unicellular, this is based on the nature of their organization. They do not form the physiological unity of a centrally controlled multicellular confederation. However, there are the first indications of some tendencies towards a cell association. Colonial associations occur in the flagellates Chrysomonadina, Phytomonadina and Protomonadina. Some genera of the Radiolaria (*Collozoum*, *Sphaerozoum*, *Collosphaera*) form 4-6 cm swimming colonies. Many peritrichous ciliates, too, form colonies. These associations generally provide greater protection to the individual cells and increase the water current and the procuring of food by means of the action of the flagella and cilia. The protective effect is increased still more when, as in

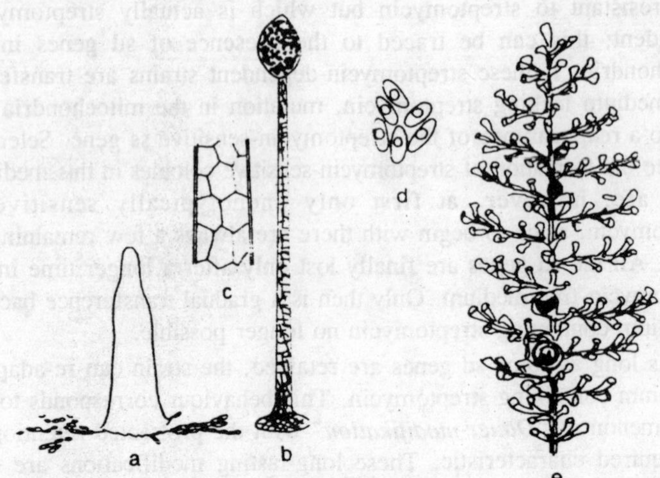

Fig. 3.30. Formation of colonies.

Carchesium and *Zoothamnium*, a contact stimulus can be transmitted by one individual to the entire colony, thereby leading to a united escape reaction by contraction.

A further advance conducing to organism survival is cell differentiation. This occurs temporarily in some amoebae, the *Acrasiae*, for example in *Dictyostelium discoideum*. When environmental conditions are unfavourable, individual amoebae secrete a hormone, acrasin, which diffuses into the medium and stimulates the amoebae to mass together into a conus structure. The conus tips over; all the constituent amoebae shift their position, and finally pile up again to form the so-called *sporophore*. Cell differentiation occurs in this structure; the amoebae constituting the stalk fuse together into a compact tissue, while the amoebae in the head of the sporophore survive as cysts. In the flagellates, the individuals forming an association of cells usually retain their capacity for independent development. However, in the Phytomonadina, some Volvocidae show differentiation.

The individual colonies contain cell which are still all the same and which can all divide, but in *Pleodorina californica* there are two types of cell which can be distinguished by their difference in size. The small somatic cells have lost the capacity to divide, so that only the larger generative cells are still capable of forming new colonies. In *Volvox aureus* and *V. globator* only a few generative individuals remain. A corresponding condition is found in the peritrichous ciliate *Zoothamnium alternans*. Only a few macrozooid individuals can detach themselves and are still able to divide, while the many somatic microzooids have lost the ability to produce daughter colonies. The *Cnidosporidia* have a particular form of cell differentiation during the production of spores.

The site formation of the spores in Myxosporidia is multinuclear at first, and becomes temporarily multicellular. In this relatively short phase of development, the physiological unity related to that of the metazoan cells is still less marked that in, for example, the colonies of the Phytomonadina. In the Phytomonadina (Volvocidae) not only agamous reproduction but also sexuality is sometimes limited to a few individuals scattered within the colony. This differentiation into pure mortal soma cells and the cells of the germ path achieves at highest development in the genus *Volvox*. A parallel intracellular development occurs in the macronuclei of the euciliates and of some Foraminifera which also have only a somatic function. These examples reveal a certain tendency in the Protozoa towards the subsequent development of the Metazoa, but the fact remains that they do not yet have the physiological unity of a multicellular organism as possessed by the true Metazoa.

4

INHERITANCE

Inheritance within the Strain

A Protozoan species is composed of strains (races, biotypes, or stocks) which differ among themselves in hereditary traits. Such races have long been known in *Paramecium*, *Difflugia*, *Arcella* and *Centropyxis*. Observations on mating types of ciliates have shown that a conventional species also may include varieties which are completely, or almost completely, unable to interbreed. This situation creates taxonomic problems which cannot be solved until more is known about mating types and the comparative characteristics of these different ciliate strains. Racial characteristics are of various kinds. Differences in size and fission-rate are well-known in various ciliates. Strains of *Tetrahymena* in pure cultures have shown minor differences in biochemical activities, and in the extent to which they can become acclimatized to salt solutions.

Differences in pathogenicity, noted among strains of parasitic species, may be paralleled by morphological differences. Strains of *Entamoeba histolytica* with relatively low pathogenicity may show a small average size. In *Plasmodium vivax*, relatively low and high degrees of pathogenicity may be correlated with slow and rapid reproduction. That these racial characteristics are inherited is indicated by their persistence in cultures or in infected animals.

Tendency Toward Genetic Unformity

Although non-herediatary differences, induced perhaps by environmental factors, may be expected within a race, a reproduction by fission or budding should insure exact duplication of genes from generation to generation, barring mutations or mitotic accidents.

Therefore, genetic constancy of the race would be expected in the absence of sexual phenomena. In general, this expectation has been realized. Such was the case in the early work of Jennings on *Paramecium*. Although separation of wild populations into several races was usually possible, selection for size within the race was no longer effective. Similar findings of Ackert (1) and Jollos and indicated that the race is relatively constant. However, certain apparent exceptions have been reported.

Apparently Spontaneous Changes

Selection within the race, continued for many generations, has produced distinct stocks of *Difflugia corona* differing in number and length of spines, diameter and height of the shell, and diameter of the mouth. Comparable effects have been observed in *Centropyxis aculeata* and *Arcella dentata*. These results remain unexplained. Although it is possible that gene mutations were involved, undetected environmental differences might have been perpetuated under the experimental conditions of continued selection. In the latter case, the different types probably could be considered results of acclimatization rather than mutations. Comparable changes within the race have been reported in a few ciliates. By opposite selection through more than 150 generations, Middleton established two strains of *Stylonychia pustulata* deffering in rate of fission.

During selection there was a gradual increase in the average difference, indicating that the effects of selection were cumulative. Since these differences persisted after conjugation and also through fassion for several months after selection was discontinued, Middleton suggested that the selection of small variations may be "an effective evolutionary procedure." Similar changes, involving size, division-rate and resistance to environmental factors, were observed by Raffel in a clone of *Paramecium aurelia*. The new types did not revert to normal, even after conjugation, and were believed to have arisen by gene mutation.

The origin of two unusual biotypes—differing from the parent stock in rarity of conjugation, lower division-rate, higher mortality, and frequency of morphological abnormalities—also has been reported by Sonneborn and Lynch in *P. aurelia*. Some of these hereditary changes in ciliates were attributed to endomixis. Diller's report of autogamy in *P. aurelia* was discounted as a possible explanation on the grounds that hids evidence for autogamy was far from convincing and that extraordinary assumptions would be necessary in relating autogamy to the appearance and subsequent disappearance of particular traits. The recent conclusion of Sonneborn, that autogam (and *not* endomixis) occurs in his strains of *P. aurelia*, evidently leaves some of these intraracial changes in ciliates unexplained for the present.

The effects of selection described in *Stylonychia pustulata* cannot be ascribed to endomixis or autogamy because neither process seems to have been observed in this species. Certain morphological changes in ciliates have been interpreted as mutations. An example is Hance's race of *Paramecium caudatum* with 2-7 extra contractile vacuoles, an abnormality inherited in fission and unaffected by selection or conjugation. Likewise, hereditary changes in number of nuclei have been observed in *P. bursaria*. A truncated type of *P. aurelia* has shown similar behaviour, persisting through more than 400 generations without being influenced by conjugation or endomixis. MacDougall's tetraploid mutant in *Chilodonella uncinatus* also bred true.

Environmentally Induced Changes

A variety of changes may be induced by modification of environmental conditions. Although some cases of acclimatization may represent merely the selection of resistant strain from a genetically mixed population, serologically distinct types evidently can arise within a pure line, as in *Trypanosoma brucei*. Changes in resistance to chemical agents have been investigated in parasitic and free-living species. Among the parasites, most of the work has been done on trypanosomes in which antigenic modifications, occurring during an infection, are especially interesting. After inoculation of a guinea pig, for example, with *Trypanosoma rhodesiense*, the flagellates increase in number for a time. Suddenly, most of them are killed by a newly developed antibody.

The survivors continue to multiply, so that the blood is repopulated by a *relapse strain*. This relapse strain in resistant to the trypanocidal antibody which is still present in the host and is still active against the original strain (*passage strain*). Two explanations have been suggested: (1) the activity of the antibody brings about selection of a resistant strain (the relapse strain) from an originally mixed population; (2) as the result of an antigenic change, the trypanocidal antibody is no longer specific for flagellates which give rise to the relapse strain. The second interpretation receives support from the fact that an infection started with a pure line of *T. brucei* showed the usual development of a relapse strain.

The treatment of different strains of *Paramecium aurelia* with homologous antisera also has induced antigenic changes which are inherited. Similar phenomena have been observed in chemotherapy of trypanosomiasis. Most of the flagellates are killed, but a few may survive to produce a resistant strain—often termed an "arsenic-fast" or "antimony-fast" strain, depending upon the type of drug, although such a designation may not be entirely accurate. In tests of several substituted phenylarsenoxides, for instance, trypanosomes have seemed to develop resistance to substituent basic or acidic groups on the phenylarsenoxide molecule rather than to the arsenoxide group as such. This drug resistance may persist for long periods. Strains of *T. rhodesiense* have remained resistant to atoxyl, tryparsamide, and acriflavine for 7.5 years through 900 mouse transfers, and to atoxyl for 12.5 years through 1,500 mouse transfers. A strain of *T. brucei* was still tryparsamide-resistant after 59 transfer through guinea pigs and four through *Glossina morsitans*.

A tryparsamide-resistant strain of *T. rhodesiense* has been produced also by repeated treatment of the flagellates *in vitro*. The trypanosomes were exposed to the drug, washed, and then inoculated into a mouse. The strain was recovered from the mouse and the procedure was repeated a number of times, with the result that the flagellates became at least 500 times as resistant as the original stock. Among the malarial parasites, *Plasmodium gallinaceum* has inherited paludrine-resistance in five cyclical transfer through mosquitoes without intervening drug treatment. The mechanism involved in development of resistance of drugs is unknown. One suggestion is that resistant trypanosomes have lost their normal ability to absorb active drugs.

In addition, differences in stain ability of normal and resistant strains have been demonstrated, and the development of resistance may accompany shifts in isoelectric points of various trypanosomal proteins. Although genetic significance has not been considered in many studies of acclimatization in free-living Protozoa, inherited modifications have been reported in a few instances. Neuschloss acclimatized *Paramecium caudatum* to quinine, arsenic and antimony compounds, and various dyes by exposing the ciliates to gradually increasing concentrations. The developed resistance was specific except for some reciprocal effects of arsenic and antimony compounds. Similar results have been obtained in *P. aurelia* and *P. caudatum* by a combination of selection and acclimatization. For example, a strain was grown in a non-lethal concentration of an arsenical and then subjected to a dosage lethal for most of the ciliates. The survivors were returned to a non-lethal arsenic medium for a time before heavy dosage was repeated. As the procedure was continued, the strain became progressively more resistant.

Resistance was inherited for long periods after a return to normal culture media. Such modifications—although inherited through hundreds of fissions, through endomixis (or autogamy?), and in rate cases through conjugation—eventually disappeared after removal of the stimulus. Accordingly, Jollos called such changes "Dauermodifikationen, " distinguishing them from true mutations. The more recent acclimatization of both amicronucleate and normal strains of *Colpoda steinii* to arsenicals indicates that the micronucleus is not necessarily involved in "Dauermodifikationen". Comparable acclimatization has been reported in *Bodo caudatus*, strains of which developed a tolerance to acriflavine in concentrations of 1:500, as compared with the normal susceptibility to dilutions of 1:50,000 to 1:10,000. This resistance was inhertied, in decreasing degree, for at least a year in drug-free media.

Morphological modifications have been reported in several cases. Loss of the kinetoplast, induced in *Trypanosoma brucei* by inoculating infected mice with certain dyes, became an apparently fixed characteristic. Loss of the parabasal body also was induced in *Bodo caudatus* by treatment with acriflavine, but no permanently abnormal strain was obtained. Various structural changes have been reported in *Chlamydomonas debaryana*. One type could be transformed into another by maintenance in an appropriate medium for a period varying with the length of time the original strain had been exposed to the conditions which produced it. In view of these findings, Moewus suggested that many of the varieties found in natural populations are merely "Dauermodifikationen" induced by specific environmental conditions. Morphologically distinct types of *Chilodonella uncinatus* have been induced by ultraviolet irradiation. These changes, believed to be mutations, persisted through fission and conjugation.

Likewise, a physiological change, expressed as a lowered fission rate, has been induced in *P. aurelia* by treatment with X-rays. Homozygous strains were obtained in autogamy, and the abnormality was transmitted through both exconjugants in matings between normal and abnormal clones. This induced changes was attributed to a micronuclear mutation. Aside from the rare cases which may have involved true mutations, the genetic significance to these induced changes remains uncertain. Jollos considered them the result of cytoplasmic modification rather than gene mutation—an interpretation with interesting implications. In reproduction by fission, an original mass of modified cytoplasm would already be diluted several milion times at the twentieth generation, and some of these induced modifications have persisted for several hundred generations after removal of the stimulus. It is inconceivable that modified cytoplasm could exert significant effects in such high dilutions. If "Dauermodifikationen" are strickly cytoplasmic, the modified cytoplasm obviously must reproduce itself in a sort of cytoplasmic inheritance.

Genetic Effects Of Syngamy

Syngamy in Haploid Flagellates

Meiosis appears to be zygotic in Phytomonadida, with the result that heterozygous vegetative stages are eliminated by persistence of the haploid chromosome number throughout most of the life-cycle. Since the genotypic composition of the flagellate is indicated by its phenotype after division of the zygote, the phytomonads may be favourable material for the study of biochemical genetics because so many species can be grown bacteria-free in media of known composition. The induction of mutations in autotrophic and heterotrophic types might produce physiological changes which could be analyzed genetically.

Experimentally induced loss of chlorophyll might make possible crosses between green and colourless strains of the same species. Such matings might supply significant data on the genetics and biochemistry of chlorophyll formation and perhaps on cytoplasmic inheritance. Although such aspects of phytomonad genetics have not been explored, the inheritance of morphological traits has been traced in a few species. The first observations were reported by Pascher in two strains of *Chalmydomonas*. In some cases, the lines derived from hybrid zygotes were essentially identical with one parental type or the other.

Occasionally, some of the lines showed combinations of parental characteristics and apparently represented new genetic combinations. Essentially the same pattern of inheritance was reported by Moewus in intraspecific and interspecific crosses of *Polytoma pascheri* and *P. uvella*. Linkage of such features as size of the body and length of flagella was described, and occasional crossing over was reported. Similar results were obtained with *Chlamydomonas eugametos*, *C. paradoxa*, *C. paupera*, and *C. pseudoparadoxa*. Although these observations are very interesting, they need confirmation because the validity of the data on crossing-over has been questioned.

Syngamy in Diploid Protozoa

A number of Protozoa undergo gametic meiosis and are diploid throughout most of the life-cycle. Except for the ciliates, which carry on conjugation instead of syngamy, the genetics of diploid species is yet to be investigated.

Genetic Effects of Conjugation

The significance of conjugation inheredity was discussed by Bütschli, R. Hertwig, and Maupas long before adequate experimental data were available. It was suggested that conjugation, in bringing about diparental inheritance, forms new combinations and thus increase variation. At the same time, conjugation was believed to level out

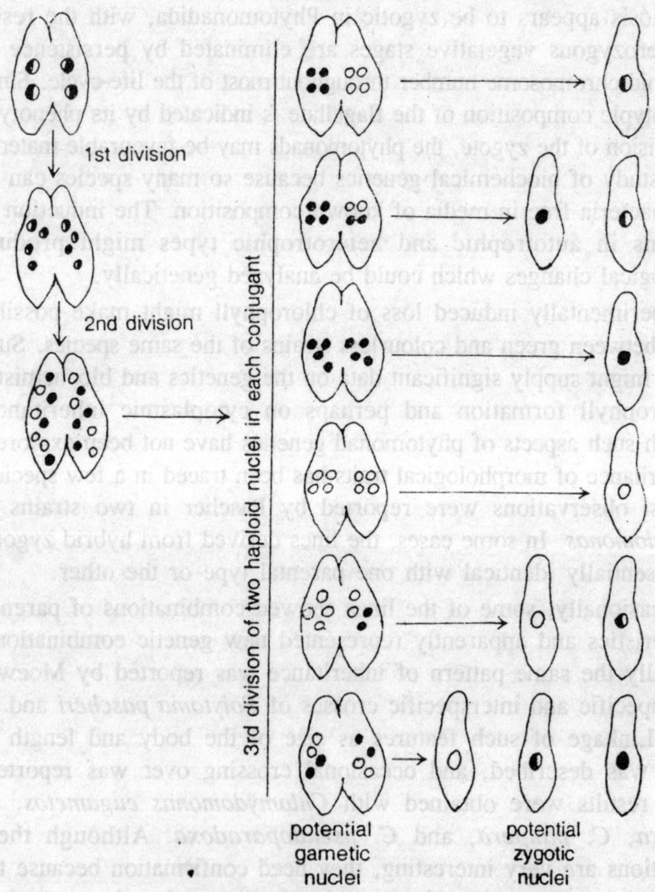

Fig. 4.1. Theoretical genetic effects of conjugation in Paramecium aurelia.

major differences arising in other ways, and in this sense, to limit the range of variation.

General Effects of Conjugation

The work of Pearl indicated that exconjugants are less variable than non-conjugants, and that conjugation tends to prevent extreme variation instead of inducing variation. However, Jennings found that exconjugants were more variable than non-conjugant lines with respect to fission-rate. Since these differences were inherited, conjugation in a population apparently gave rise to new biotypes, although the descendants of a single pair were closely similar as a result of biparental inheritance.

The appearance of new combinations after conjugation within a population was reported also in later investigations. The effects may vary with the strain of *Paramecium aurelia*, variation being increased in some strains but not in others. The other general effect of conjugation is the production of similarities through biparental inheritance in single pairs of exconjugant lines. In tracing the effects of hybridization on viability, body-length, and fission rate-of *Paramecium aurelia*, Sonneborn and Lynch found that some lines resembled one parental type, some resembled the second, and others were intermediate. Inbreeding showed that the intermediate types wre heterozygous; the others were apparently homozygous. It was concluded that the inheritance of these traits in *P. aurelia* is basically mendelian.

Cytoplasmic Lag in Biparental Inheritance

An unusual feature of hybridization has been the occurrence of a "cytoplasmic lag" in the exconjugant phenotypes of *P. aurelia*. In the experiments of Sonneborn and Lynch, the two lines from each pair of conjugants did not become phenotypically identical unit ten generations or so had passed. A similar lag characterizes inheritance of body-size in *P. caudatum*, the original size being retained in a hybrid exconjugant line for 10-36 generations. Since the two lines derived from a pair of conjugants were considered genotypically identical, this lag in appearance of the new phenotypes supposedly represented the time required for elimination of the old cytoplasm and the production of new cytoplasm under the influence of the heterozygous synkaryon. Assuming that the volume of old cytoplasm is halved at each fission, a dilution of at least 1:1,000 would seem to be required in these cases before the new zygotic nucleus can assert itself by producing a new phenotype.

Micronucleus in Conjugation

The behaviour of the micronucleus and its derivatives must be considered in relation to the potential genetic effects of conjugation. For instance, it is often assumed that the two gametic nuclei in a conjugant are derived from the same parental haploid nucleus. If this is the case, the nuclear contributions of a heterozygous conjugant to the two zygotic nuclei of the conjugating pair would be identical. So far as cytological evidence goes, this is not necessarily true in *Paramecium aurelia* because "two to five products of the second division continue to divide", and thus produce a number of potential gametic nuclei. Therefore, it is possible that the two successful gametic nuclei of a heterozygous conjugant could originate from different nuclei and thus be genetically different.

In *P. caudatum* also, a variable number of nuclei undergo the third pregamic division to produce more than two potential gametic nuclei, and both cross-fertilization and self-fertilization (cytogamy) are believed to occur in conjugation. Two products of the second maturation division normally undergo the third division in *Euplotes* so that there are four potential gametic nuclei. Are the functional gametic nuclei derived from one second-division nucleus or from two? Genetic data indicate that both methods of origin occur in *Euplotes*.

Behaviour of Mating Types in Conjugation

The effects of conjugation on mating types apparently vary with the species and the variety of ciliates. In variety I of *Paramecium bursaria* the descendants of each pair of conjugants have belonged to the same mating type in most cases. The few exceptional pairs show various results. In some cases, the two exconjugants may produce clones of different mating types. In other pairings, a single exconjugant has produced two different mating types. In some cases, these two types were parental types; in others, they were not. Five crosses, of the six possible for the four types in variety. I, have produced descendants belonging to all four mating types. Jenningss concluded that mating types are controlled by the genetic composition of the synkaryon and that the appearance of non-parental types might represent new nuclear combinations.

Chen has suggested that inheritance of mating types in *P. bursaria* is probably independent of the chromosomes count, which may vary as much among races within the same type as it does among different types. Inheritance of mating types in group A of *P. aurelia*—types I and II (variety 1), V and VI (variety 3) and IX and X (variety 5)—

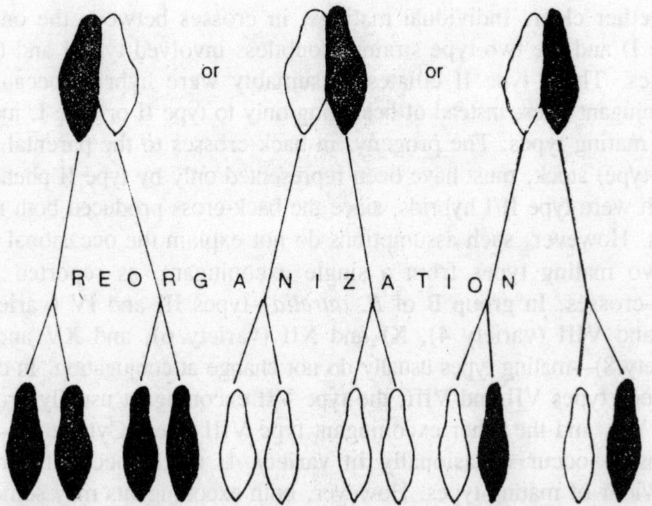

Fig. 4.2. Inheritance of mating types in conjugation of Paramecium aurelia group A, mating types I (solid black) and II.

may be illustrated by crosses between types I and II. Three results are possible. All of the exconjugant lines may be type I; all may be type II; or an individual exconjugant may differentiate into types I and II, usually at the first exconjugant fission. The third type of result was believed to indicate that mating types are controlled by the macronucleus, although it is not clear how this can explain the production of two types from one exconjugant.

The possible action of cytoplasmic factors (*plasmagenes*) in variety 1 seems to have been dismissed in the statement that "mating is determined by the genes and is not affected by whatever initial differences in cytoplasm may have existed". Mendelian inheritance has been reported in matings between a permanently type I strain and a "two-type" strain in which both types I and II appear within a clone. The hybrid exconjugant lines usually showed the two-type condition, indicating its dominance over the one-type trait. In back-crosses between the recessive (one-type) and the hybrid (two-type) progeny, about half of the exconjugant lines showed the one-type and the rest the two-type condition, as in a mendelian back-cross.

Certain exceptional results were later attributed to cytogamy, which may occur in about 60 per cent of the conjugating pairs at 27°, a temperature within the optimal range (25-30°) for conjugation of variety I. The method of inheritance, in this "first discovery of inheritance in Mendelian ratios in the ciliate Protozoa", is not altogether clear. Individual matings, in crosses between the one-type (type I) and the two-type strains, doubtless involved type I and type II ciliates. These type II ciliates presumably were hybrids because the exconjugant lines, instead of belonging only to type II or type I, included both mating types. The progeny, in back-crosses to the parental type I (one-type) stock, must have been represented only by type II phenotypes which were type II/I hybrids, since the back-cross produced both mating types. However, such assumptions do not explain the occasional origin of two mating types from a single exconjugant, as reported in the back-crosses. In group B of *P. aurelia*—types III and IV (variety 2), VII and VIII (variety 4), XI and XII (variety 6), and XV and XVI (variety 8)—mating types usually do not change at conjugation. In crosses between types VII and VIII, the type VII exconjugant usually produces type VII, and the other exconjugant type VIII lines. Cytogamy, which seems to occur occasionally in variety 4, might account for such behaviour of mating types. However, both exconjugants may sometimes produce type VII lines, or both may give rise to type VIII lines.

At present, the combined results cannot be explained logically on the basis of nuclear behaviour in conjugation. Consequently, Metz has insisted that cytoplasmic factors afford the only mechanism which can account for the behaviour of mating types in group B. The inheritance of mating types in *Euplotes patella* (types I-VI) has been explained by

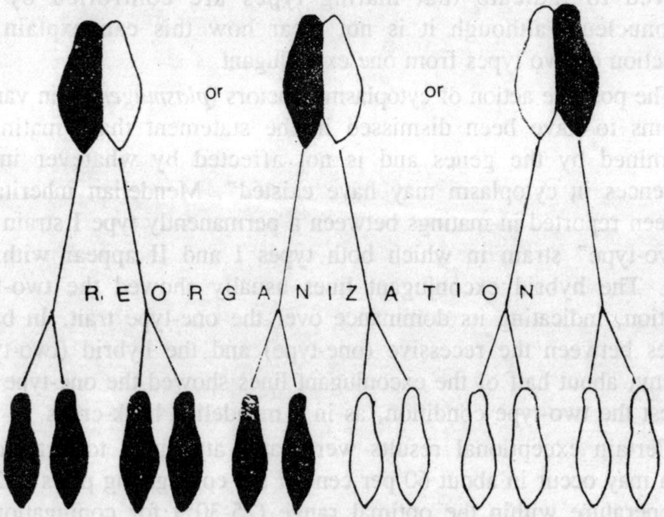

Fig. 4.3. Inheritance of mating types in conjugation of Paramecium aurelia group B, mating types VIII (solid black) and VIII.

a system of triple alleles in which the genotypes are represented as follows: type I, $mt_1\ mt_2$; type II, $mt_1\ mt_3$; type III, $mt_3\ mt_3$; type IV, $mt_1\ mt_1$; type V, $mt_2\ mt_3$; type VI, $mt_2\ mt_2$. Crosses between types I and II yield types I, II, IV, and V, while IV x VI crosses yield only type I. The crosses I x II, II x VI, and III x IV, also have produced the expected results. An interesting feature of *E. patella* is that amicronucleate strains have retained their type characteristic and have shown mating reactions with other amicronucleate *E. patella*.

Mating types in this ciliate are correlated with the production of specific substances which are released into the culture fluid and can induce conjugation in certain combinations. Induction of conjugation by a given substance is restricted to strains which cannot produce that substance, and may occur within a clone previously showing only one mating type. For instance, mating-type substance I, from a type IV culture, induces conjugation within a clone of type III, V, or VI. The relation of this phenomenon to the results obtained in crosses of two mating types is not yet clear.

Behaviour of Antigenic Types in Conjugation

At least four, and possibly six, antigenic types have been identified within killer stock 51 of *P. aurelia* and its variants. In crosses between

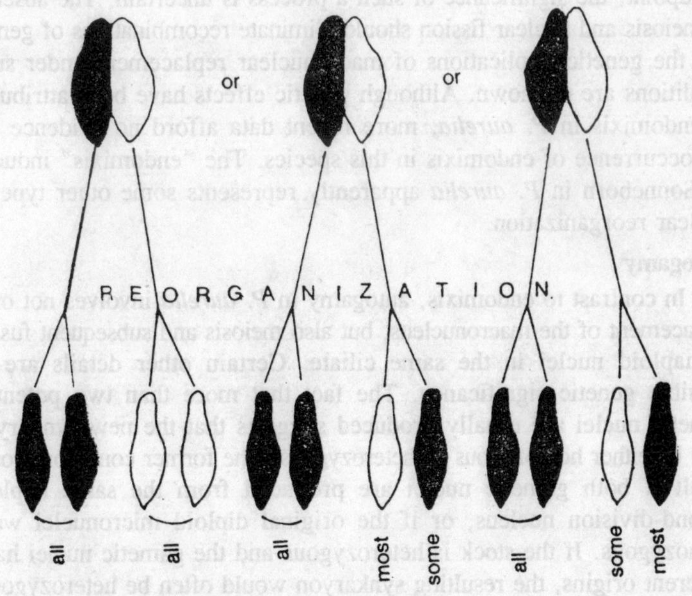

Fig. 4.4. Inheritance of antigenic types A (solid black) and B in conjugation of Paramecium aurelia.

antigenic types A and B, three different results may be expected. (1) The progeny of the type A conjugant remain type A and those of the other remain type B, even through autogamy. (2) The type A conjugant gives rise to type A lines, while the type B conjugant produces some type A, but mostly type B. (3) Descendants of the type A conjugant remain type A, while the other conjugant produces mostly type A and few or no type B lines. If a different type B strain is mated with type A, the proportions may be exactly reversed. The behaviour of micronuclei affords no apparent basis of all these results.

GENETIC SIGNIFICANCE OF ENDOMIXIS, AUTOGAMY AND CYTOGAMY

Endomixis

As described in *Paramecium aurelia*, endomixis involves: (1) a periodic disintegration and resorption of the macronucleus; (2) division of the micronuclei to produce eight daughter nuclei, all but two of which usually disintegrate; (3) a fission which produces two ciliates, each with a functional nucleus; (4) division of this nucleus to produce four, two of which become macronuclei; (5) division of both micronuclei at the next fission to restore the normal equipment. From the genetic standpoint, the significance of such a process is uncertain. The absence of meiosis and nuclear fission should eliminate recombinations of genes, and the genetic implications of macronuclear replacement under such conditions are unknown. Although genetic effects have been attributed to endomixis in *P. aurelia*, more recent data afford no evidence for the occurrence of endomixis in this species. The "endomixis" induced by Sonneborn in *P. aurelia* apparently represents some other type of nuclear reorganization.

Autogamy

In contrast to endomixis, autogamy in *P. aurelia* involves not only replacement of the macronucleus, but also meiosis and subsequent fusion of haploid nuclei in the same ciliate. Certain other details are of possible genetic significance. The fact that more than two potential gametic nuclei are usually produced suggests that the new synkaryon may be either homozygous or heterozygous. The former condition would result if both gametic nuclei are produced from the same haploid second-division nucleus, or if the original diploid micronuclei were homozygous. If the stock is heterozygous and the gametic nuclei have different origins, the resulting synkaryon would often be heterozygous. In other words, there is no cytological assurance that autogamy invariably results in a homozygous synkaryon. In contrast to this lack

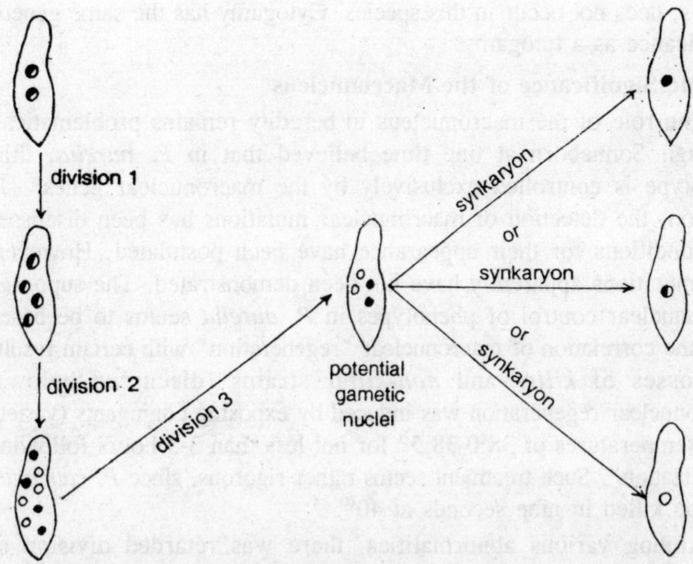

Fig. 4.5. Theoretical genetic effects of autogamy in heterozygous ciliate, based on Diller's description of autogamy in Paramecium aurelia.

of cytological evidence, it has been concluded from genetic data that the two gametic nuclei in *P. aurelia* are always genotypically identical. Genetic changes in autogamy have been reported in *Paramecium*. A clone of *P. bursaria* may differentiate into two mating types after autogamy. After autogamy in variety 1 of *P. aurelia*, all of the progeny usually belong to either mating type I or mating type II, and heterozygous strains apparently become homozygous. A lag has been noted in the transformation of heterozygous type II into type I. The change occurs at different times in different ciliates, so that cultures come to contain organisms of both types and may show mating reactions. Such as lag was not observed in the change from heterozygous type I to type II. In addition to the usual production of all type I or all type II, both mating types sometimes arise after autogamy. These unusual cases are not readily explained since the lines arising from one ciliate all contain micronuclei and macronuclei derived from the same synkaryon.

Cytogamy

In addition to autogamy, cytogamy is believed to occur in *P. aurelia*. Cytogamy is essentially incomplete conjugation in which the exchange of pronuclei is inhibited and each "conjugant" undergoes autogamy. The occasional occurrence of cytogamy in *Euplotes patella* also is suggested by genetic data; autogamy, as described for other ciliates, does not occur in this species. Cytogamy has the same genetic significance as a tutogamy.

Genetic Significance of the Macronucleus

The role of the macronucleus in heredity remains problematical, although Sonneborn at one time believed that in *P. aurelia*, "the phenotype is controlled exclusively by the macronuclear genes". In addition, the detection of macronuclear mutations has been discussed and conditions for their appearance have been postulated. However, such mutations apparently have not been demonstrated. The supposed macronuclear control of phenotypes in *P. aurelia* seems to be based upon the correlation of macronuclear "regeneration" with certain results in crosses of *killer* and *non-killer* strains (discussed below). Macronuclear regeneration was induced by exposing conjugants (variety 1) to temperatures of 38.0-38.5° for not less than 3-5 hours following "fertilization". Such treatment seems rather rigorous, since *P. caudatum* may be killed in nine seconds at 40°.

Among various abnormalities, there was retarded division of differentiating macronuclei. In the postconjugant fissions, some ciliates received new macronuclei and others only the fragments of the old macronuclei. The latter developed macronuclei from the fragments, each of which became a complete nucleus. The resulting macronuclei were distributed in subsequent fissions until the normal nuclear situation was restored. In applying this process to the study of genetic problems, Sonneborn crossed non-killer (*kk*) with homozygous killers (*KK*). Macronuclear regeneration was induced in the exconjugants derived from the non-killers. These were supposed to have received from their mates an excess of *kappa*, a cytoplasmic factor essential to development of the killer condition.

The zygotic nucleus of each exconjugant was a *Kk* genotype, and the macronuclei derived from the synkaryon had the same genotype. The non-killers which regenerated their macronuclei had to use fragments of the old *kk* macronucleus. Only ciliates with new *Kk* macronuclei developed into killers. Furthermore, non-killers (*Kk* micronuclei and *kk* macronuclei) produced no killers after autogamy, although some must have developed *KK* micronuclei and macronuclei. Their supply of kappa presumably was exhausted before autogamy occurred. Accordingly, the presence of gene *K* in the macronucleus was considered essential to the continued production of kappa in the cytoplasm.

It might be interesting to extend these observations to such a combination as *Kk* micronuclei and regenerated *KK* macronuclei, or *kk* micronuclei and regenerated *Kk* macronuclei, since the production of kappa is inhibited at temperatures above 33.5° It has been suggested that the macronucleus controls mating types in group A of *P. aurelia*. This assumption does not explain the origin of two mating types from one ciliate after autogamy, a phenomenon implying formation of two genotypically different macronuclei from one zygotic nucleus. The least improbable explanation, according to Sonneborn, involves macronuclear mutation, an assumption which cannot be tested experimentally at present.

Cytoplasm in Inheritance

There is a growing tendency to relate inheritance of certain traits in *Paramecium aurelia* to cytoplasmic factors which may, in different cases, be dependent upon or independent of nuclear genes. There characteristics include the *killer* trait, mating types, and antigenic varieties. Cytoplasmic inheritance is, in a rather real sense, a phenomenon familiar to all protozoologists in the self-perpetuation of blepharoplasts in flagellates and basal granules in ciliates. Therefore, there is nothing very startling in the possibility that self-perpetuating cytoplasmic particles may induce the appearance of specific substances which show physiologial activity without assuming the concrete form of new organelles. The current investigations on ciliates are being followed with much interest and with the hope that future developments may furnish logical explanations for some of the unsolved puzzles in protozoan genetics.

Killer* Trait in *Paramecium Aurelia

A *killer* strain of *P. aurelia*, as described in variety 4, gives off into the culture fluid a substance, *paramecin*, which is lethal to sensitive strains but without effect on killers. As an antibiotic substance, paramecin is interesting in that it shows differential effects within a single variety. Different strains of *P. aurelia* seem to produce different quantities and different kinds of paramecin. A single particle of paramecin (stock 51, variety 4), produced by a killer about once every five hours, may be enough to kill a sensitive ciliate (2), whereas Sonneborn has found that as many as half of the sensitive ciliates may survive when exposed singly or in small numbers to 10,000 or more particles of paramecin.

Paramecin seems to be a deoxyribonucleoprotein which is inactivated by pepsin, chymotrypsin, and desoxyribonucleases and shows a sensitivity to high temperatures comparable to that of various enzymes. The ability to produce paramecin is said to depend upon the presence of a dominant gene *K* and a self-reproducing plasmagene, *kappa*, Non-killer strains may be genotypes *KK, Kk* or *kk* which lack the factor kappa. Although gene *K* may occur in the absence of kappa, maintenance of kappa in the cytoplasm depends upon the presence of gene *K*, perhaps in the macronucleus as well as the micronucleus. Sonneborn suggested that kappa particles are distributed as single molecules throughout the body. Later calculations of Preer, however, indicate a size of 0.3-3.0μ for kappa particles.

Comparable particles, present only in killer strains, have been identified as granules stainable by Feulgen and Giemsa techniques and containing desoxyribonucleic acid. Treatment of *P. aurelia* with nitrogen mustard inactivates kappa and at the same time reduces the number of these granules. Kappa particles apparently lose the power to reproduce at temperatures above 33.5°. Some of these characteristics suggest close chemical similarity between kappa and paramecin. Under optimal conditions, kappa in variety 2 of *P. aurelia* is quadrupled daily. Consequently, it is possible to decrease the kappa content by increasing fission-rate, or even to eliminate kappa completely and permanently from strains of variety 2. This apparently is not possible for the kappa of variety 4 which increases fast enough to keep up with rapid fission. By depressing the fission-rate, the kappa content of a law-kappa strain may be increased progressively, at a rate which apparently varies with the strain or with the kind of kappa.

Mutations of kappa have been reported in variety 4 of *P. avrelia*. The mutant plasmagenes stimulated production of a new kind of paramecin with a different lethal action on sensitive ciliates. The behaviour of kappa in autogamy and conjugation has been described. After autogamy in a *Kk*-kappa line, the persistence of the killer trait for a few generations in homozygous recessive (*kk*-kappa) lines suggests that kappa can maintain the production of paramecium even without gene *K*. Sooner or later, however, the recessives become non-killers, presumably because kappa cannot increase in the absence of gene *K*. The disappearance of kappa thus shows a "lag" analogous to that described for inheritance of size in conjugation.

If gene *K* is reintroduced, by crossing the recessive with a homozygous killer (*KK*) strain, before all the kappa has been lost, the hybrid (*Kk*-kappa) descendants remain killers. If the cross, *kk* x *KK*-kappa, is made after the recessive has become a non-killer strain upon exhausting its original supply of kappa, the heterozygous (*Kk*) descendants of the non-killer conjugants remain non-killers. Such results are said to demonstrate that gene *K* cannot initiate the production of kappa after it has disappeared from the cytoplasm. Therefore, cytoplasmic inheritance is very important in transmission of the killer trait. It has been reported more recently that, in certain crosses of *KK*-kappa x *kk*, both exconjugant lines become killers (*Kk*-kappa). This result is said to depend upon transfer of cytoplasm from the killer to the non-killer conjugant.

When autogamy occurs in such lines, some of the descendants remain killers, presumably as *KK*-kappa genotypes, while the rest become non-killers after kappa has disappeared from the cytoplasm. The status of the killer and non-killer traits is further complicated by the conclusion that another pair of genes, *S* and *s,* and a corresponding plasmagene, *sigma*, are involved. The relationship between sigma and its homologous alleles are comparable to those between kappa and its related genes. Sigma supposedly has the ability to compete with kappa and to replace it under certain conditions, but is "not actually a factor for sensitivity". It now appears that a homozygous killer (*KKSS*) strain can yield pure killer and pure sensitive strains differing only in cytoplasmic factors.

Mating Types and Cytoplasmic Inheritance

The difficulty of explaining inheritance of mating types in variety 4 of *P. aurelia* is responsible for the conclusion that plasmagenes are involved. The usual appearance of the original mating type in each exconjugant line would have to be explained on the basis of cytogamy rather than interchange of gametic nuclei, if nuclear genes are responsible. The production of type VIII from conjugants belonging originally to types VII and VIII would imply that type VIII is dominant to VII, if nuclear control exists.

Other matings, in which the results are exactly reversed, would indicate that type VII is dominant to type VIII. Furthermore, the origin of both mating types from one exconjugant can scarcely be explained on a micronuclearr basis, although a similar phenomenon has been attributed tentatively to macronuclear mutation in group A. These peculiarities in the inheritance of group B mating types are attributed to the occurrence or the lack of cytoplasmic transfer during conjugation, although such a process has not been detected in cytological studies on *P. aurelia* and *P. bursaria*.

According to this hypothesis, no transfer to cytoplasm has occurred when type VII conjugants produce only type VII descendants, and type VIII only type VIII. If the exconjugant lines are all type VIII, in a VII x VIII mating, cytoplasm has been transferred from the type VIII to the VII conjugant. If the results are reversed, cytoplasm has been transferred from type VII to type VIII. If one exconjugant produces two mating types, interchange of cytoplasm has been followed by segregation of plasmagenes in post-conjugant fissions. Sonneborn assumes that there are two kinds of plasmagenes in variety 4, one controlling type VII and the other type VIII. The same explanation is believed to hold for other varieties of group B.

Antigenic Types and Cytoplasmic Inheritance

Hereditary antigenic variations in *Paramecium aurelia* have been reported by several workers. According to Harrison and Fowler, such variations may arise spontaneously. The stability of the variant differs with the strain and has ranged from less than three months to about four years. Similar variations have been induced in *P. aurelia* (stock 51) by exposure to X-rays and to homologous antisera. Temporary changes, lost after 1-15 fissions, also have been induced by exposure to trypsin and by maintenance of cultures at 14° for a number of generations (39). Antigenic types may be modified in different ways by different experimental methods. Types B, C and D may be converted into A by incubation at 32°, and types A, C and D may be changed to B by incubation at 12°.

All the various types remain stable at 27° if the food supply is controlled to maintain one fission a day. The antigenic type in variety 4 of *P. aurelia* is said to depend upon competition between plasmagenes. In the only experiments with bacteria-free cultures, an antigenic modification, exhibited as insensitivity to antiserum, has been produced by exposure of *Tetrahymena* to homologous antiserum, but the ciliates reverted to the original type after two transfers in normal culture medium. The status of antigenic varieties in *P. aurelia* is not yet settled. Since there is no evidence for micronuclear control, it has been suggested that cytoplasmic inheritance determines the observed behaviour in conjugation and autogamy.

Results obtained in conjugation of types A and B are attributed to a lack of cytoplasmic transfer in some cases, and to the transfer to large or small amounts of cytoplasm in others. After an A x B cross, the inbreeding of type A exconjugant lines yields only type A; that of type B exconjugants, only type B. Cytoplasmic inheritance is believed to offer the most logical explanation for these various results. Possible mechanisms involved in the antigenic transformations of *P. aurelia* have been discussed by several workers, and Beale has suggested that the postulated plasmagenes are the antigens themselves.

5

Living Activities

Nutritional Requirements of Protozoa

The establishment of various species in cultures free from other microorganisms opened a new era in the study of protozoan nutrition, making possible for the first time a realistic approach to the determination of basal food requirements. As a result, the investigation of metabolic activities in Protozoa is steadily expanding, and is leading to increasingly precise interpretations. One of the important factors in this rapidly developing field has been the availability of many phytoflagellates in pure culture. For this, much credit is due E.G. Pringsheim whose unfailing determination was responsible for maintaining in invaluable collection intact through a number of trying years.

The phytoflagellates are particularly favourable material for investigation because their absolute requirements are much less extensive than those of higher Protozoa. Consequently, the afford they most direct routes to the determination of mineral requirements, and also the need for certain vitamins and organic foods. The relatively simple requirements of certain phytoflagellates, in contrast to the complex brews needed by higher Protozoa, also should expedite the study of metabolic pathways. Furthermore, the diversity of phytoflagellates, with respect to the possession or lack of chlorophyll and the presence of absence of holozoic habits, encourages consideration of the evolutionary and taxonomic aspects of protozoan nutrition. Until recently, ciliates and Zoomastigophorea had been grown only in broths of unknown chemical composition which endangered the validity of conclusions concerning basal food requirements.

The development of almost completely defined media for at least a few of these higher Protozoa insures much the same results as those now obtainable with many phytoflagellates. Aside from the intrinsic interest to protozoologists, the study of protozoan nutrition promises significant contributions to the general fields of biochemistry and physiology. In the study of plant nutrition, the rapidly growing chlorophyll-bearing flagellates are readily adaptable to the investigation of various fundamental problems. The study of animal nutrition might be served in the same way by the typical animals among the Protozoa. As microscopic animals, which approach bacteria in rates of growth and ease of handling, Protozoan in pure cultures also offer material for studying animal metabolism under conditions controllable to a degree not attained with tissues of higher animals.

Table 5.1. Constituents of Culture Media for the Colourless Phytoflagellate, Chilomonas Paramecium, and the Ciliate, *Tetrahymena Pyriformis*, Strain E (121)

Chilomonas paramecium		*Tetrahymena*
NH_4Cl	Arginine	riboflavin
$(NH_4)_2SO_4$	histidine	pteroglyglutamic acid
KH_2PO_4	isoleucine	biotin
$MgCl_2 . 6H_2O$	leucine	thiamine
$CaCl_2 \; 2H_2O$	lysine	choline
$FeCl_3 . 6H_2O$	methionine	yeast nucleic acid
Acetate (or ethyl alcohol)	phenylalanine	protogen
Thiamine	serine	$MgSO_4 . 7H_2O$
	threconine	K_2HPO_4
	tryptophane	$CaCl_2 . 2H_2O$
	valine	$FeCl_3 . 6H_2O$
	dextrose	$Fe(NH_4)_2(SO_4)_2 . 6H_2O$
	sodium acetate	$CuCl_2 . 4H_2O$
	pantothenate	$MnCl_2 . 4H_2O$
	nicotinamide	$ZnCl_2$
	pyridoxine	

In addition, precise control of the food supply favours use of these organisms in the search for new vitamins, as in the discovery of protogen, as well as in micro-biological assays of known growth-factors and amino acids. To the parasitologist and the pathologist, parasitic

Protozoa in pure cultures offer unique opportunities for correlating metabolic activities of parasites with susceptibilities to chemotherapeutic agents and with reactions of the host's tissues of infections. To the explorer, these organism extend a challenge to trace nutritional evolution from the possibly complete synthesis of needed vitamins to an essentially complete dependence upon external sources. Did plant-like Protozoa suddenly become 'animals, with wholesale loss of synthetic abilities, or did they lose their original abilities one by one as evolution tempted them toward the animal kingdom? Or have the phytoflagellates arisen from more animal-like Protozoa, acquiring in their evolution various synthetic powers unknown to their ancestors?

General Types of Nutrition

For many years, general types of nutrition were classified merely as *autotrophic* (or holophytic), *saprozoic*, and *holozoic*, the last two representing varieties of heterotrophic nutrition. By definition, autotrophs could grow in inorganic media while heterotrophs required organic foods. With the exception of saprozoic types, Protozoa probably are not limited to one method in a natural environment. Chlorophyll-bearing species are often saprozoic and some can grow in darkness. A number of the green flagellates also ingest solid food, while typically holozoic organisms also may carry on saprozoic nutrition.

The relative importance of one method or another depends largely upon environmental conditions. The ecological classification of Kolkwitz and Marsson (301) divides Protozoa into *katharobes*, living in water containing almost no organic matter; and *saprobes*, found in water containing appreciable quantities of organic matter. Saprobes are divided into *oligosaprobes*, *mesosaprobes*, and *polysaprobes*, living in the presence of small, moderate, and large amounts of organic matter. This classification involves oxygen relationships as well as food supply, since katharobes are typically aerobic, while polysaprobes are more probably anaerobes or facultative anaerobes. The earlier results with pure cultures soon demonstrated that these older concepts were inadequate. For instance, it became clear that the terms, *autotroph*, could no longer be applied automatically to any green flagellate that occurs naturally in fairly pure water exposed to light.

In fact, it is not yet certain that the existence of complete autotrophs, as as originally defined, has been demonstrated. Furthermore, some of the "saprozoic" flagellates proved to be "autotrophic" with respect to nitrogen sources, although needing organic foods as a source of energy. This situation furnished the stimulus for several more modern classifications of protozoan nutrition. Although such classifications were a distinct improvement over the older systems and were a convenience in discussions, it now appears that even the more modern classifications have a tendency to oversimplify protozoan nutrition. From present indications, food and vitamin requirements of Protozoa will show many variations from species to species, so that much more information will be needed before definitive classification can be attempted.

Determination of Food Requirements

Two general methods have been followed in the study of protozoan food requirements. In one procedure the experimental media have been the simplest ones which would support growth in serial transfers. In such media, growth is often at a minimum and the organisms do little more than maintain themselves in successive transfers. In the other general procedure, media have been devised to maintain growth at a maximum and any constituent which cannot be omitted is considered essential to growth. This method is based upon the generalization the growth of any species will reach a maximum when all conditions are optimal—qualitative and quantitative aspects of the substrata, concentrations of stimulatory and essential growth-factors, concentrations of essential minerals, and non-dietary environmental factors. On a theoretical basis, the first procedure might seem to offer the more direct analysis.

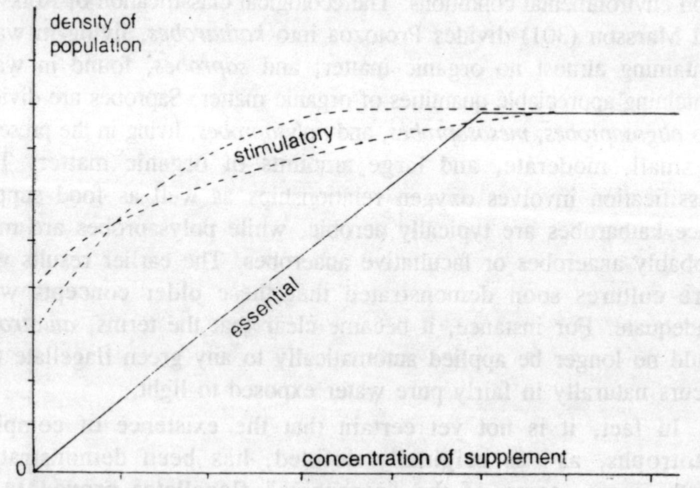

Fig. 5.1. Hypothetical growth responses of a test organism to essential and stimulatory growth-factors.

In a medium reduced to bare essentials, it might be possible to recognize species capable of synthesizing required materials at a rate so slow that growth could never reach the maximum attainable in a rich medium. In the second procedure, with maximal growth as the goal, slow synthesis of a particular factor might conceivably be overlooked. In practice, however, the first methods has certain limitations. Validity of the results obviously depends upon purity of the reagents and cleanliness of the culture vessels. In addition, contamination of the medium with dust or with volatile materials from the atmosphere of a laboratory could be a possibly serious source of error. Even minute contaminations might turn the balance in favour of slight growth, with resulting faulty interpretations of experimental data. Hence, it is essential, is following the first procedure, to take all possible precautions. In the use of media which support maximal growth, in influence of minute contaminations would be less likely to account for positive instead of negative results.

Furthermore, the response of graded increments of a given growth-factor can be traced over a wide range of growth. An approximately linear growth-response to a vitamin or a mineral in concentrations ranging from zero to an optimum would indicate that the factor is essential. Omission of stimulatory substance, on the other hand, would decrease growth but not prevent it completely.

Autotrophic Nutrition

This general variety of nutrition, in which inorganic nitrogen is adequate for growth, is sometimes considered a primitive type which was gradually lost during the "regressive" evolution of heterotrophs. Another view is that the evolution of autotrophic organisms has involved the acquisition of synthetic abilities lacking in more primitive ancestral types which were dependent upon the environment for critical organic materials. Whether autotrophic nutrition should be considered primitive or not, the food requirements of the supposedly autotrophic phytoflagellates differ considerably from those of other Protozoa. Chemoautotrophic, photoautotrophic, and heteroautotrophic nutrition have been reported in various phytoflagellates. Three species have been grown under conditions suggesting chemoautotrophy—two of the three in glass-covered culture vessels.

The investigation of chemoautotrophic nutrition illustrates, to an extreme degree, the difficulties inherent in the use of media supplying a bare minimum for growth. After an adequate number of transfers, it may be concluded that the food supply of the organism is limited to constituents of the test medium—*plus* any contaminants absorbed from glassware, introduced by way of distilled water, stock solutions and dust or absorbed from the air. The first obvious refinement in technique is the elimination of cotton plugs from culture tubes or flasks. Although bleached cotton may be inactive as a source of thiamine or its components plugs of this material contribute significantly to the growth of *Chilomonas paramecium* in an inorganic medium and the effect is not eliminated by the addition of thiamine in excess.

Table 5.2. Reported Cases of "Autotrophic" Nutrition in Flagellates

Species	*"Chemoauto-trophic"*	*"Photoauto-trophic"*	*Heteroauto-trophic*
Cryptomonadida			
Chilomonas paramecium	NC (390)		NC (390)
Chilomonas paramecium			CP (220)
Chilomonas paramecium			NC (354)
Chilomonas paramecium	NC (73)		NC (73)
Phytomonadida			
Chlamydomonas agloëformis			CP (340)
Chlorogonium elongatum		CP (321)	
C. euchlorum		CP (321)	
Eudorina elegans		NC (108)	
Haematococcus pluvialis		CP (340)	
Haematococcus pluvialis		CP (420)	CP (420)
Lobomonas piriformis	CPP (422)	CP (422)	CP (422)
Polytoma caudatum			CP (352)
P. obtusum			NC (355)
P. ocellatum			NC (353)
P. uvella			CP (340, 452)
P. uvella			NC (355)
Polytomella caeca			NC (353)
Euglenida			
Astasia longa (Jahn strain)	NC (505)		NC (503)
Euglena anabaena		CP (105, 175)	
Euglena anabaena		NC (109)	
E. gracilis		CP (104, 189)	NC (504)
E. klebsii		CP (105)	
E. stellata		CP (105)	
E. viridis		CP (177), NC (508)	

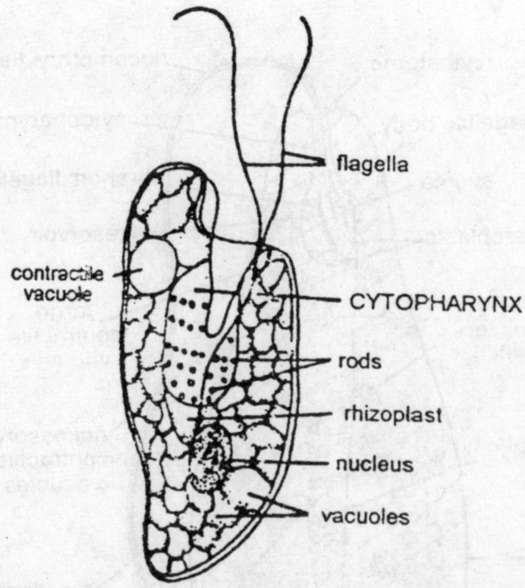

Fig. 5.2. Chilomonas.

Organic pollution of distilled water and stock solutions by bacterial growth can be avoided by aseptic techniques or by the use of a volatile preservative. However, other difficulties remain. Certain inorganic salts used for culture media contain organic contaminants, which in terms of carbon balance could account for the observed growth in an experimental medium. Furthermore, there must be a demonstrable in organic source of energy. In the case of *Chilomonas paramecium*, there is no significant oxidation of ammonia, although tests have not been reported for several other "chemoautotrophs." Accordingly, it must be concluded that the evidence for chemoautotrophy in Protozoa is still inadequate, and that under experimental conditions now attainable it is seemingly impossible to limit a flagellate to chemoautotrophic nutrition. So far as photoautotrophic nutrition is concerned, it should be much less difficult to prove that organic contaminants are an unimportant factor in growth.

In this connection, the failure of *Chlamydomonas moewusii* to grow without both light and carbon dioxide—even in inorganic media supplemented with a variety of nitrogen sources, vitamins and oxidizable carbon sources—seems very significant. Investigations are still handicapped, however, by inadequate knowledge of qualitative and quantitative mineral requirements. "Photoautotrophic" nutrition has been reported in several Phytomonadida and Euglenida. In most cases, culture vessels have been plugged with cotton, but *Eudorina elegans* has been maintained in all-glass vessels. Such results also have supported earlier

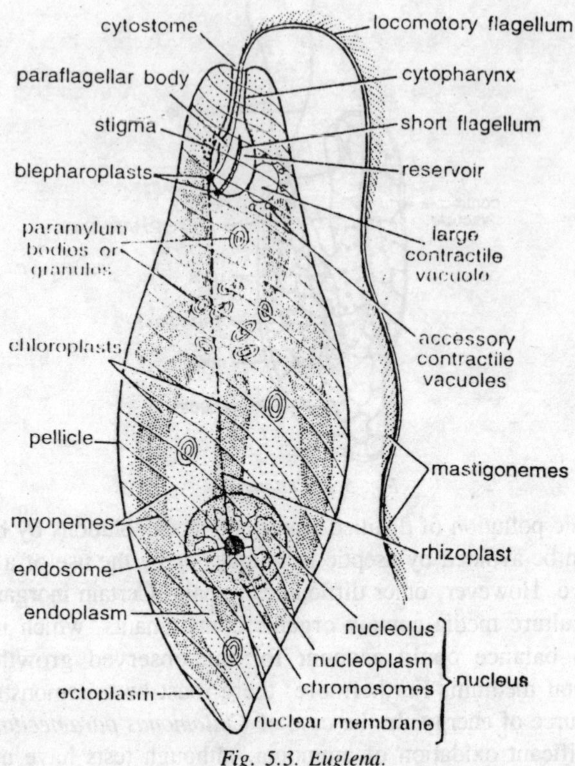

Fig. 5.3. Euglena.

observations on *Euglena anabaena*. Similar confirmation is needed for other reports of photoautotrophy.

All-glass vessels do not eliminate all potential organic contaminants, but it is possible that such impurities could not account for the observed growth without cotton plugs. In contrast to *Euglena anabaena*, *E. gracilis* is said to require the pyrimidine component and *E. pisciformis*, both the pyrimidine and thiazole components of thiamine for growth as "photoautotrophs." Whether or not any or all of the currently reported photoautotrophs will eventually prove to be such, it appears at present that certain green flagellates may not be photoautotrophs. For example, failures to demonstrate "photoautotrophy" have been reported for *Euglena deses*. Such failures could have been caused by inadequate culture media. On the other hand, *E. deses* may represent a more advanced stage in the type of regressive evolution suggested for *E. gracilis* and *E. pisciformis*. Heteroautotrophic nutrition was first noted by Pringsheim in *Polytoma uvella*. These observations have been repeated and comparable findings have been reported for *P. obtusum*. The related flagellates, *Polytoma caudatum*, *P. ocellatum* and *Polytomella caeca*, have been grown in inorganic salt and acetate media with supplementary growth-factors.

More recently, *Astasia longa*, *Euglena gracilis* and *Lobomonas piriformis* have been maintained as heteroautotrophs, the green species being incubated in darkness. *Chilomonas paramecium* also is a facultative heteroautotroph, although there is one report that supplementary growth-factors are required under such conditions. At present, it appears that certain flagellates are facultative heteroautotrophs for which thiamine and possibly other vitamins are stimulatory but not essential, while other species require one or more supplementary growth-factors under such conditions. This possibility needs further investigation with careful attention to mineral requirements and with a variety of substrates.

Mineral Requirements

In contrast to older beliefs that some ten or eleven elements are essential of life, modern investigations have detected about fifty elements in the tissues of different animals and plants. It remains to be determined just how many are essential and how many are chance accumulations conditioned by a particular chemical environment. So far as the Protozoa are concerned, little is known about qualitative and quantitative mineral requirements.

For microorganisms in general, certain metal requirements are related to particular enzyme systems. Such metals may be integral parts of enzymes or may serve as "activators" whose exact functions are not yet understood. Consequently, it is at least conceivable that certain mineral requirements may vary quantitatively, and possibly even qualitatively, in the presence of different substrates. Growth in organic media cannot be expected to yield many clues because such materials contain quite a variety of trace elements. In yeast extract, for example, barium, bismuth, boron, calcium, chromium, copper, gold, iron, magnesium, manganese, phosphorus, potassium, silicon, silver, and strontium have been demonstrated by spectroscopic analysis, and a preparation of asparagine has contained as many as 21 trace metals. Such difficulties have been illustrated in the investigation of mineral requirements in *Tetrahymena pyriformis*.

For various phytoflagellates, culture media have been prepared from analyzed reagents and used in all-glass culture vessels. Within such limits of accuracy, the composition of the medium is known except for possibly significant contaminations from glassware or other sources. Some of these media contain, as impurities in the salts, barium, cobalt, chromium, copper, gold, iron, manganese, sodium, zinc, and other metals, in addition to the intentionally supplied calcium, magnesium, phosphorus, potassium, sulfur, chlorine, ammonium salts, and carbon dioxide. It has been possible, from such starting points, to detect a few trace-mineral requirements.

The determination of others, merely by selecting different salts so as to exclude particular elements, has been impossible. The purification of reagents permits some further progress. The removal of certain trace metals has been accomplished with chelating agents, such as 8-hydroxyquinoline, which form metal complexes soluble in chloroform or ethyl acetate. In addition, several other methods for purifying important components of culture media are available and a number of purified elements and reagents can be obtained commercially. Additional advantages are promised by the use of non-toxic chelating agents in culture media. Such substances as citrate and ethylenediamine-tetraacetic acid, which form soluble and fairly stable metal complexes, minimize and tend to eliminate certain technical difficulties. A metal-chelate complex forms a sort of metal 'buffer" with an action somewhat analogous to that of pH-buffers.

Since the precipitation of metals as hydroxides, phosphates, or sulfates in prevented, it is possible to add quantities sufficient for heavy growth. Metals which are toxic above certain concentrations also can be supplied at levels favouring heavy growth without danger of toxic effects. Aside from such general improvement of the simpler culture media, chelating agents can be of assistance in analyzing mineral requirements. The addition of a chelating compound to a medium containing essential trace metals in minimal amounts may induce a metal deficiency which will prevent growth. Incidentally, materials often supplied as substrates—a-amino acids, glycerol, malate, ammonium salts—also may be active enough as chelating agents to induce metal deficiencies. Therefore, they might be considered of little value to the organism unless allowance is made for chelating activity. Intentional induction of an inhibitory metal deficiency makes it possible, by trial and error, to identify various trace elements which seem to be essential.

After preliminary qualitative observations, individual requirements can be analyzed quantitatively by adding an excess of all needed elements except one and then determining the amounts of this element which will compensate for graded increases in the chelating agent. A curve plotted from such data and then extrapolated to zero chelate should give a fairly good approximation of the basal requirement for a particular metal. There remains to be considered the situation in which a trace element minutely contaminating a supposedly required metal may actually be the essential factor. This complication can be eliminated, if elimination is possible, only by the use of highly purified metal sources. However, the failure of progressive purification to alter the apparent requirements quantitatively would suggest that the original indications were valid. The older techniques were adequate only to the extent of indicating qualtitative requirements for certain metals in a few species. Calcium—apparently needed by *Euglena anabaena*, *E. stellata*, *Hyalogonium klebsii*, *Chilomonas paramecium*, *Oikomonas termo* and *Tetrahymena pyriformis*—may prove to be a general requirement. Magnesium, necessary for *Chilomonas paramecium* and *Tetrahymena pyriformis*, is a component of carboxylase and should be a general requirement.

As a constituent of chlorophyll, magnesium also is obviously essential to green flagellates. Iron was found to be a requirement of *Chilomonas paramecium*, *Eudorina elegans*, *Euglena anabaena*, *Polytoma obstusum* and *P. uvella*, and *Tetrachymena pyriformis*. Since this metal is a constituent of cytochromes, cytochrome oxidase, catalase, and peroxidase, it may be impossible to find Protozoa which do not need iron unless it can be shown that obligate anaerobes have absolutely no iron requirements. Phosphorus, essential for *Chilomonas paramecium* and *Tetrahymena pyriformis*, is obviously a general requirement for phosphorylation of metabolites and vitamins.

Observations on the phosphate cycle in *T. pyriformis* indicate that during the lag phase of growth there is a rapid liberation of inorganic phosphate from organic sources, whereas uptake of inorganic phosphate is more characteristic of logarithmic growth. Manganese, which favours growth of *Euglena anabaena* in inorganic media, and apparently participates in oxidation of pyruvate and other metabolites by *Plasmodium gallinaceum*, may be generally needed as an activator of phosphorylases and peptidases and possibly other enzymes. Potassium, required by *Chilomonas paramecium* and *T. phriformis*, seems to be needed in certain phosphorylations and probably is a general requirement.

The possible significance of sodium, apparently needed by *C. paramecium*, is uncertain. Sulphur, as a constituent of several vitamins and amino acids, is presumably essential. A variety of inorganic salts, cystine, glutathione, and cysteine are satisfactory sources for *C. paramecium*. Silicon stimulates growth of *C. paramecium* and prolongs life in a phosphate-deficient medium. Whether this effect is attributable to silicon or to impurities in the silicate used, remains to be determined. Vanadium also seems to accelerate growth of *C. paramecium*. A need for copper, apparently a component of ascorbic and phenol oxidases, became apparent in *Tetrahymena pyriformis* when natural products were replaced by purified constituents of culture media.

Zinc, apparently involved in aldolase, carbonic anhydrase, and uricase activity, has often been included in culture media on the assumption that it is essential to protozoan growth. Cobalt, as a constituent of vitamin B_{12} (cyano-cobalamin), is required by *Euglena gracilis* and probably various other Protozoa. Molybdenum, which accelerates nitrogen-fixation by bacteria and seems to be essential for certain molds, needs investigation as a protozoan requirement. In summary, fragmentary evidence now indicates that, in addition to carbon, hydrogen, oxygen, and nitrogen, at least twelve other elements—calcium, cobalt, copper, iron, magnesium manganese, phosphorus, potassium, silicon, sodium, sulfur, and vanadium—are either stimulatory or essential to growth of certain Protozoa.

In addition, there are reasons for believing that others, such as molybdenum and zinc, may be important. Highly purified chemicals and the newer techniques of investigation may expand the list of required trace elements, and should clarify the status of some of them in protozoan metabolism. Even so, a complete list of the basal requirements apparently remains unobtainable with the inorganic materials and culture vessels now available.

Vitamin Requirements

The requirements of many Protozoa, although incompletely known, are probably comparable to those of Metazoa. *Colpoda steinii* (*duodenaria*) needs more than five vitamins, *Tetrachymena pyriformis* needs at least nine or ten, and not less than six are important in the metabolism of malarial parasites. At the other extreme, a few phytoflagellates have been grown in media apparently free from vitamins. There is every reason to believe that such differences depend upon the ability or inability to synthesize particular vitamins. Among the phytoflagellates, *Chilomonas paramecium*, in an acetate, inorganic salt and thiamine medium, synthesize nicotinic acid and the diphosphopyridine nucleotide (DPN, or coenzyme I) which contains nicotinamide and adenine.

Microbiological assays of comparable cultures have confirmed the synthesis of nicotinic acid and demonstrated that of pyridoxal and riboflavin. In addition, growth of *Tetrahymena pyriformis* on *C. paramecium* and on *Polytoma ocellatum* in similar culture media suggests that these flagellates are able to synthesize a variety of vitamins needed by the ciliate. Therefore, it may be assumed, in the absence of evidence to the contrary, that metabolic activities of the phytoflagellates involve essentially the same vitamins as do those of the higher Protozoa. Under favourable conditions, which must include a medium satisfying essential mineral requirements, some species may be able to synthesize all of their needed vitamins.

Present indications, that certain other phytoflagellates cannot synthesize at least one or two vitamins from simple materials, raise interesting possibilities. Perhaps it will be feasible, in this group to trace a series of stages in the development of multiple vitamin requirements (or multiple losses in synthetic powers) as represented by ciliates, for example. As the scope of the pure-culture techniques is broadened, the ability to visualize vitamin requirements on a taxonomic framework may prove very interesting—possibly to the extent of furnishing clues to the phylogeny of the higher Protozoa. From the practical standpoint, the determination of vitamin requirements for many different Protozoa may reveal unsuspected new vitamins and may also furnish additional tools for microbiological assays. Both possibilities have already been realized to a limited extent. Some information on vitamin requirements is now at hand for a number of species. Although the present data may be definitive for a few phytoflagellates, this is far from true for nearly all of the other Protozoa which have been investigated.

Thiamine

This vitamin is an absolute requirement for certain strains of ciliates and parasitic flagellates and probably for malarial parasites. The case of *Chilomonas paramecium* is still puzzling. Certain strains apparently require either thiamine or its thiazole and pyrimidine components, while others have been grown in all-glass vessels without added thiamine on acetate as a substrate. Under such conditions, supplementary thiamine markedly increases growth on acetate and becomes essential instead of stimulatory when pyruvate is substituted for acetate.

Thiamine is stimulatory for *Polytoma obtusum* and *P. uvella* in simple media, although both will grow without the added vitamin. In the same types of media, certain other colourless phytomonads need thiazole or both the pyrimidine and thiazole components of thiamine. Several substituted thiazoles and pyrimidine also are active for *Polytomella caeca* and *Chilomonas paramecium*. In addition, heavy growth of *Euglena gracilis* var. *bacillaris*, in amino acid and inorganic salt medium containing vitamin B_{12}, depends upon an adequate concentration of thiamine.

Riboflavin

Earlier reports of growth-acceleration in ciliates were soon followed by evidence that this vitamin is essential for *Colpoda steinii* (*duodenaria*) and *Tetrahymena pyriformis* and probably for malarial parasites. The synthesis of riboflavin by *Chilomonas paramecium* suggests the probable importance of this vitamin in phytoflagellate metabolism.

Pyridoxine complex

In the earlier investigations a stimulation of growth by pyridoxine was noted in several ciliates. A ciliate, *Colpoda steinii*, also was the first protozoon shown to need pyridoxine. *Tetrahymena pyriformis* has since been found to required pyridoxine, pyridoxal, or pyridoxamine, the two derivatives being 100-500 times as active as pyridoxine, a relationship similar to that previously reported for certain bacteria. Pyridoxine proved to be a component of 'Factor II", a concentrate of natural origin previously found essential to growth of *T. pyriformis*. Since pyridoxine inhibits the action of quinine and atebrin against *Plasmodium cathemerium* and *P. lophurae* in ducklings, the vitamin probably is a requirement of malarial parasites. Among the phytoflagellates, *Chilomonas paramecium* synthesizes pyridoxal.

Pantothenic acid

In the first tests on Protozoa, Elliott found that growth of *T. pyriformis* was accelerated, within the pH range 5.5-6.5, by a concentrate of pantothenic acid. Garnjobst, Tatum and Taylor next found pantothenate essential for *Colpoda steinii*, and it now appears that *Tetrahymena pyriformis* has the same requirement. Supplementary evidence involves inhibiton of growth of *T. pyriformis* by a-methyl-pantothenic acid and reversal of the effect by pantothenic acid. This vitamin also favours survival of *P. lophurae in vitro*. Furthermore, a pantothenate deficiency in chickens decreases the severity of infections with *P. gallinaceum*, and dosage with certain analogues is more effective than quinine therapy. Pantothenate analogues are active likwise against *Trichomonas foetus*, *T. gallinae*, and *T. vaginalis* in pure cultures.

Supplementary pantothenate in the diet of rats also increases the population of *Eimeria nieschulzi*.

Table 5.3. Vitamin Requirements of Various Protozoa

Species	*Vitamins*
Phytomastigophorea	
Chilomonas paramecium	A, B (353); none (73)
Euglena gracilis	B (353); N (230)
E. pisciformis	A, B (107)
Haematococcus pluvialis	C, K (420); non (340)
Polytoma caudatum	A (352, 353)
P. obtusum	none (355)
P. ocellatum	A (353)
P. uvella	none (355)
Polytomella caeca	A, B (351)
Zoomastigophorea	
Eutrichomastix colubrorum	K, L, Q (49)
Leishmania agamae	M, Q (366)
L. ceramodacyli	M Q (366)
L. donovani	K, M, Q (366)
L. tropica	K, M, Q (366)
Leptomonas ctenocephali	M (366)
L. pyrrhocoris	M (366)
Strigomonas culicidarum	C, M (366); C, D, E, M (566)
S. fasciculata	C, M (366)
S. muscidarum	M (366)
S. oncopelti	C (366)
Trichomonas columbae	K (50), L (47)
T. foetus	K, L, Q (48)
T. gallinarum	K, Q (52)
T. vaginalis	F (303), L (537)
Trypanosoma cruzi	K, M, Q (366)
T. lewisi	M (432)
T. rabinowitchi	M (410)
Sarcodina	
Acanthamoeba castellanii	A, B (345)
Sporozoa	
Plasmodium gallinaceum	c, g, j (535)
P. knowlesi	d, j, p (11)
P. lophurae	f, h (556)

Ciliatea	
Colpidium camphylum	Q (437)
Colpoda duodenaria	C, E, G (545); D, F, Q (139)
Pleurotricha lanceolata	D, E, Q (317)
Stylonychia pustulata	C (120, 122, 179, 13, 357);
Tetrahymena pyriformis	D (179, 29, 290); E, F, G (121, 289, 290); I (122, 275); J (284); O (121, 540)

Key : A, thiazole component of thiamine; B, pyrimidine component of thiamine; C, thiamine; D, riboflavin; E, pyridoxine; f, pantothenic acid; G, nicotinic acid or nicotinamide; H, biotin; I, pteroylglutamic acid; J, nucleic acid components (purines, pyrimidines); K, ascorbic acid; L, sterols; M, hematin; N, cyanocobalamin (vitamin B_{12}); O, protogen; P, *p*-aminobenzoic acid; Q, unidentified growth-factors. Small letters indicate *probable* requirements.

Nicotinic acid

Colpoda steinii was the first protozoan species shown to require nicotinic acid. Later on, *Tetrahymena pyriformis*, at first believed to

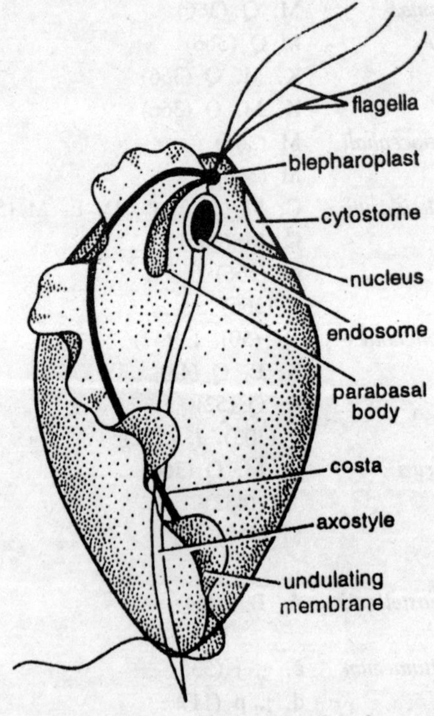

Fig. 5.4. Trichomonas.

grow without nicotinic acid, was found to need the vitamin. Diphosphopyridine nucleotide (DPN), which contains nicotinamide, also has been demonstrated in *T. pyriformis*. As a component of DPN and TPN, nicotinamide also is involved in oxidative metabolism of *Plasmodium gallinaceum* and *Trypanosoma hippicum*. Among the phytoflagellates, growth of *Eugena viridis* in an asparagine medium and that of *Chilomonas paramecium* as a heteroautotroph are stimulated by nicotinic acid. The latter also synthesized this vitamine.

Biotin

Although a biotin deficiency has decreased division-rate and reduced the density of populations evidence that this vitamin is essential for *Tetrahymena pyriformis* is still lacking. However, it seems to be important in the metabolism of malarial parasites.

Pteroylglutamic acid and p-aminobenzoic acid

In the first report on Protozoa, Kidder found this vitamin essential to growth of *T. pyriformis*. Calculated on the basis of free pteroylglutamic acid, the vitamin has about the same activity as it conjugates, pteroylglutamylglutamic acid and pteroylhexaglutamic acid. Apparently, *p*-aminobenzoic acid cannot be substituted for folic acid. This ability to use conjugates and the holozoic nature of *T. pyriformis* suggest the probable value of this ciliate in assays of natural products. The action of sulfadiazine against *P. gallinaceum* in chickens is reversed by pteroylglutamic acid; *p*-aminobenzoic acid has the same effect on sulfonamides used against *P. lophurae* and *P. gallinaceum*. Some of these sulfonamides, such as sulfanilamide and sulfathiazole, inhabit oxygen consumption of malarial parasites. In addition to the evidence obtained with analogues, growth of *P. knowlesi* is stimulated *in vitro* by *p*-aminobenzoic acid. For the phytoflagellates, a reversal of sulfanilamide action by *p*-aminobenzoic acid has been reported in *Polytomella caeca*.

Nucleic acid derivatives

Ribonucleic acid contains certain purines (adeninge, guanine), pyrimidines (cytosone, uracil) and D-ribose; in desoxyribonucleic acid, uracil is replaced by thymine and D-ribose by D-2-desoxyribose. Several nucleic acid derivatives have been tested on *Tetrahymena pyriformis*. Together with folic acid, the purines (guanine apparently being essential) form the active components of "Factor I," an undefined concentrate previously found essential to growth of *T. pyriformis*. Although adenine and hypoxanthine show a guanine-sparing action, neither can replace guanine. However, the inhibitory action of an adenine analogue (adenazolo) on growth of *T. pyriformis* is reversed specifically by adenine. Among various substituted purines, I-methyl-guanine is about 75 per cent as active as guanine, several are inert, and others are inhibitory. "Factor III", another concentrate which appeared necessary to growth of *T. pyriformis*, has been resolved into the pyrimidine derivatives, uracil and cytosine, or their ribosides or ribonucleotides. *T. pyriformis* is believed to synthesize thymine from non-pyrimidine precursors in reactions involving pteroylglutamic acid as reported previously for bacteria.

Ascorbic acid

Although a need for this vitamin has been attributed to *Haematococcus pluvialis* and several parasitic flagellates, there is at present no conclusive evidence that ascorbic acid is essential to growth of Protozoa.

Sterols

Several flagellates require sterols, a requirement whih may be satisfied by cholesterol or certain other sterols. Among 66 different sterols tested on *Trichomonas gallinae* (*T. columbae*), comparable activity was shown by cholesterol, cholestanol, sitosterol, and several others. Ergosterol was moderately active if not heated, whereas irradiated ergosterol ("vitamin D") was inactive. Cholesterol also seems to be required by *Entamoeba histolytica* and is a possible requirement of *Trichomonas vaginalis*. Growth of *Colpidium campylum* is slower with certain concentrations of cholestrol but reaches a greater density than in the control medium.

Hematin

That species of *Trypanosoma* and related flagellates need blood in culture media was first noted many years ago. Later on Salle and Schmidt found that, for *Leishmania tropica*, blood could be replaced by hemoglobin, which they suggested as a probable growth-factor. This question has been investigated extensively by M. Lwoff, who has shown that certain Trypanosomidae can grow in ordinary peptone media while others require supplementary blood or a more active substitute, hematin. The latter are unable to synthesize porphyrin groups in the production of cytochrome, cytochrome oxidase, and related enzymes. *Strigomonas fasciculata* apparently can combine iron and exogenous protoporphyrin to produce heme. On the other hand, certain Trypanosomidae and free-living Protozoa containing the cytochrome system apparently can synthezise porphyrins from simpler materials.

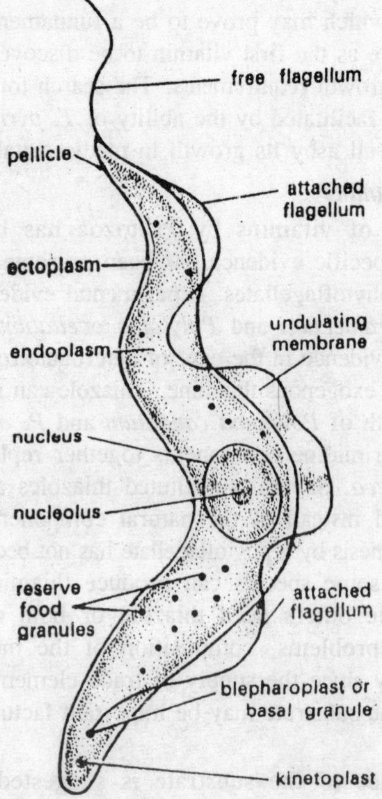

Fig. 5.5. Trypanosoma.

Vitamin B_{12} (cyano-cobalamin)

It is interesting that the first evidence for protozoan requirements has been obtained with a phytoflagellate. Vitamin B_{12}, or "cyanocobalamin", tremendously stimulates growth of *Euglena gracilis* var. *bacillaris* in the presence of adequate thiamine and is believed to be an absolute requirement. These findings have extended earlier observations that heavy growth of *E. gracilis* depends upon certain factors present in crude casein. This growth-response of *E. gracilis* has been applied to microbiological assay of cyano-cobalamin.

Protogen

A previously undefined "Factor II", a concentrate of natural origin essential for *Tetrahymena pyriformis*, has been resolved into fractions IIA and IIB. The name, *protogen*, was proposed for Factor IIA, which is not identical with any known vitamin or with the "animal protein factor." Protogen, which may prove to be a fundamental requirement of animals, is unique as the first vitamin to be discovered through the study of protozoan growth requirements. The search for natural sources of protogen will be facilitated by the ability of *T. pyriformis* to digest complex foods as well asby its growth in media suitable for assays.

Biosynthesis of vitamins

The synthesis of vitamins by Protozoa has been suggested Occasionally, but specific evidence has been presented in only a few cases. Among the phytoflagellates, experimental evidence is available for *Chilomonas paramecium* and *Polytoma ocellatum*. There is also some presumptive evidence in the case of heteroautotrophs which have been grown without exogenous thiamine. Thiazole can replace thiamine in stimulating growth of *Polytoma caudatum* and *P. ocellatum*, while the thiazole and pyrimidine components together replace the vitamin for *Polytomella caeca*. Several substituted thiazoles and pyrimidines also can be utilized instead of the natural components of thiamine. Although actual synthesis by a phytoflagellate has not been demonstrated, it is assumed that some species can produce thiamine from simple raw materials while others need thiazole or both components. In investigating such problems, composition of the medium must be considered carefully since the supply of trace elements, such as iron and the nature of the substrate may be important factors in a potential synthesis.

The significance of the substrate is suggested by failure of *Chilomonas paramecium* to grow on pyruvate without added thiamine, although the flagellate grows slowly on acetate in a thiamine-free medium. This is an interesting parallel to *Prototheca zopfi* which can oxidize acetate in a thiamine-deficient medium but apparently requires thiamine for utilization of pyruvate. In general, the burden of proof seems to rest upon those who would deny that phytoflagellates can synthesize a variety of vitamins. For the higher Protozoa, such assumptions are not justified because these organisms have been grown almost exclusively in chemically undefined media.

So long as material of natural origin are included, it is unsafe to assume that a particular vitamin has been eliminted from a culture medium. Synthesis of thiamine from its thiazole and pyrimidine components has been reported for *Acanthamoeba castellanii*, and from unspecified intermediates in the case of *Tetrahymena pyriformis*. However, *A. castellanii* was grown in peptone media of unknown vitamin content, and the interpretation of the earlier data for *T. pyriformis* has been questioned. Although more recent data have been supplied for the ciliate, there is no conclusive evidence that either *A. castellanii* or *T. pyriformis* can synthesize thiamine. The Trypanosomidae which need exogenous hematin presumably are unable to synthesize the porphyrins necessary to the formation of heme.

Others, which possess the cytochrome system but do not require ready-made porphyrins, obviously synthesize heme from simpler constituents of culture media. The problem of obtaining suitable raw materials is a minor one because such a substrate as acetate may serve as a starting point. Syntheses of this nature may be assumed for *Chilomonas paramecium*, *Polytoma uvella*, *Astasia klebssi*, *Euglena gracilis*, and *Tetrahymena pyriformis*, for example. As a source of direct evidence, the hematin-requiring Trypanosomidae should be useful for microbiological assays. The synthesis of nicotinic acid, adenine, pyridoxal, and riboflavin has been demonstrated in *Chilomonas paramecium*. Synthesis of riboflavin, pantothenic acid and probably of biotin has been reported for *Tetrahymena pyriformis* on the basis of microbiological assays, but these conclusions were later withdrawn. The synthesis of *p*-aminobenzoic acid and inositol by *T. pyriformis* has been reported on the basis of *Neurospora* assays. Synthesis of the former would seem to be no advantage to the ciliate since *p*-aminobenzoic acid apparently cannot replace pteroylglutamic acid as an absolute requirement.

Requirements of Various Groups

At present, little has been published on two major groups of the phytoflagellates, the Chrysomonadida and Dinoflagellida. Earlier work on *Oikomonas termo* and dinoflagellates was interrupted, and although extensive investigations are in progress, the two orders offer quite a variety of unsolved nutritional problems. Both groups include colourless and chlorophyll-bearing species and a number of holozoic types, and both are represented in fresh and salt water. Representatives of two other orders. Heterochlorida and Chloromonadida, apparently are not yet available in pure cultures.

Cryptomonadida

So far, the chlorophyll-bearing cryptomonads have been neglected in favour of *Chilomonas paramecium*, several strains of which have been investigated. As nitrogen sources, ammonium salts are satisfactory, nitrate is inadequate and utilization of nitrate has not been demonstrated. Reported chemoautotrophy had not been confirmed. Although *C. paramecium* has been grown in glycine and acetate medium, little is known about amino acids as nitrogen sources or their possible value as sources of both carbon and nitrogen. Excellent carbon sources, added to a basal inorganic medium supplemented with thiamine (or its components) include acetate, ethanole, lactate, and pyruvate. In such a medium about 45 per cent of the available acetate is oxidized while the rest is assimilated.

Without added thiamine, acetate has supported growth in all-glass culture vessels—about one per cent of the growth obtained with thiamine. In thiamine-supplemented acetate medium, growth of *C. paramecium* has been tripled by raising the carbon dioxide concentration from that of the atmosphere to 100 mm Hg at atmospheric pressure. Similar stimulation by carbon dioxide had previously been detected in peptone media. Certain alcohols, fatty acids and carbohydrates also accelerate growth in peptone media, although a number of these supplements have not yet been tested as carbon sources in heteroautotrophic nutrition.

Phytomonadida

As inorganic nitrogen sources in "photoautotrophic" nutrition, ammonium salts have been more satisfactory than nitrates for *Haematococcus pluvialis* and *Lobomonas piriformis*. No appreciable difference have been reported for *Chlamydomonas agloeformis*, *Chlorogonium elongatum*, and *C. euchlorum*. However, comparative tests of nitrates and ammonium salts over an adequate range of salt and hydrogen-ion concentrations have not been reported. In heteroautotrophic nutrition *Polytoma ocellatum* has been grown in a nitrate medium, but species of *Polytomella* and other species of *Polytoma* apparently are limited to an ammonium-N source. This situation deserves further investigation in view of Lwoff's characterization of an autotroph as an organism which can reduce nitrate in an inorganic medium.

Organic nitrogen sources have not been investigated extensively but growth on asparagine or a single amino acid has been reported for species of *Chlamydomonas* and *Haematococcus*, *Chlorogonium*, *Lobomonas* in darkness, *Polytoma* and *Polytomella*. Acetate, butyrate, and lactate are good carbon sources in heteroautotrophic nutrition. A number of other substrates probably would be satisfactory since salts of additional acids, including propionic, valerianic, and caproic and also certain alcohols accelerate growth of various colourless species in peptone media. Certain carbohydrates also have stimulated growth of *Polytoma* and *Polytomella* but may or may not be adequate substrates in heteroautotrophic nutrition. The ability to use an amino acid, as the sole source of nitrogen, carbon, and energy, has not yet been demonstrated. Several green species—*Chlorogonium elongatum*, *C. euchlorum*, *Lobomonas piriformis*, *Chlamydomonas agloeformis*, and *Haematococcus pluvialis*—have been grown in darkness, particularly in peptone media supplemented with acetate, and are obviously facultative heterotrophs. Under such conditions, acetate is a rather satisfactory substitute for photosynthesis. On the other hand, *Chlamydomonas moewusii* is an obligate phototroph in a wide variety of media.

Euglenida

Further study of these flagellates should prove interesting because the order includes chlorophyll-bearing species, colourless saprozoic types (*Astasia*, *Menoidium*, etc.), and various holozoic genera (*Heteronema*, *Peranema*, etc.) furthermore, such species as *Euglena anabaena*, *E. deses*, *E. klebsii*, *E. pisciformis* and *E. stellata* have failed to grow in darkness, whereas *Euglena gracilis* grows well under such conditions. Although investigations on holozoic types are in progress, previous reports are limited to *Euglena* and *Astasia*. Inorganic sources of nitrogen have been tested for several species of *Euglena*. The available data indicate that ammonium salts are generally more satisfactory than nitrates. However, nitrates apparently are adequate for *Euglena anabaena*. Among the colourless Euglenida, *Astasia longa* has been maintained on ammonium-N. Organic sources of nitrogen have been investigated for several species of *Euglena*.

Asparagine and various amino acids have supported growth of one species or another, but several species have shown interesting differences in their apparent abilities to utilize particular amino acids. Possible relations of mineral requirements and pH of the basal media to the utilization of amino acids have not been investigated adequately. The available information on carbon sources is based mainly on growth of Euglenida in peptone media, although acetate has supported slow growth of *Astasia longa* and *Euglena gracilis* in heteroautotrophic nutrition. A number of organic acids—including acetic, propionic, butyric, valerianic, caproic, *iso*-caproic, octylic, nonilic, lactic, and pyruvic—have stimulated growth of *Astasia* and *Euglena* in peptone medium. Growth of *Euglena* and *Astasia* also is stimulated by certain alcohols, including ethyl, propyl, butyl and hexanol.

Protomastigida

Pure cultures of parasitic Protomastigida have been available for many years but most investigators have been interested in these flagellates as parasites rather than in their nutrition. However, the investigations of Marguerite Lwoff on several parasites of insects showed that *Strigomonas oncopelti* grows well in peptone media while certain other species have more complex requirements. *S. fasciculata*, from mosquitoes, requires a small amount of blood or hematin as a supplement to peptone medium. *Leptomonas ctenocephali*, from the dogflea, requires such as supplement in higher concentrations. Specific requirements of *S. culicidarum* include at least nine amino acids, hematin, thiamine, riboflavin, pyridoxamine, trace minerals, and possibly two or three additional vitamins.

Certain Trypanosomidae of vertebrates require, in addition to hematin, other growth-factors not supplied by peptone solutions. *Trypanosoma cruzi*, *Leishmania brasiliensis*, *L. donovani*, and *L. tropica* need factors in serum other than thiamine, *p*-aminobenzoic acid,

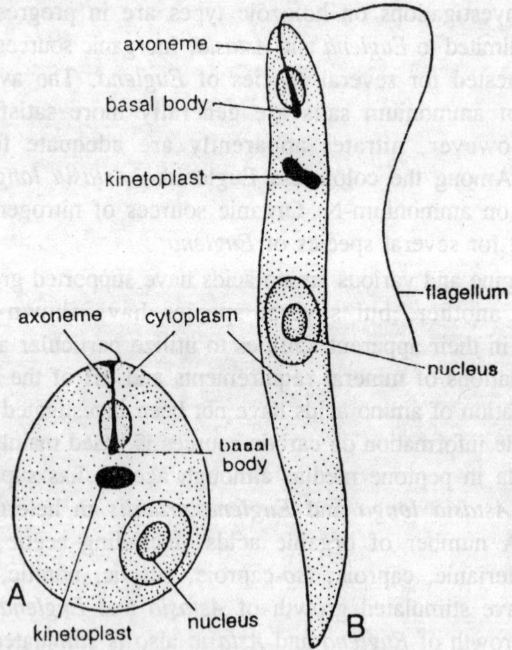

Fig. 5.6. Two forms of Leishmania. A—Amastigote or leishmanial form in man. B—Promastigote or leptomonad form in sandfly.

pyridoxine, or nicotinic acid. Hence, it appears that one aspect of physiological specialization among the trypanosomes and their relatives involves an increasing dependence upon the host for essential growth-factor.

Trichomonadida

Relatively little work has been reported on these flagellates. Peptone solutions enriched with whole blood, serum and fragments of liver have supported growth of *Eutrichomastix colubrorum*, *Trichomonas gallinae* (*T. columbae*), *T. foetus*, and *T. vaginalis*. Liver-infusion agar slants, overlaid with serum-enriched Ringer's solution, also have been satisfactory for *T. vaginalis*. The investigations of Cailleau have shown that the first three species need certain growth-factors apparently not required by free-living flagellates. More recently, *T. vaginalis* have been grown in a peptone solution supplemented with acetate, moltose, about 15 growth-factors, and after sterilization, diluted serum and ascorbic acid. In such a medium, two fractions of human serum—an ether-soluble and an ether-insoluble aqueous fraction—are essential. Linoleic acid seems to be the most active component of the ether-soluble fraction and serum albumin of the ether-insoluble one. The first fraction could be replaced by a mixture of linoleic and oleic acids, cholesterol, ergosterol, lecithin, α-estradiol, α-tocopherol, vitamin A, and β-carotene. *Trichomonas foetus* has been grown in a mixture of thirteen amino acids, various vitamins, and minerals.

Sarcodina

Published reports on the Sarcodina include little more than the development of suitable media for pure cultures, although detailed investigations are in progress. *Acanthamoeba castellanii* has been grown in a medium containing serum and liver fragments and in simpler peptone and inorganic salt medium. A medium containing peptone, dextrose, and inorganic salts also is satisfactory of *Mayorella palestinensis*. Progress is also being made toward pure-culture techniques for parasitic amoebae. *Entamoeba invadens* has been maintained bacteria-free for several transfers after elimination of a single bacterial contaminant by treatment with penicillin. In addition, *E. histolytica* has been grown on non-viable bacteria for more than 200 transfers, and also in a non-particulate medium without bacterial growth.

Ciliates

Under natural conditions free-living ciliates feed mainly upon ingested microorganism. In the investigation of such natural foods, strains of ciliates have been grown in cultures with other living or killed microorganisms. Such "species-pure" cultures with living bacteria involve complex relationships, and the ciliates tend to be swamped unless the initial proportions between ciliates and bacteria are satisfactory. Such relationships, particularly important when the medium supports growth of the bacteria, may be controlled by using non-nutrient basal media.

A wide variety of bacteria may serve as food for particular ciliates. About 20 species, as individual suspensions in salt solutions, were each adequate for growth of *Tetrahymena pyriformis*. Killed yeasts and washed and killed suspensions of green flagellates also were satisfactory, although living flagellates failed to support growth in serial transfers. On the other hand, *Perispira ovum* thrives on living *Euglena gracilis*, and *Tetrahymena pyriformis* grows on either *Chilomonas paramecium* or *Polytoma ocellatum*. For *Colpidium colpoda*, species of Enterobacteriaceae are more satisfactory than Bacillaceae. Killed bacteria seem to be an indequate diet for *Colpoda duodenaria*. Likewise, *Pleurotricha lanceolata* and *Stylonychia pustulata* can be grown on living *Tetrahymena geleii* but not on killed ciliates. Just what heat-labile factors, supplied by living organism, are significant in such cases is still unknown. However, *Didinium nasutum* is said to have lost the ability to synthesize peptidases and must obtain these enzymes from the living *Paramecium* which it ingests.

Definitive observations on food requirements of ciliates awaited, first of all, the establishment of pure cultures. This step was taken some thirty years ago when A. Lwoff isolated *Tetrahymena* (*Glaucoma*) *pyriformis* in a peptone medium. Comparative data on various culture media were published later. These observations furnished a timely stimulus, and within a relatively few years, additional bacteria-free strains—referred to the genera *Colpidium*, *Colpoda*, *Glaucoma*, *Tetrahymena*, *Loxocephalus*, and *Paramecium*—were isolated by other workers.

With a few exceptions, culture media included solutions of commercial peptones, yeast-extract or yeast autolysates, usually supplemented with inorganic salts. Such media as those Giaser and Coria were more complex. Although *Paramecium bursaria* has been maintained in peptone media. Cultivation of other species of *Paramecium* has proven more difficult. However, *P. aurelia* and *P. multimicronucleatum* are now in pure culture and investigations on their food requirements are in progress. The next progressive step led towards the development of chemically defined culture media. The first apparently successful results were those of Kline with *Colpidium striatum*. Unfortunately, Kline's strain of *C. stratum* seems to have been lost and several other strains have failed to grow in the medium.

Much better results have been obtained with somewhat similar media developed for *Tetrahymena geleii* W. The simpler of these media contained eleven amino acids (arginine, histidine, isoleucine, leucine, lysine, methionine, phenylalanine, serine, threonine, tryptophane, valine), glucose, eleven known vitamins, several inorganic salts, and in the proportion of 1:10, a plant or animal tissue extract apparently containing a number of amino acids, certain unidentified growth factors and known vitamins and several minerals. Further progress has made possible the gradual substitution of known growth-factors for supplements of natural origin. This general type of medium, which has proven satisfactory for several strains of *Tetrahymena*, is illustrated, although nucleic acid may be replaed by known purines and pyrimidines. Such media are now almost completely defined in a chemical sense and are potentially useful in the assay of certain vitamins and also such amino acids a histidine, isoleucine, lysine, and tryptophane.

Oxygen Relationships And Oxidations

Ecological Distribution

The distribution of Protozoa suggests that some species are obligate aerobes, that others are microaerophiles (requiring only a little oxygen), and that many intestinal parasites and some free-living types may be obligate or facultative anaerobes. Natural waters containing much putrefying material are anaerobic at their lower levels, and such artificial environments as Imhoff sewage tanks also insure anaerobiosis beneath the surface. Species characteristic of such a fauna, sometimes termed polysaprobic or sapropelic, are at least facultative anaerobes. They include Ctenostomina and scattered species of other ciliates, as well as a few flagellates and Sarcodina.

Another practically anaerobic environment is found near the bottom of deep fresh-water lakes, but very little is known about this fauna. Other Protozoa, typical of clean waters with a high oxygen content, are aerobes and some may be obligate aerobes. However, such ciliates as *Coleps hirtus* and *Frontonia leucas*, which are not sapropelic, may survive anaerobically for several weeks. Oxygen relationships of parasites doubtless vary with the usual site of infection. Species which invade the blood and other tissues probably have access to about as much oxygen as the surrounding tissue cells and may be predominantly aerobic. Oocysts of *Eimeria stiedae* and *E. magna*, for instance, apparently cannot sporulate under strictly anaerobic conditions. On the other hand, conditions in the vertebrate intestine suggest that intestinal parasites are anaerobes. Experimental evidence indicates that the rumen ciliate, *Eudiplodinium neglectum* is an obligate anaerobe. Such is true also for flagellates of termites *Entamoeba histolytica*, in contrast to many other intestinal parasites, normally invades the wall of the colon. Yet this species grows as an anaerobe in cultures. Relationships between oxygen tension and growth of laboratory populations have been investigated in a few cases. Aeration of flask cultures increases growth of *Tetrahymena pyriformis* reduction of the oxygen supply (pyrogallol technique) decreases populations of about half the normal density and complete anaerobiosis prevents growth.

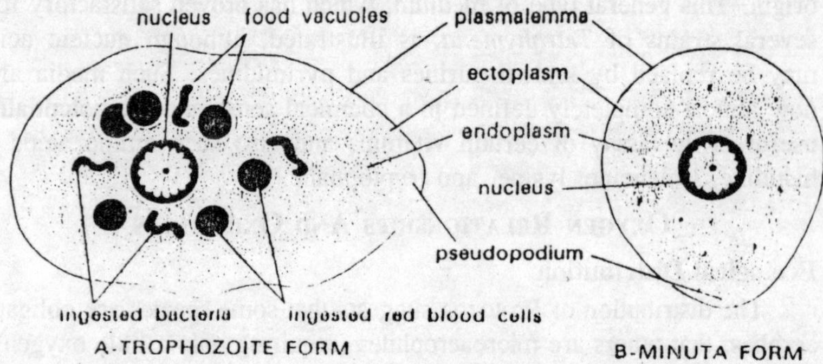

Fig. 5.7. Entamoeba histolytica. The trophozoite and minuta form.

Growth of *Chilomonas paramecium* on the other hand, is retarded by aeration of cultures. Quantitative data also have been reported for *Trichomonas vaginalis* and for *C. paramecium* and *T. pyriformis*. Growth of *T. vaginalis* is heaviest in complete anaerobiosis and is inhibited progressively by increasing oxygen tensions. Oxygen pressures of 0.5 to 500 mm Hg permit growth of *C. paramecium*, with an optimum at about 75 mm Hg (about half the normal atmospheric concentration of oxygen), while pressure of 600 mm Hg and higher are lethal. Growth of *T. pyriformis* increases from 10 mm Hg to a maximum (about twice the growth with atmospheric concentrations of oxygen) in an atmosphere of pure oxygen (739 mm Hg).

Oxidation-reduction Potentials

The oxidation-reduction potential of the culture medium is another factor to the growth of microorganism. In a general sense, this potential is a measure of the reducing intensity of oxidizing intensity of a given system. Examples of such systems—each of which consists of a more reduced and a less reduced substance—are leuco-methylene blue/methylene blue, lactate/pyruvate, and reduced cytochrome *a*/oxidized cytochrome *a*. If a platinum electrode and a calomel electrode, in a potentiometer hookup, are immersed in such a system, a potential difference can be measured. Since the calomel electrode is standardized against the hydrogen electrode, measurements are expressed in millivolts in terms of the hydrogen electrode potential. The more negative the potential, the greater is the reducing power of the system; the more positive, the greater the oxidizing power (and the lower the reducing power). Each system has characteristic E_0' value at which it is half reduced at particular temperature and pH. Consideration of pH is necessary because the potential varies with pH. For the methylene blue system, E_0' at pH 7 is 11 mv; for the cytochrome *a* system. 290 mv at pH 7.4; for lactate/pyruvate,–180 mv at pH 7.

If two systems with different E_0' values are mixed together, a reaction, which may be considered the transfer of electrons from one system to the other, continues unit equilibrium is reached. The higher potential is lowered and the lower potential raised to a common level; or reduction of the first system (gain of electrons) and oxidation of the second (loss of electrons) take place. The potential of the culture medium undoubtedly influences the growth of microorganisms. For bacteria, it is possible to lower the potential of a liquid medium with a suitable reducing agent so as to inhibit growth of aerobes and permit growth of anaerobes. In the case of *Chilomonas paramecium*, appropriate additions of the sulfhydryl radical (-SH) both lower the oxidation-reduction potential of the medium and stimulate growth.

Influence of the potential on *Entamoeba histolytica* apparently varies with the type of medium. According to one worker, growth of *E. histolytica* decreases from a maximum, at a potential below -300 mv, to almost none at -200 mv; unencysted amoebae die after an hour or more at -50 mv (56). Jacobs on the other hand, found that the potential of the medium was about -25 mv while *E. histolytica* was growing most rapidly in cultures containing "*organism*." In contrast to *E. histolytica*, *Trypanosoma cruzi* and several species of *Leishmania* grow best in cultures at a potential of about 330 mv. Growth of microorganisms themselves also may modify the potential. For instance, a drop of about 290 mv has been traced in cultures of *Chilomonas paramecium*. It is uncertain just how extensively the potential of the medium influences internal oxidations and reductions, although the 'internal potential" of *Amoeba proteus*, as measured by injected indicators, may be changed from about -70 mv (pH 7) under aerobic conditions to -143 mv in anaerobiosis.

Oxygen Consumption

Measurements of oxygen consumption make it possible to trace effects of environmental factors on metabolic rates, to investigate the utilization of particular substrates, and to correlate stage in the life-cycle with metabolic activity. Such measurements are necessary in the study of oxidative mechanisms by the use of poisons or stimulants, and may indicate the relative importance of particular systems, such as the cytochrome system, in the metabolism of particular species. Manometric techniques can be used also in the estimation of specific enzyme systems and metabolites. Comparative data on oxygen consumption of different species should be interpreted cautiously, since extensive variations apparently occur even in single species. For example, oxygen consumption of *Paramecium caudatum* has been recorded as 0.00014, 0.0004 and 0.0052 mm^3/hour/organism. In addition to differences attributable to different manometric techniques, the physiological condition of the test organisms may be a significant factor.

Starvation significantly reduces oxygen consumption of *Paramecium caudatum* and *Pelomyxa carolinensis*. Likewise a marked decrease occurs in old cultures of *Colpidium colpoda*, *Bodo caudatus*, *Chilomonas paramecium*, *Tetrahymena pyriformis*, *Trichomonas foetus* and *Trypanosoma cruzi*. In. *T. pyriformis* the change occurs after the logarithmic phase of growth and is not correlated with any decrease in cytochrome content. Changes in consumption also have been traced during conjugation of *Paramecium caudatum*.

Environmental conditions also may influence oxygen consumption. For *Tetrahymena pyrimformis* consumption is at a maximum in media at pH 5.5 and is distinctly lower on each side of the optimum. Increasing of temperatures, within physiological limits, stimulate oxygen consumption ciliates and *Strigomonas fasciculata*. The oxygen consumption of *Spirostomum ambiguum* increases with increasing oxygen concentration of the atmosphere to which cultures are exposed. The maximum, observed with pure oxygen, was about 50 per cent higher than for ciliates exposed to air. On the other hand, changes in oxygen tension within fairly wide limits have produced little effect on *Paramecium caudatum*.

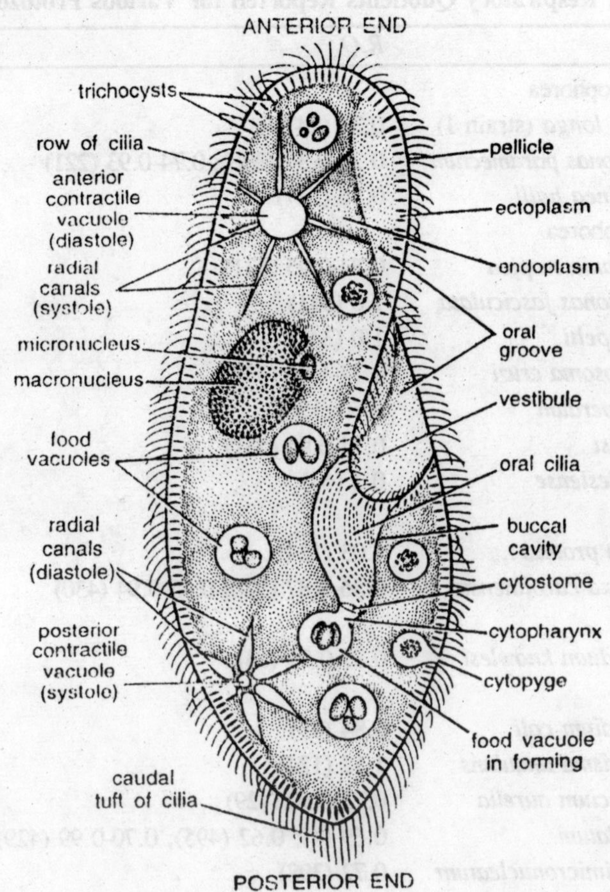

Fig. 5.8. Paramecium.

Respiratory Quotients

The ratio of the carbon dioxide produced to the oxygen consumed the respiratory quotient (R.Q.)—has interested physiologists as a theoretical index to the type of material being consumed. The R.Q for complete oxidation of carbohydrate is 1.0 and is about the same for acetate; for butyrate, about 0.8; for fats, approximately 0.7; for proteins, about 0.8 (urea as the nitrogenous waste) or about 0.9 (ammonia as the nitrogenous waste). Quotients well above 10.0 may indicate synthesis and storage of fat produced from carbohydrate. Low values (0.4-0.6) might indicate conversion of protein to carbohydrate, or incomplete oxidation of carbohydrate.

Table 5.4. Respiratory Quotients Reported for Various Protozoa

Species	*R.Q.*
Phytomastigophorea	
Astasia longa (strain J)	0.34 (247)
Chilomonas paramecium	0.28-0.37 (398); 0.74-0.93 (221)
Khawkinea halli	0.56 (247)
Zoomastigophorea	
Leishmania tropica	0.84-0.95 (531)
Strigomonas fasciculata	1.0 (341)
S. oncopelti	1.0 (341)
Trypanosoma cruzi	0.74-1.06 (33)
T. equiperdum	0.60 (126)
T. lewisi	0.74-0.94 (531)
T. rhodesiense	0.2 (64)
Sarcodina	
Amoeba proteus	1.03 (124)
Pelomyxa carolinensis	0.56-0.87 (526) 0.45-0.94 (430)
Sporozoa	
Plasmodium knowlesi	0.87-0.93 (63)
Ciliatea	
Balantidium coli	0.84 (85)
Blepharisma undulans	1.12 (124)
Paramecium aurelia	0.73-0.90 (429)
P. caudatum	0.69 (4); 0.62 (495); 0.70-0.99 (429)
P. multimicronucleatum	0.72 (398)
Spirostomum ambiguum	0.84 (532)
Tetrahymena pyriformis	0.81-1.27 (431)

Most of the R.Q. values reported for Protozoa fall within the usual range. This is especially true of Trypanosomidae, some of which have shown a higher R.Q. with glucose than without, as would be expected. Unusually low value for several phytoflagellates have been attributed to synthesis of carbohydrates from carbon dioxide and to the conversion of protein into carbohydrate or the incomplete oxidation of carbohydrate. Varying quotients for a species may reflect differences in condition of the organisms. An R.Q. of 0.87 has been reported for well-fed *Pelomyxa carolinensis* and one of 0.56 for starved specimens. Age of the culture also is a factor. The R.Q. of *Chilomonas paramecium*

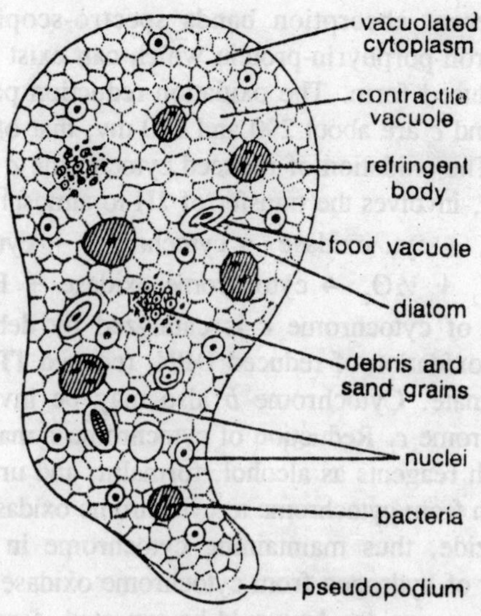

Fig. 5.9. Pelomyxa.

drops from 0.91-0.93 in 24-hour cultures to 0.75 at 72 hours that of *Tetrahymena pyriformis* from 1.21-1.27 at three days to 0.81 after seven days; that of *Trypanosoma cruzi*, from a maximum of 0.06 to a low of 0.74 after the population reaches its peak.

In the last case, the lower quotient is attributed to exhaustion of carbohydrates and subsequent utilization of proteins. The R.Q. also may vary with temperature—0.73 at 20° to 0.90 at 30° for *Paramecium aurelia*; 0.70 at 15° to 0.99 at 35° for *P. caudatum* 0.45 at 10° to 0.94 at 30° for *Pelomyxa carolinensis*.

Oxidations

Most biological oxidations consist of series of oxidations and reductions catalyzed by a variety of enymes, and maybe pictured as involving the transfer of hydrogen step by step from one acceptor another. Each step involves an oxidation reduction system and each dehydrogenation yields energy for anabolism. For cells in general, a number of enzymes and oxidation-reduction systems are known to be involved in metabolism. Gradually accumulating evidence indicates that at least some of these are operative in Protozoa, as would be expected.

Cytochrome system

In aerobes the final stages of oxidative reactions—the transfer to hydrogen to oxygen (2H to O_2)—involve the cytochrome system. This system includes several cytochrome pigments which, in their reduced forms, show different absorption bands spectro-scopically. Each cytochrome is an iron-porphyrin-protein which can exist in either the oxidized or the reduced form. The oxidation-reduction potential (E_o') of cytochromes *a* and *c* are about 290 and 270 mv; that of cytochrome *b* about —40 mv. The oxidation of reduced cytochrome *c* catalyzed by cytochrome oxidase, involves the transfer of 2H to atmospheric oxygen:

cytochrome-H_2 + cyt. oxidase → cytochrome + cyt. oxidase-H_2

cyt. oxidase-H_2 + ½O_2 → cytochrome oxidase + H_2O

The reduction of cytochrome *c* is catalyzed by dehydrogenases which bring about oxidation of reduced DPN, reduced TPN, and such substrates as succinate. Cytochrome *b* also may be involved in the reduction of cytochrome *c*. Reduction of cytochrome *c* may be blocked by heat and by such reagents as alcohol, formalin, and urethanes. The transfer of hydrogen from cytochrome to cytochrome oxidase is inhibited by cyanide and azide, thus maintaining cytochrome in the reduced condition. Transfer of hydrogen from cytochrome oxidase to oxygen is inhibited by carbon monoxide. As would be expected, aerobic Protozoa resemble other aerobic microorganisms in possessing cytochrome pigments. Cytochromes *a, b* and *c* have been reported in *Astasis klebsii* and *Tetrahymena pyriformis*; cytochromes *b* and *c,* in *Euglena gracilis*, *Tetrahymena* (*Glaucoma*) *pyriformis*, *Polytoma uvella*, *Strigomonas fasciculata*, and *S. oncopelti*; cytochrome *c*, in *Colpidium campylum*, and with cytochrome oxidase, in *Chilomonas paramecium*. Cytochrome oxidase is said to occur in the pigment granules (mitochondria) of *Stentor coeruleus*. In contrast to the typical aerobes, *Trichomonas foetus* apparently contains no cytochrome.

Poisoning techniques have supplied additional information. The respiration of *Tetrahymena pyriformis* is decreased by 9.57 per cent in different concentrations of methyl-ethyl, and propylurethanes (364). Cyanide decreases oxygen consumption about 61-64 per cent in *Astasia longa* and *Khawkinea halli*, about 90 per cent in *Polytoma uvella* and about 95 per cent in *Astasia klebsii*, for which azide is almost as inhibitory. A number of Trypanosomidae also are susceptible to such poisons. *Leptomonas ctenocephali*, *Strigomonas fasciculata*, and *S. oncopelti* show 83-95 per cent inhibition with cynide and are about as sensitive to carbon monoxide.

Cyanide in certain concentrations inhibits respiration about 90 per cent in *Trypanosoma cruzi*, 11-82 per cent in *T. congolense* 97-98 per cent in *Leishmania tropica*, *L. rasiliensis*, and *L. donovani*, 85-88 per cent in *T. lewisi* from cultures, and 66-69 per cent in *T. conorhini*. Certain other Trypanosomidae are relatively insensitive to cyanide-*T. equiperdum*, *T. brucei*, *T. hippicum*, *T. rhodesiense*, *T. evansi*, and *T. equinum*. Although the oxygen consumption of *T. gambiense* from blood is not decreased by cyanide, flagellates from cultures are moderately sensitive. Among the Sarcodina, *Pelomyxa carolinensis* is sensitives to cyanide, and sensitivity increases with temperature in the range, 10-35°. Respiration of *Plasmodium knowlesi* also is inhibited by cyanide, and carbon monoxide. Earlier reports that free-living ciliates are insensitive to cyanide, are contradicted by later observations.

Respiration of *Tetrahymena pyriformis* is sensitive both to carbon monoxide and to cyanide while that of well-fed, but not of starved specimens, also is cyanide-sensitive in *Paramecium aurelia* and *P. caudatum*. The results obtained with poisoning techniques indicate that in general, aerobic Protozoa are cyanide-sensitive and presumably oxidize substrates mainly through the cytochrome system. On the other hand, some questions are unanswered. Why are starved ciliates, in contrast to well-fed ones, relatively insensitive to cyanide? What converts insensitive *T. gambiense* from the blood into cynide-sensitive flagellates in culture media? It is necessary for these organisms to oxidize certain substrates only partially ("anaerobically") in the blood but completely, or nearly so, in culture media? And what are the biochemical differences between the cyanide-sensitive "*lewisi* group" of trypanosomes and such insensitive species as *T. brucei* and *T. evansi*? Such problems are of practical as well as theoretical interest, since the response of parasites of chemotherapeutic agents may depend to an important extent upon the oxidative mechanisms of particular species.

Pyridine nucleotide enzymes

The pyridine nucleotides are coenzymes for a number of important oxidative enzymes. Diphosphopyridine nucleotide (DPN), or coenzyme I, contains nicotinamide, D-ribose, adenine, and two phosphoric acid groups. Triphosphopyridine nucleotide (TPN), or coenzyme II, contains a third phosphoric acid group. Both coenzymes are involved in protozoan metabolism, *Chilomonas paramecium* and *Tetrahymena pyriformis* S contain DPN, while *Trypanosoma hippicum* requires DPN for glycolysis *in vitro*. In addition, supplementary DPN and TAP accelerate oxidation of pyruvate by *Plasmodium gallinaceum* in the presence of dicarboxylic acids.

Diphosphothiamine enzymes

This phosphoric acid ester of thiamine is the coenzyme of carboxylase which catalyzes the decarboxylation of pyruvic acid and probably certain other α-keto monocarboxylic acids. Supplementary thiamine is necessary for the oxidation of pyruvate by *Tetrahymena pyriformis*, is essential to growth of *Chilomonas paramecium* on pyruvate and accelerates oxidation of pyruvate by *Plasmodium gallinaceum*. The thiamine content of *Tetrahymena pyriformis* is at least 60 per cent that of yeast and the vitamin is essential to growth of this and certain other Protozoa. Therefore, thiaminoprotein enzymes are probably of general importance in protozoan metabolism.

Flavoprotein enzymes

In these enzymes the prosthetic groups contain riboflavin, either as riboflavin-phosphate or a flavin dinucleotide (a union of riboflavin-phosphate and adenylic acid). Enzymes of the first type apparently catalyze the oxidation of reduced TPN and oxidation of L-amino acids (L-amino acid dehydrogenase). Enzymes of the second type include xanthine oxidase, D-amino acid dehydrogenase, glycine dehydrogenase, and apparently "diaphorase I" (catalyzing oxidation of reduced DPN). These flavoprotein enzymes, which are not significantly affected by cyanide poisoning, are probably present in Protozoa. Riboflavin occurs in high concentration in *Tetrahymena pyriformis* and is essential to growth of several ciliates although synthesized by *Chilomans paramecium*.

Pyridoxine enzymes

Pyridoxal phosphate appears to be the coenzyme for transaminases, tryptophanase, and certain amino-acid decarboxylases. The presence of comparable enzymes in Protozoa may be suspected since pyridoxine is essential to growth of certain ciliates and inhibits the antimalarial action of quinine and atebrine against *P. lophurae*, and also since pyridoxal is synthesized by *Chilomonas paramecium*.

Peroxidase and catalase

These are iron-porphyrin-protein enzymes. Catalase probably catalyzes coupled oxidations by means of the hydrogen-peroxide formed in some primary reaction, the peroxide being decomposed to water and molecular oxygen. Peroxidase catalyzes the oxidation of such substrates as tyrosine, adrenaline, bilirubin, pyrogallol, and various other phenols in the presence of hydrogen peroxide. Peroxidase has been demonstrated in *Tetrahymena pyriformis* and catalase in certain related ciliates, but the activities of neither enzymes in protozoan metabolism have been investigated.

Glutathione

In the reduced form, this is a tripeptide of glycine, cysteine, and glutamic acid. Although reduced glutathione has been demonstrated in

Tetrahymena pyriformis and may play a part in respiration of this ciliate, its functions in protozoan metabolism are still unknown.

Pantothenic acid enzymes

Pantothenic acid is a component of *coenzyme A* which may be involved in acetylation reactions in general and perhaps in the utilization of acetylmethylcarbinol by certain bacteria. Although pantothenate is essential to growth of certain ciliates and malarial parasites, its possible functions, in protozoan metabolism have not been investigated.

Adenosine phosphate system

The adenosine phosphate system includes *adenylic acid* (adenosine monophosphate), *adenosine diphosphate* (ADP), and *adenosine triphosphate* (ATP). Each contains adenosine (a riboside of adenine) and one, two, or three phosphoric acid groups, respectively. The system functions in phosphorylation of metabolites and enzymes and especially in the transfer of high-energy phosphate bonds. This system apparently makes available for anabolic activities the energy derived from oxidation of metabolites. Essentially, TPN serves as a participant common to a variety of exergonic and endergonic reactions, making it possible for reactions of the first type to drive those of the second variety. Little is known about the adenosine-phosphate system in Protozoa. However, adenylic acid, ADP, and ATP have all been demonstrated in *Euglena gracilis* (I). In addition, *Trypanosoma hippicum* needs ATP in the formation of hexose-phosphates and *Tetrahymena pyriformis* contains adenosine triphosphatase.

Tricarboxylic acid cycle

This so-called cycle involves the oxidation of various metabolites through a common catalytic system to carbon dioxide and water under aerobic conditions. The cycle is fed by glycolysis, leading to pyruvate and thence to acetyl; by the breakdown of fats, yielding acetyl from fatty acids; and by the breakdown of proteins, through the deamination of certain amino acids to α-keto acids which enter the cycle. At each turn of the oxidative cycle, CO_2 and H_2O are produced as end-products in certain reactions, and energy is made available by the generation of high-energy phosphate bonds in several dehydrogenations.

Aside from its importance in the oxidation of substrates, the tricarboxylic acid cycle may also be considered a basic reservoir of important materials which can be drawn upon for the synthesis of amino acids, carbohydrates, and fatty acids. The tricarboxylic acid cycle has been studied both by the use of isotopes and by the control

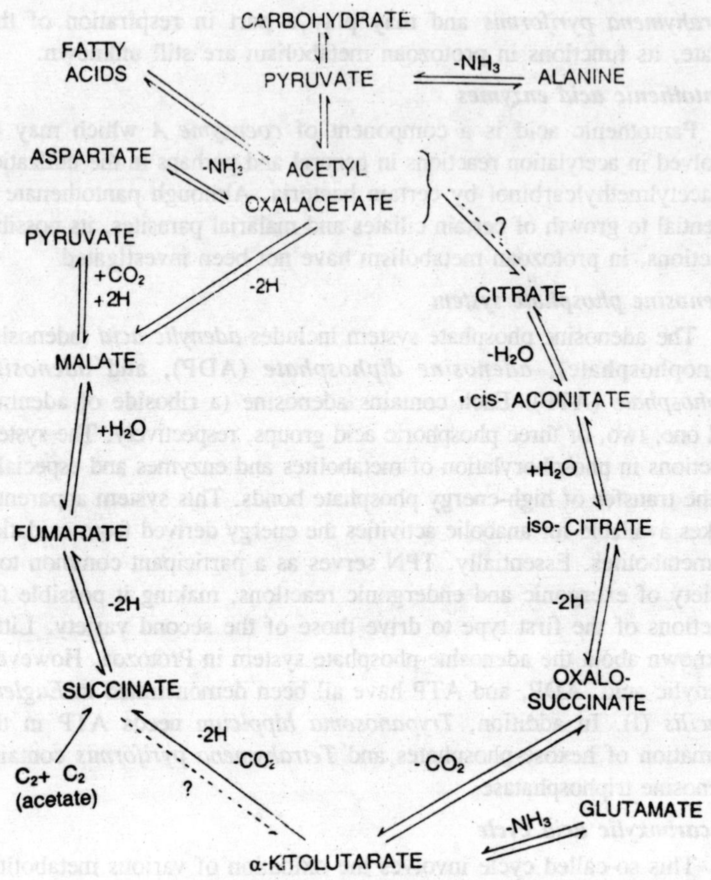

Fig. 5.10. The tricarboxylic acid cycle.

of enzyme systems with blocking reagents. In the presence of cyanide at a suitable concentration, oxalacetate is trapped; in the presence of sufficient malonate, which inhibits succinic dehydrogenase, the cycle stops with the accumulation of succinate. Arsenite checks the cycle by inhibiting the oxidation of malate to oxalacetate.

In addition, the ability of a species to use components of the cycle may be tested by measuring oxygen consumption with each as a substrate or by determining possible stimulation of growth. Little work has been done on the tricarboxylic acid cycle in protozoan metabolism. There is no reason a *priori*, for suspecting that all aerobic Protozoa must complete the oxidation of metabolites through a typical tricarboxylic acid cycle; there may be some species which do not.

There apparently are such reactions as C_2 + C_2 condensations which skip the C_6 acids of the typical cycle. *Rhizopus nigricans* can carry out this condensation, producing succinate and fumarate from labelled acetate or ethanol with essentially quantitative recovery of radioactive carbon, indicating no decarboxylation of intermediate C_6 acids.

In relation to heteroautotrophic nutrition of phytoflagellates, it is interesting that a mutant form of *Azotobacter agilis* has lost the ability to use glucose, lactate, pyruvate, and vaiious Krebs-cycle acids, but retain the ability to use acetate and ethanol. Evidence for the occurrence of the tricarboxylic acid cycle in certain ciliates seems conclusive. *Tetrahymena pyriformis* takes up carbon dioxide with formation of succinate in the anaerobic dissimilation of glucose (561). Studies on oxygen consumption show that pyruvate, succinate, α-ketoglutarate, fumarate, malate, and oxalacetate are utilized, and that malonate produces typical inhibition. It is interesting that malonate in low concentrations (5 μg/ml) serves as a substrate for *T. pyriformis* (528), although it is distinctly inhibitory at high concentrations.

With the exception of citrate and *cis*-aconitate, the various intermediates of the cycle are readily utilized by *Plasmodium gallinaceum* and fumarate, pyruvate, and succinate are known to be oxidized by *P. lophurae*. The evidence for such a cycle in phytoflagellates is not yet conclusive. Added to a peptone medium, malate stimulates growth of *Astasia longa*, *Euglena gracilis* and *Polytoma ocellatum* while succinate stimulates growth of these and six other species (565), and also accelerates growth of *E. gracilis* in darkness. In addition, fumarate, malate, and succinate are satisfactory substrates for *E. gracilis* var. *bacillaris* at pH 3.0-3.6. Occasional failures to use Krebs-cycle acids have been observed but these cases probably should be reconsidered. Experimental data for *Tetrahymena pyriformis* indicate that permeability of the surface to the substrate is a major factor to be considered. Such a factor may explain the reported inability of *Polytoma uvella* to use malate and pyruvate and that of *Astasia klebsii* to oxidize succinate or citrate at a significant rate.

The need for carbon dioxide, established for several phytoflagellates, might suggest that carboxylation occurs as an essential part of the cycle but such an assumption is yet to be confirmed experimentally. The data for Zoomastigophorea also are fragmentary. *Trypanosoma lewisi* oxidizes several of the intermediates rather slowly (403), although the basal "medium" used for such tests may not be the most favourable one for reactions which depend upon a variety of growth-factors.

Trypanosoma cruzi and the species of *Leishmania* from man form succinate as one of the products in oxidation of glucose and levulose.

Digestion

Digestion in holozoic species occurs typically within vacuoles which enclose the food after ingestion. The mechanical features of ingestion vary with the species and with the type of food. Ingestion in Amoebida commonly involves extension of pseudopodia or formation of "food cups" to engulf the food. A food cup may be quite deep, as in *Amoeba vespertilio* in which the food is taken in through a temporary "cytostome" similar to the permanent structures in more specialized Protozoa. In shelled types ingestion is limited to an area of naked protoplasm. If there is a large enough opening, ingested particles may be passed into the shell, as in *Arcella*. In the shell contains only small pores, as in many Foraminiferida, the fusion of already extended pseudopodia to enclose trapped food is the essential feature.

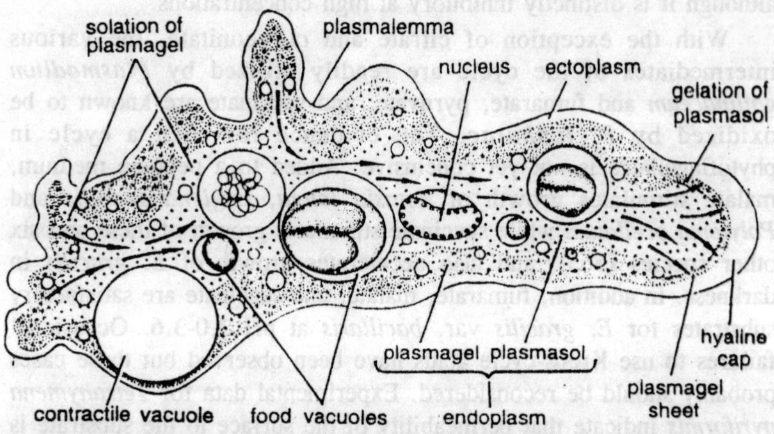

Fig. 5.11. Amoeba.

A comparable process often follows the adherence of food particles to axopodia in Heliozoida and Radiolarida. Ingestion by pseudopodial activity has been reported in various holozoic phytoflagellates and other simple flagellates, while fairly large particles are ingested without marked pseudopodial activity in *Lophomonas* and *Trichonympha*. In certain flagellates and in typical ciliates, ingestion is limited to a cytostome. In simple cases, this appears to be merely a thin region of the cortex. More often, the cytostome lies at the base of a groove or pit. The oral groove, or the peristomial area, of ciliates is often equipped with strong cilia, membranelles, or undulating membranes which drive particles into the cytostome.

Food Vacuoles

The wall of the vacuole in *Amoeba* and similar types is derived from the surface layer of the body. In ciliates feeding on small particles, the vacuole develops at the inner end of the cytopharynx as an enlarging bulb which is eventually pinched off. Formation of the vacuole apparently is stimulated by the passage of solid particles through the cytostome into the cytopharynx, since ciliates in a non-particulate medium contain few, if any, food vacuoles. The ingestion of large masses, as in the engulfment of *Paramecium* by *Didinium nasutum*, is a less simple process. In Suctorea, a food vacuole is formed at the base of a tentacle which is usually attached to the prey and sucks its protoplasm into the captor's body.

Food vacuoles apparently may fuse or divide. Fusion of small vacuoles and the division of large vacuoles into several smaller ones have been noted in *Amoeba proteus*. Also, the collection of small ingested particles into one mass, which becomes surrounded by a common vacuolar membrane, has been described in *Ichthyophthirius multifliis*. A continuous "digestive tract," in which successively formed food vaculoes remain joined with one another by slender tubes, has been described in *Paramecium* and *Vorticella*. In addition, a coiled "canal," extending from cytostome to cytopyge, has been described in *Colpidium* cyclosis of food vacuoles being merely an optical illusion caused by movement of food along this canal. Such a canal resembles to some extent the long sausage-shaped food vacuoles formed by *Paramecium* in certain salt solutions, but other workers have failed to find a tubular digestive system in ciliates.

After formation of the food vacuole, the contents apparently become acid sooner or later. As reported in *Paramecium caudatum* and *Actinosphaerium eichorni*, a drop to pH 4.0-4.3 occurs after a time. In *Amoeba*, the pH of the vacuolar fluid falls to some point between 4.0 and 6.5. The acidity of the vacuole in *P. caudatum* is said to approach that of 0.8N hydrochloric acid; in certain other ciliates, less than that of 1×10^{-4} N acid. The origin of this acid is uncertain, although Mast attributed it to the respiration of ingested organisms and their later autolytic changes. During later stages of digestion and absorption, there is a gradual rise in pH, sometimes to about pH 7.0 in old vacuoles containing undigested residues. Undigested materials are usually eliminated through a definite area (*cytopyge*) in Protozoa with a well-developed cortex. In various Peritrichida the contents of the old vacuole are discharged into the vestibule. In many other ciliates, the cytopyge lies at the surface somewhere in the posterior half of the body. A similar differentiation also may be found in such flagellates as *Peranema trichophorum*, in which the supposed cytopyge is a small area in the postero-lateral body wall lacking the usual cortical inclusions.

Digestion of Proteins

Protozoa which ingest solid food presumably are equipped with digestive enzymes, and would thus be expected to produce both endopeptidases (proteinases) and exopeptidases (peptidases). *Didinium nasutum* seems to be an interesting exception which depends upon its ingested prey for a supply of peptidases. *Glaucoma pyriformis* produces an endopeptidase active in the pH range, 2.2-9.6, with an optimum at about pH 6.0; comparable enzymes of *Plasmodium gallinaceum* are more active at pH 6.5 than at 7.4. Exopeptidases have been reported in *Paramecium caudatum*, *Frontonia* sp., and *Amoeba proteus*. Certain trypanosomes likewise produce exopeptidases and also an endopeptidase of the kathepsin type, but no enzymes of the pepsin or trypsin types. The ability to digest proteins also has been reported for *Euglena gracilis* and for such a saprozoic flagellate as *Leishmania tropica*. Digestion in such cases is presumably extracellular but it is uncertain whether the enzymes are eliminated during life or are released by disintegration of the flagellates.

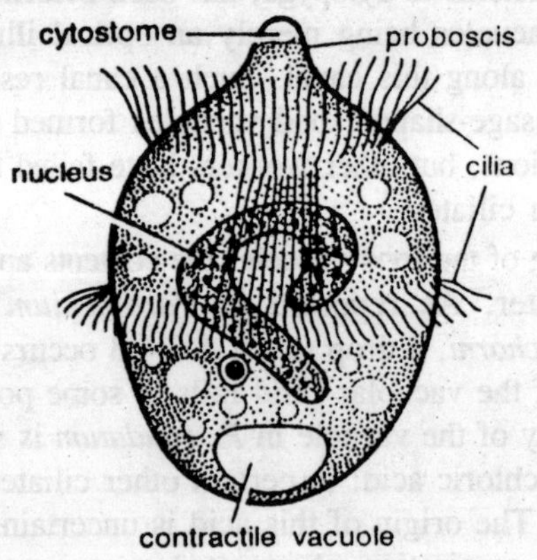

Fig. 5.12. Didinium.

Digestion of Carbohydrates

Enzymes catalyzing the digestion of carbohydrates are of two general types—*polysaccharidases*, or polysaccharases, acting on cellulose, starch, and similar large molecules, and the *glycosidases*, acting on such small molecules as disaccharides and trisaccharides. The digestion of cellulose has been reported for several Protozoa and the same ability may be assumed for species which feed on plant materials or those which invade plant tissues.

Table 5.5. Utilization of Polysaccharides

Species	*Cellulose*	*Starch*	*Insulin*	*Dextrin*	*Glycogen*
Mastigophora					
Eutrichomastix colubrorum	?	?	–	–	–
Leishmania tropica	?	?	+	–	–
Strigomonas media	?	?	+	–	?
S. muscidarum	?	?	+	+	?
S. porva	?	?	–	–	?
Trichomonas columbae	?	+	+	+	?
T. foetus	?	+	+	+	?
T. termopsidis	+	?	?	?	?
flagellates of termites	+	?	?	?	?
flagellates of wood roach	+	?	?	?	?
Sporozoa					
Plasmodium knowlesia	?	?	–	–	–
Ciliatea					
Chilodon cucullus	+	+	?	?	?
Colpidium campylum	?	–	?	?	?
Diplodinium denticulatum	+	?	?	?	?
D. maggii	+	?	?	?	?
D. multivesiculatum	+	?	?	?	?
Entodinium caudatum	–	?	?	?	?
Eudiplodinium neglectum	+	?	?	?	?
Glaucoma scintillans	?	+	?	?	?
Paramecium caudatum	+	+	?	?	?
P. multimicronucleatum	+	+	?	?	?
Saprophilus oviformis	+	+	?	?	?
Tetrahymena pyriformis	?	+	+	+	?
T. pyriformis	?	+	–	+	+
T. pyriformis	?	+	–	+	?
T. pyriformis	?	+	–	+	?
T. pyriformis	?	+	–	+	?
T. pyriformis	?	+	+	+	?
T. pyriformis	?	+	–	+	+
T. pyriformis	?	–	–	?	?
T. pyriformis	?	+	?	?	?
T. pyriformis	?	+	–	+	+
T. vorax	?	+	–	+	+
Trichoda pura	+	+	?	?	?

The demonstration of cellulases in certain flagellates of termites and the wood roach, and in certain ciliates of ruminants, presumably qualifies these organisms as symbiotes of their respective hosts. In addition to the Protozoa in which other polysaccharidases have been reported the many species which store and utilize starch and glycogen doubtless have enzymes capable of splitting these reserve foods into simple sugars. However, the utilization of stored polysaccharides may involved phosphorolysis rather than digestive hydrolysis. Starch, for example, would yield α-glucose-I-phosphate instead of maltose or glucose. This situation raises the possibility that some of the phytoflagellates which store starch may be unable to use exogenous starch as a substrate.

Table 5.6. Utilization of Disaccharides

Species	*Cellobiose*	*Lactose*	*Maltose*	*Sucrose*
Mastigophora				
Eutrichomastix colubrorum	?	+	+	+
Leishmanis tropica	?	?	±	+
Strigomonas media	?	–	–	+
S. muscidarum	?	+	+	+
S. parva	?	–	–	+
Trichomonas columbae	?	+	+	+
T. foetus	?	+	+	+
T. termopsidis	+	?	?	?
flagellates of termites	+	?	?	?
flagellates of wood roach	+	?	?	?
Sporozoa				
Plasmodium knowlesi	?	–	–	–
Ciliatea				
Colpidium campylum	–	–	+	+
Diplodinium denticulatum	+	?	?	?
D. maggii	+	?	?	?
D. multivesiculatum	+	?	?	?
Eudioplodinium neglectum	+	?	?	?
Glaucoma scintillans	+	–	+	–
Saprophilus oviformis	+	?	+	?
Tetrahymena pyriformis E, GF-J, GP, H, I	?	–	+	–
T. pyriformis E, GHH, H, T, T-P	–	–	+	–
T. pyriformis W, *T. vorax*	+	–	+	–
T. pyriformis W, *T. vorax*	–	–	+	–

The utilization of disaccharides has been reported on the basis of fermentation reactions, the activity of extracts prepared from Protozoa, or the effects of sugars on oxygen consumption. Such abilities also may be inferred for species which digest polysaccharides.

Digestion of Lipids

Little is known about the utilization of lipids by Protozoa, although the production of lipases by many species seems probable in view of the common storage of cytoplasmic fats and oils, and the absence of evidence that different enzyme systems are involved in storage and utilization of lipids, *Amoeba proteus* apparently hydrolyzes several animal and vegetable oils after their injection into the cytoplasm and also digests fats and food vacuoles. The products of digestion pass into the cytoplasm and are combined there to form droplets of neutral fat. Unlike amoebae, certain trypanosomes apparently do not produce lipases.

Nitrogen Metabolism

For certain phytoflagellates no amino acid need be supplied from external sources. Therefore, the major feature of nitrogen metabolism presumably is the assimilation of ammonium-N in synthesis of the amino acids needed for growth. In *Chilomonas paramecium* and *Polytoma ocellatum*, these syntheses apparently include all the amino acids which are absolute requirements for *Tetrahymena geleii*. Such flagellates may have some promise in tracing the intermediate stage and the growth-factors involved in synthesis of amino acids. Perhaps the general techniques of "inhibition analysis" will prove applicable here. The fact that certain phytoflagellates can grow on a single amino acid indicates that transaminations may be as effective, in a general way, as the assimilation of ammonium-nitrogen from an inorganic source. Utilization of amino acid as the sole source of energy has not been demonstrated and the dissimilation of amino acids has not yet been traced. A little more is known about the metabolism of amino acids in other Protozoa.

Strains of *Tetrahymena pyriformis* seem to need eleven amino acids, ten of which are considered irreplaceable for higher animals. *T. pyriformis* undoubtedly synthesizes additional amino acids. In a chemically defined medium striped to essentials, these syntheses must involve transaminations with at least certain number of the eleven serving as nitrogen-donors. However these reaction have not yet been traced. The production of ammonia—reported for *Bodo caudatus*, *Leishmania tropica*, *Acanthamoeba castellanii*, *Plasmodium gallinaceum*, *Didinium nasutum*, *"Glaucoma" pyriformis*, *Paramecium caudatum*, *Spirostomum ambiguum*—indicates that such species can deaminate amino acids but nothing is known about the specific dehydrogenases involved. For Protozoa in pure cultures there is little critical information on nitrogenous excretory products. The rather general production of ammonia, and also the failure of tests for urea and uric acid in cultures of *Tetrahymena pyriformis* and in washed suspensions of *Paramecium caudatum* fed powdered fibrin, have suggested ammonia as the probable excretory product.

Carbohydrate Metabolism

The utilization of various carbohydrates by ciliates, trichomonad flagellates, trypanosomes and related forms is well known. On the other hand, the significance of sugars in the metabolism of phytoflagellates remains problematical. Quantitative techniques have not demonstrated utilization of glucose by *Chilomonas paramecium* and *Chlorogonium euchlorum*. Furthermore, sugars do not accelerate growth of *Polytomella agilis* and *Polytoma ocellata* in peptone media and sugars apparently cannot replace acetate in the synthesis of reserve carbohydrates by phytoflagellates.

In a few cases, stimulation of growth has been reported in peptone media supplemented with certain sugars. However, such effects have been minor ones and cannot, with any assurance, be attributed to utilization of sugars as substrates. This situation has been puzzling in view of the fact that these flagellates store carbohydrates and evidently utilize such reserves. Perhaps the difficulty lies in some fundamental deficiency such as the lack of an adequate phosphorylating mechanism for utilizing exogenous carbohydrates. Or possible the permeability of the body wall is too low for effective absorption. Low rates of absorption presumably would not be a hindrance in holozoic species. Consequently, the investigation of polysaccharides and disaccharides as substrates for holozoic Euglenida and Chrysomonadida might yield significant information. Utilization of monosaccharides has been demonstrated by consumption, and by quantitative sugar determinations.

The decomposition of a monosaccharide involves a series of reactions catalyzed by a number of enzymes, the initial step being phosphorylation of the sugar to glucose-6-phosphate. This reaction precedes the dissimilation of exogenous sugar and often its storage as polysaccharide. In this connection, it is interesting that hexokinase has not been found in *Polytomella caeca* and also that hexosediphosphate but not glucose can be oxidized by *Astasia klebsii*. Although utilization of glucose also has not been demonstrated for *Euglena gracilis*, this species does contain the following compounds which appear in

Table 5.7. Utilization of Monosaccharides

Species	*Sugars*
Mastigophora	
Eutrichomastix colubrorum	a, B, D,E, I, j, l
Leishmania donovani	B,E
Leishmania tropica	a, B, D, E, G, I, l
Leptomonas ctenocephali	B, E
Strigomonas media	a, B, D, E, G, I, j, l
S. muscidarum	A, B, D, E, G, I, j, l
S. parva	a, B, D, E, G, I, j, l
Trichomonas columbae	*A*, B, D, *E*, *I*, j, *L*
Trichomonas foetus	a, B, D, E, I, j, l
Trichomonas vaginalis	a, B, D, E, j, l
Trypanosoma brucei	B, D, E, G
Sarcodina	
Acanthamoeba castellanii	b
Sporozoa	
Plasmodium knowlesi	a, B, d, E, G, h, i, j, l
Ciliatea	
Colpidium compylum	a, B, d, E, l
Glaucoma scintillans	a, B, d, E, l
Tetrahymena pyriformis E	a, B, d, E, G, h, j, l
T. pyriformis E	a, B, c, D, E, f, G, h, i, j, k, l
T. pyriformis GF-J	a, B, D, e, g, h, j, l
T. pyriformis GHH	a, B, c, d, E, f, g, h, i, j, k, l
T. pyriformis GP	a, B, D, E, g, h, j, l
T. pyriformis H	a, B, c, d, E, f, G, h, i, j, k, l
T. pyriformis L	a, B, D, E, l
T. pyriformis T, T-P	a B, c, d, E, f, G, h, i, j, k, l
T. pyriformis W	a, B, c, d, E, f, G, h, i, j, k,
Tetrahymena vorax D	a, B, d, E, l
T. vorax PP, V	a, B, c, D, E, f, G, h, i, j, k, l
Trichoda pura	B

A, arabinose; B, dextrose; C, fucose; D, galactose; E, levulose; F, lyxose; G, mannose; H, melizitose; I, raffinose; J, rhamnose; K, ribose; L, xylose. Bold-face capitals indicate utilization; italicized capitals, slow utilization; small letters, no utilization.

glycolysis; glucose-l-phosphate, fructose-6-phosphate, fructose-1, 6-diphosphate, and glycerophosphoric acid (l). In organisms equipped with hexokinase, glucose-6-phosphate is produced and also may be stored, presumably by conversion into glucose-l-phosphate and thence into polysaccharide. Or, glucose-6-phosphate may undergo dissimilation, the next step being conversion into fructose-6-phosphate.

Phosphorylation of this ester yields fructose-1, 6- diphosphate. The diphosphate then undergoes cleavage into two interconvertible triose-phosphates. The series of reactions up to this point yields a certain amount of utilizable energy. Aerobically, pyruvate may be oxidized through the tricarboxylic acid cycle, with more efficient utilization of the original free energy in the glucose molecule. Anaerobically, pyruvate may be converted into lactate or into ethanol.

Glycolysis has been traced, at least to some extent, in a number of Protozoa. Trypanosomes apparently vary in their methods of attacking glucose. In the earlier work, no evidence was obtained for phosphorylation. More recently, *Trypanosoma equiperdum* has been shown to phosphorylate glucose to fructose-1, 6-diphosphate, which is split into triose-phosphate. Glycolysis is similar in *T. evansi* and *T. hippicum*. In the latter, hoxokinase, aldolase, triose-phosphates dehydrogenase, glycerol dehydrogenase, and glycerophosphate dehydrogenase have been demonstrated. The activity of hexokinase in certain trypanosomes is inhibited significantly by arsenicals and in malarial parasites by quinacrine.

The products of dissimilation vary in different species of *Trypanosoma*. Certain species decompose glucose mainly, or even quantitatively to pyruvate. *Trypanosoma equiperdum* produces pyruvate and glycerol; glycerol accumulates under anaerobic condition but is converted almost completely to pyruvate aerobically. Sugar metabolism of *T. evansi* and *T. hippicum* is essentially similar to that of *T. equiperdum*. It has been suggested that *T. hippicum*, which lacks cytochrome oxidase, is dependent upon the host for removal of waste pyruvate. However, it might be interesting to test *T. hippicum* and *T. equiperdum* under conditions which would insure an adequate supply to thiamine and other growth-factors *in vitro*. Such species as *T. lewisi* and *T. rhodesiense* produce several intermediates. Succinic forms about 40 per cent of the acids recovered from *T. rhodesiense* suspensions in glucose-Ringer's solution, and is also a major product for *T. lewisi*. Both species also produce acetic, lactic, pyruvic, and formic acids, ethanol and CO_2 and *T. rhodesinese* produces glycerol in addition. Formate and CO_2 appear only under aerobic conditions in *T. lewisi* suspensions. Whether succinate is produced through the tricarboxylic acid cycle is uncertain. Since succinic dehydrogenase is cytochrome-linked and *T. rhodesiense* presumably lacks the cytochrome system, this trypanosome may be unable to oxidise succinate after producing it. This may not be true for *T. lewisi* which is rather sensitive to cyanide poisoning and presumably contains the cytochrome system.

Certain flagellates of termites decompose glucose anaerobically to carbon dioxide, hydrogen, acetic acid, and certain unidentified products. Lactic and pyruvic acids, acetaldehyde, methyl glyoxal, and ethanol have not been detected in significant amounts.

In *Plasmodium gallinaceum*, hexokinase has been demonstrated and glucose, lactate, and pyruvate all seem to be oxidized through the tricarboxylic acid cycle. The oxidation of pyruvate is inhibited by malonate, with accumulation of succinate. Parasitized erthrocytes oxidize pyruvate almost completely to CO_2 and H_2O, by way of the Krebs cycle and accumulate very little acetate. Cell-free suspension of *P. gallinaceum* produce considerable acetate, as well as CO_2 and H_2O, under aerobic conditions and the acetate is not further decomposed; anaerobically, pyruvate does to disappear and acetate is not formed. In *P. knowlesi*, lactate is produced from glucose and can be oxidized and the increase in lactate is more or less parallel to the production of pyruvate.

Little is known about sugar metabolism of ciliates. *Paramecium caudatum* decomposes glucose to unidentified organic acids which account for about a third of the sugar utilized. *Tetrahymena pyriformis* produces lactic, acetic, and succinic acids from glucose under anaerobic conditions. When *T. pyriformis* was supplied with glucose and radioactive CO_2, all the radioactive carbon appeared in the carboxyl groups of succinic acid, indicating that CO_2 is assimilated in the production of succinate as previously reported for *Trypanosoma lewisi* under anaerobic conditions. The oxidation of substrates through the tricarboxylic acid cycle in ciliates is indicated by the presence of succinic dehydrogenase in *Tetrahymena pyriformis* and *Paramecium caudatum* and by the stimulatory effects of fumarate, succinate, and a-ketoglutarate on oxygen consumption of the former.

Synthesis of Carbohydrates and Lipids

Many Protozoa synthesize and store carbohydrates and lipids as visible deposits. Little is known about the relations of particular substrates and other factors to such syntheses. Photosynthesis makes an important contribution in many phytoflagellates, but those without chromatophores also store carbohydrates. In pure cultures, lipids may accumulate as the cultures grow older, whereas carbohydrates may be predominant in young cultures.

The lipids synthesized by *Tetrahymena pyriformis* have been estimated quantitatively sterols make up about 0.05 per cent of the total. A mixture of fatty acids extracted from *T. pyriformis* has shown bacteriostatic activity against several Gram-positive bacteria *in vitro* but not *in vivo*. Similar material from *Chilomonas paramecium* showed activity against pneumococccus type III *in vitro*. Acetate is on effective substrate for the synthesis of lipids and carbohydrates by *T. pyriformis*. Although arsenite and malonate inhibit oxidation of acetate, they do not influence synthesis of either carbohydrates or lipids.

Contractile Vacuoles in Hydrostatic Regulation

The major function contractile vacuoles seems to be that of hydrostatic regulation. Although they probably do eliminate some soluble wastes, their excretory function is of doubtful importance. The many Protozoa which lack contractile vacuoles must carry on excretion through the general body surface or some permeable portion, and the same mechanism probably is operative in species with contractile vacuoles.

The nature of the excretory products is uncertain for most Protozoa. So-called excretion-crystals have been described in various species, but the chemical nature of these inclusions has been disputed and their excretory significance has not been demonstrated satisfactorily. In attempts to identify less problematical waste products, Howland was unable to demonstrate uric acid in the vacuoles of *Amoeba*, *Paramecium* and *Vorticella*, but did detect it in fluid from cultures of *Amoeba* and *Paramecium*. Weatherby found urea in culture fluid but not in the contractile vacuole or cytoplasm of *P. caudatum*. However, urea has been reported in the vacuolar fluid of *Spirostomum*, the low concentration suggesting that only about 1.0 per cent of the theoretical urea production could be eliminated by the contractile vacuole. Ammonia, rather than urea, seems to be the nitrogenous waste product for a number of species.

The assumption that the contractile vacuole is a hydrostatic regulator is based upon the fact that, in a system involving two fluids of different densities separated by a semipermeable membrane, water should pass from the less dense into the denser medium until equilibrium is reached. The cytoplasm would represent the denser medium in fresh-water Protozoa, and the occurrence of endosmosis would necessitate a mechanism for preventing excessive dilution of the cytoplasm. The general occurrence of contractile vacuoles in fresh-water Protozoa and the absence of such structures in many marine and parasitic species support this assumption. An osmoregulatory function also is indicated by certain experimental data. Injection of distilled water into *Amoeba dubia* increases rate of pulsation and water output of the contractile vacuole. A decrease in frequency of contraction with increasing salinity of the medium has been observed in *Amoeba verrucosa*, species of *Paramecium*, *Gastrostyla steinii* and *Blepharisma undulans*. In *A. verrucosa*, pulsation ceases at a salt concentration of 1.5-2.5 per cent; in *G. steinii*, at 1.25 per cent. Conversely, the rate of pulsation in certain marine and parasitic species rise with decreasing salinity, as in *Amphileptus guttula*, *Nyctotherus cordiformis* and *Balantidium entozoon*. Under similar conditions, appearance *de novo* of contractile vacuoles has been described for *Amoeba biddulphiae* and *Vahlkampfia calkinsi*. However, *Flabellula mira* develops no contractile vacuoles even in a 1:20 dilution of sea water. This species seems to eliminate water by way of large food vacuoles which are emptied at intervals.

The water eliminated by the contractile vacuole may be traced to several sources. Endosmosis may account for much of it in fresh water species. Such a process demands the maintenance of a difference in osmotic pressure across a selectively permeable membrane. The internal electrolyte concentration of *Amoeba proteus* and various ciliates, determined with microelectrodes for measurement of intracellular conductivity, is equivalent to 0.01-0.68N KCl. The internal osmotic pressure of *Spirostomum ambiguum*, determined by the vapour pressure method, is equivalent to that of 0.15 per cent NaCl, and the difference of osmotic pressure across the body wall of fresh water peritrichs approximates that of a 0.05M sucrose solution. Formation of food vacuoles is another source of water in holozoic Protozoa, although there is some compensation in the evacuation of old vacuoles. This source accounts for 8.20 per cent of the water eliminated by contractile vacuoles of marine ciliates. Another source of water is that arising in metabolism, but the relative amount has not been estimated.

Vacuolar Cycle

In the simplest cases, small vacuoles appear in the cytoplasm and fuse to form a new vacuole which increases in volume (diastole) and then collapses (systole) in discharging its contents to the outside. Canal-fed vacuoles receive fluid during diastole from feeder canals which may persist throughout the cycle. As described by Lloyd and Beattie in *Paramecium caudatum*, diastole involves: (1) an early rapid phase, coinciding with contraction of the canals to force fluid into the vacuole; and (2) a slow phase, in which further distension involves diffusion of water into the vacuole from the cytoplasm. In systole there are: (1) a preliminary slow phase, in which fluid passes from the vacuole into the canals, distending them; and (2) a rapid phase, in which the remaining fluid is expelled from the vacuole to the outside. Fluid of relatively high osmotic pressure-that derived from the vacuole at the beginning of systole—supposedly remains in the canals and facilitates withdrawal of water from the cytoplasm in the next cycle. On the other hand, Gelei believed that connections between the vacuole and the canals are closed before systole. This is also the case in *Paramecium multimicronucleatum*.

The frequency of pulsation, in general, is greater in fresh-water species than in marine or parasitic forms. Cycles range from 6 seconds to 20 minutes for fresh-water species, 45 seconds to 32 minutes for marine and brackish water types, and 72 seconds to 16 minutes for endoparasitic forms. Fresh water species eliminate a volume of water equivalent to body volume in 4-45 minutes, whereas marine ciliates require 2.75-4.75 hours. In a given species, frequency of pulsation increases as the temperature rises within non-injurious limits. Temperature characteristics (μ values), calculated from the equation of Arrhenium, have been reported for *Spirostomum ambiguum*, *Blepharisma undulans*, and four species of *Paramecium* over the range, 16-26.8°.

According to the *osmotic theory* of diastole, water passes into the contractile vacuole by osmosis from the cytoplasm. This mechanism would require an osmotic gradient favouring the contents of the contractile vacuole. Since it is not clear just how such a gradient would be maintained, it is difficult to account for diastole on this basis alone. The *filtration theory* holds that hydrostatic pressure forces water through the vacuolar membrane. Haye and Kitching have pointed out that hydrostatic pressure would not be relieved by passage of water into the contractile vacuole, since this organelle is surrounded by cytoplasm during diastole. The *secretion theory*, favoured by Kitching, postulates secretion of water into the contractile vacuole by the membrane. This assumption seems logical enough and it conflicts with no available data.

The discharge of the contractile vacuole has been explained in two general ways: (1) that the process involves contraction of the vacuolar membrane; and (2) that systole is produced by cytoplasmic pressure on the vacuole. Although the change from a sol to a gel in the vacuolar membrane might exert enough contraction to initiate systole, this mechanism could not in itself bring about complete discharge. Cytoplasmic pressure theories maintain that contraction of the vacuole is brought about by pressure of the cytoplasm against the vacuolar wall. Observations on *Amoeba dubia* and *A. proteus*, in which the vacuole becomes embedded in a zone of gelated cytoplasm just before systole, suggest that pressure from this zone bring about systole. For such ciliates as *Paramecium*, it appears that hydrostatic pressure is exerted by a more or less fluid cytoplasm and that systole may be initiated by some other factor, such as adjustment of the vacuole to the excretory pore. Adjustment to the pore, as a preliminary step, would presumably insure discharge of a full vacuole rather than a partially filled one.

Growth Of Protozoa

Individual and Populations

The growth of individual Protozoa has been traced in very few cases. In some species, a constant growth-rate has been reported, as in *Plasmodium praecox* and *Ichthyophthirius multifiliis*; in others, a decreasing rate which may or may not follow a sigmoid curve. Investigations on this phase of protozoan growth have been reviewed by Richards.

Cultures of Protozoa are essential for investigation a variety of problems *Pure cultures*, containing one species of Protozoa and no other microorganisms, are necessity for tracing many biochemical and physiologic activities and for determining basal food requirements. *"Species-pure" cultures*, containing one strain of Protozoa and one or more strains of other microorganisms, also have been used to advantage in many investigations. In maintenance of such cultures with known bacteria in known cencentrations has insured reproducible experimental conditions. In the case of *Entamoeba histolytica*, cultures of this type show promise for preliminary investigations on growth requirements and amoebacidal drugs. *Mixed cultures*, containing two species of Protozoa either bacteria-free or with bacteria, also have been used in a few investigations. An interesting example is the growth of *Entamoeba histolytica* with *Trypanosomma cruzi*. Cause has been interested in the struggle for existence between competitors for the some food supply (e.g, *Paramecium caudatum* and *Stylonychia mytilus* feeding on bacteria) and in captor and prey relationships (*Didinium* and *Paramecium*). In the competition between *S. mytilus* and *P. caudatum*, mutual inhibition was noted, *S, mytilus* being the stronger competitor. The observation of Brown, Dewey and Kidder, and Provasoli involve the captor-prey relationship in bacteria free media. Loefer was dealing with analogous problems in his bacteria-free cultures of *Paramecium bursaria*, in which conditions optimal for growth of *Chlorella paramecii* were not those most favorable to growth of the ciliate-algal partnership.

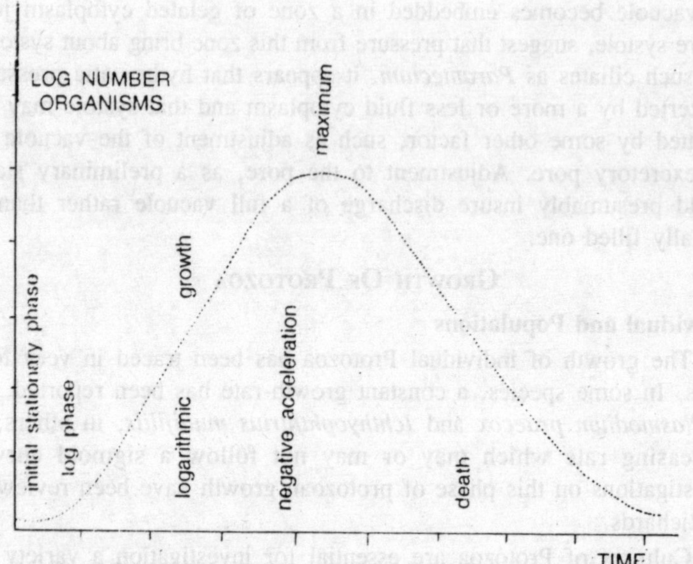

Fig. 5.13. Generalized growth curve for populations of microorganisms.

Experimental data based upon cultures are to be interpreted in terms of protozoan populations. Growth of populations in microorganisms may consists of several phases: an *initial stationary phase*, with no increase in number of organisms; a *lag phase*, during which the rate of population-growth increase to a maximum; a *phase of logarithmic growth*, during which the population increases at a constant rate; a *phase of negative growth acceleration*, in which the growth-rate decreases progressively; a m*aximal stationary phase*, in which the population remains essentially constant; and various *phases of death*, in which the density of population decreases. The first two phases are sometimes lumped together under the one term, *lag*.

The early phase in growth of populations have been investigated in *Euglena* and *Tetrahymena*, while more extended curves have been traced for *paramecium bursaria*, *Polytoma*, *Astasis longa* and *Tetrahymena pyriformis*. The histories of such populations show essentially the same phases as those reported for bacteria.

Initial stationary phase

The occurrence of this phase in cultures of *T. pyriformis* is related to the age of the inoculum. No stationary phase follows inoculation of fresh media from cultures in logarithmic growth, but with older inocula, length of this phase varies with age of the stock. A similar relationship has been observed in *Chilomonas paramecium*. The responsible factors remain unknown, although organisms in old cultures show lower reproductive ability than those in young cultures. This difference in *C. paramecium* has been attributed to storage of an "X-substance" in excessive amounts. After inoculation of fresh media with old flagellates, fission is delayed until the excessive X-substance diffuses into the medium. Another possibility is that the activity of important enzymes is impaired in old cultures, perhaps by progressive vitamin or mineral deficiency or by the accumulation of injurious substances. Upon inoculation of fresh medium, the regeneration or reactivation of essential enzymes would have to proceed growth. A change to a radically different medium high demand the development of "adaptive" enzymes before growth could occur, or perhaps the environmental selection of types adapted to growth in thc new medium. If the fresh medium contains mainly complex foods, some preliminary digestion might be a prerequisite for growth. Under certain conditions, changes in the organism must be the major factor, since no relationship between length of the initial stationary phase and size of the inoculum has been observed in *Tetrahymena pyriformis*.

Lag phase

The occurrence of lag has been explained in various ways. According to one view, favourable changes are produced by the organisms in a "biological conditioning" of the medium. This possibility is supported by the stimulatory effects of old culture fluid added to fresh media for *Chilomonas paramecium*, *Colpidium striatum* and *Tetrahymena pyriformis* in pure cultures. An analogous "conditioning" has been noted in bacterized cultures of ciliates. Another explanation is that the inoculated organisms are still recovering from damage suffered in the stock culture. Therefore, lag is a period of physiological recovery leading to the accumulation of metabolic intermediates essential for synthesis of protoplasm. During lag in cultures of *Tetrahymena pyriformis* there is marked phosphatase activity, with liberation of inorganic phosphate into the medium.

The condition of the organisms evidently in the important factor in some cases, since inocula from cultures of *T. pyriformis* in the logarithmic phase show no lag while those from older stock cultures usually do. Such a relationship has been noted also in *Chilomonas paramecium*. The duration of lag in bacterized cultures of ciliates also may increase, up to a maximum, with age of the inoculum. In addition to changes in fission-rate, changes in individual size may occur during lag. Inocula containing small *T. pyriformis* show as gradual increase to a mean size which is later maintained during the logarithmic phase. If the inoculated ciliates are large, the size decreases to about the same average as that reached by small ciliates.

Phase of logarithmic growth

In late lag the rate of growth increases to a maximum as the population enters the logarithmic phase. During this period the average size of individual organisms as well as the growth-rate may remain essentially constant, as in *T. pyriformis*. Length of this phase is influenced by various factors, and within such a genus as *Leishmania*, may vary with the species in a particular medium. The initial concentration of food is a major influence in cultures of *Astasia klebsii*, *Glaucoma scintillans*, and *T. pyriformis*, although the rate of fission may be independent of food concentration within wide limits. Supplementary thiamine extends the logarithmic phase for *Chilomonas paramecium* in an acetate and ammonium-N medium, and any essential vitamin, food, or metal presumably could become a limiting factor during this phase of growth. A pure culture also may accumulate waste products or undergo other unfavourable changes which bring the logarithmic phase to an end.

Phase of negative growth acceleration

Such unfavorable changes as depletion of the food supply or marked changes in pH of the medium sooner or later become significant and the rate of fission decreases, as in *Euglena*. Lower rates of oxygen consumption in this phase have been reported for *Chilomonas paramecium*, *Trypanosoma cruzi*, and *Tetrahymena pyriformis*. Changes in the respiratory quotient for *T. pyriformis* also indicate qualitative changes in oxidative metabolism. There is also a gradual increase in size of individual ciliates in populations of *T. pyriformis*.

Phase of maximal density

Progressive changes in the culture medium finally check increase in number and the population reaches its maximum. Maximal density has been correlated with initial concentration of food in *Astasia klebsii*, *Glaucoma scintillans*, *Mayorella palestinensis*, *Paramecium bursaria* and *T. pyriformis*.

For certain organisms at least, the vitamin supply may be a more critical factor than the total amount of food. *T. pyriformis* shows almost no growth in a filtered and autoclaved peptone medium which has previously supported a population of the same species. With added thiamine and riboflavin, however, this used medium supports populations approximating those obtained with fresh peptone. Maximal density also may be limited by adverse changes in pH, as noted for *Chilomonas paramecium* in an acetate and inorganic-salt medium. As the medium becomes increasingly alkaline, growth ceases and death of the flagellates soon follows. Periodic addition of acetic, hydrochloric, or lactic acid increases maximal density two to four-fold.

Duration of the stationary phase may depend upon a variety of factors. The thiamine content of the medium is important for *T. pyriformis* and the pH of the medium is a limiting factor for *C. paramecium*. It is somewhat uncertain just how the population is maintained during this phase. Fission may continue at a rate which balances the losses from death, or the life of individual organisms may be prolonged.

Phases of death

Little is known about this phase in protozoan populations. Morphological changes often accompany the decline in population, and a gradual decrease in individual size to about half the maximum, observed in the maximal stationary phase, has been traced in *T. pyriformis*. Death may be accelerated by a sharp drop in pH, related to thiamine deficiency in a medium containing sugar. For some species the decline in numbers in described by a fairly smooth curve; in other cases, the curve is more or less irregular. Populations of *Paramecium bursaria* show a steady decline over a period of three weeks or more in certain media. Populations of *T. pyriformis*, in a casein-peptone medium, have decreased in two major steps separated by a period of several weeks in which the population remains almost constant. Following the second step, in which most of the ciliates die, a small population may persist at least six months longer. The longevity of such small populations is related to the available thiamine. *T. pyriformis* lives for about four months in a certain gelatin medium, while added thiamine extends life of the populations to 11-12 months. With peptone culture fluid which has previously supported growth, supplementary thiamine extends life of the cultures from a maximum of one week to a minimum of at least nine months.

Size of the inoculum in relation to growth

There are three possible relationships between the initial density of population and the rate of growth. (1) The rate of growth may be independent of the initial density under a given set of conditions. (2) The growth-rate may be higher with large than with small inocula. (3)

The growth-rate may vary inversely with initial density of population. A relationship of the first type has been noted in pure cultures of *T. pyriformis*. With optimal bacterial concentrations, a similar relationship has been observed in species-pure of *Stylonychia pustalata*.

A relationship of the second type involves the so-called *allelocatalytic effect* of Robertson. According to Robertson's view, fission in stimulated by a nuclear autocatlyst which is liberated only during fission. One fission has occurred, the autocatalyst which reaches the cytoplasm during nuclear division soon passes into the culture medium, thereto accelerate later fissions. When the inoculum contains more than one organism, the liberation of more catalyst would cause mutual stimulation of fission, or *allelocatalysis*. Accordingly, the fission-rate varies more or less directly with size of the inoculum. An apparent allelocatalytic effect has been reported for bacterized cultures of certain ciliates, *Chilomonas paramecium* and *Mayorella palestinensis*. The case of *C. paramecium* has been questioned because interpretations were based upon terminal counts without any information concerning the earlier history of populations.

Various explanations have been proposed for the Robertson effect. Cutler and Crump believed that Robertson's findings resulted from failure to control the bacterial flora of his cultures. The importance of the bacterial concentration also has been stressed by Johnson who showed that, in cultures of *Oxytricha fallax*, the initial concentration of bacteria may determine whether a culture is to show a Robertson effect. Another possibility is that the initial pH of Robertson's poorly buffered medium was not optimal for his ciliates, which could pH toward the optimum.

On this basis, two ciliates should produce such a change more rapidly than one and cause an allelocatalytic effect. Jahn (240) has suggested that, in similar fashion, the oxidation-reduction potential of the medium might be responsible for an allelocatalytic effect. An inverse relationship between growth-rate and initial density was observed by Woodruff in *Paramecium aurelia* and *P. caudatum*. The more rapid reproduction with lower initial densities was attributed to less rapid accumulation of waste products. A comparable relationship has since been reported of cultures of *P. aurelia*, *P. caudatum*, and *Pleurotricha lanceolata*, *Styloncychia pustulata*, *P. caudatum*, and *Euglena* sp. This so-called Woodruff effect is variously attributed to the accumulation of waste products in the medium, exhaustion of a scanty food supply, and changes in oxygen tension and pH away from the optimum.

Initial pH of the Culture Medium

The observed relations to growth indicate that pH of the medium influences utilization of food and synthesis of protoplasm, perhaps through effects on solubility and ionization of substrates and on permeability of the organism to components of the medium. Activities of extracellular enzymes also may be influenced by pH of the medium. The "internal" pH may be relatively independent of environmental pH, since immersion of *Amoeba dubia* in liquids at pH 5.5 and 8.0 induces no change in cytoplasmic pH from the normal level of about 6.9. However, the activity of a proteinase from *Tetrahymena pyriformis* varies with pH of the medium. Likewise, the rate of oxygen consumption by *Trypanosoma rhodesiense* decrease as pH of the medium falls, and both changes can be prevented by buffering the medium. Accelerating effects of carbohydrates on growth of *Tetrahymena pyriformis* are marked below pH 7.0, but are insignificant in alkaline media.

The effect of plant auxins on growth of *Euglena gracilis* varies with pH of the medium and stimulation is greatest at pH 5.6. The rate of locomotion in *amoeba proteus* is related to pH of the medium and pseudopodial activity in ingestion of food may be influenced likewise. Also, the rate at which food vacuoles are formed in *Colpidium* increases from pH 4.5 to 6.0, and then decreases to zero at pH 8.0. For Protozoa in general growth in pure cultures has been reported between the pH limits 2.1 and 9.9. Survival for at least short periods may be possible within a wider range, such as pH 2.3-11.0 for *Euglena gracilis*, 2.0-9.65 for *E. gracilis* var. *bacillaris* and 1.4-9.6 for *Polytomella caeca*. *Euglena mutabilis* apparently can survive in polluted waters at pH 1.8, and in pure cultures, for at least 12 days within the range, 1.4-7.9. Growth throughout most of the general range seems to have been observed only in *Euglena gracilis* and *Polytomella caeca*, and the specific range varies considerably in other species.

The pH optimum also varies from species to species and within one species under different conditions. Unfortunately, it is sometimes uncertain just what a reported "optimum" means in terms of protozoan growth. The apparent optimum may depend upon the time of observation, as in *Euglena gracilis* which showed heaviest growth at pH 6.6 after 8-9 days, but at pH 7.7-7.4 after seven weeks. Present knowledge of growth pH relationship should be extended by tracing growth curves in media at different pH levels. Most of the available information does not eliminate the possibility that within reasonable limits, a pH above or below an apparent optimum may retard growth without modifying the eventural density of population.

Table 5.8. Growth-pH Relationships of Various Protozoa in Pure Cultures

Species	*pH Range*	*Optimum*
Mastigophora		
Astasia klebsii, peptone	3.2-8.2	4.2-6.0
Chilomonas paramecium, peptone	4.2-8.4	4.8-5.1; 6.8-7.1
C. paramecium, peptone, acetate	5.5-8.4	7.0
C. p aramecium, heteroautotrophy	5.7-6.7	—
Chlorogonium elongatum, peptone	4.9-8.7	7.6
C. elongatum, heteroautotrophy	5.7-8.5	—
C. euchlorum, peptone	4.9-8.7	7.4
C. euchlorum, heteroautotrophy	5.7-8.5	-
Euglena anabaena, peptone	4.5-8.3	6.9
E. deses, peptone	5.3-8.0	7.0
E. gracilis, peptone	3.0-7.7	6.7
E. gracilis, peptone	3.5-9.0	—
E. gracilis, peptone	3.9-9.9	6.6
E. gracilis var. *bacillaris*	2.5-8.8	—
E. klebsii, peptone	5.5-7.5	6.5
E. mutabilis, peptone	2.1-7.7	3.4-5.4
E. pisciformis, peptone	6.0-8.0	—
E. stellata, peptone	4.5-8.0	5.5
E. viridis, inorganic	4.0-7.2	—
Polytoma uvella, heteroautotrophy	7.1-8.5	—
Polytomella caeca, peptone	2.2-9.2	—
Trichomonas vaginalis	4.9-7.5	5.4-5.8
Sarcodina		
Mayorella palestinensis	6.4-7.2	6.8
Ciliatea		
Colpidium campylum	—	5.4
Glaucoma scintillans	—	5.6-6.8
Paramecium bursaria	4.9-8.0	6.8
Tetrahymena pyriformis E	4.5-8.5	5.5; 7.4
T. pyriformis GF-J	4.9-9.5	5.1-6.0
T. pyriformis GP	4.0-8.9	4.8-5.3
T. pyriformis H	4.5-8.5	5.5; 7.4
T. pyriformis W	—	5.6-8.0
Tetrahymena vorax	—	6.2-7.6

The growth of *Astasia longa* in acid media throws some light on such questions. Growth in peptone medium at pH 3.7, for example, is rapid for the first few days and then ceases for a period of 3-5 weeks. Later, a second period of growth produces populations comparable in density to those obtained much sooner at higher pH levels. This

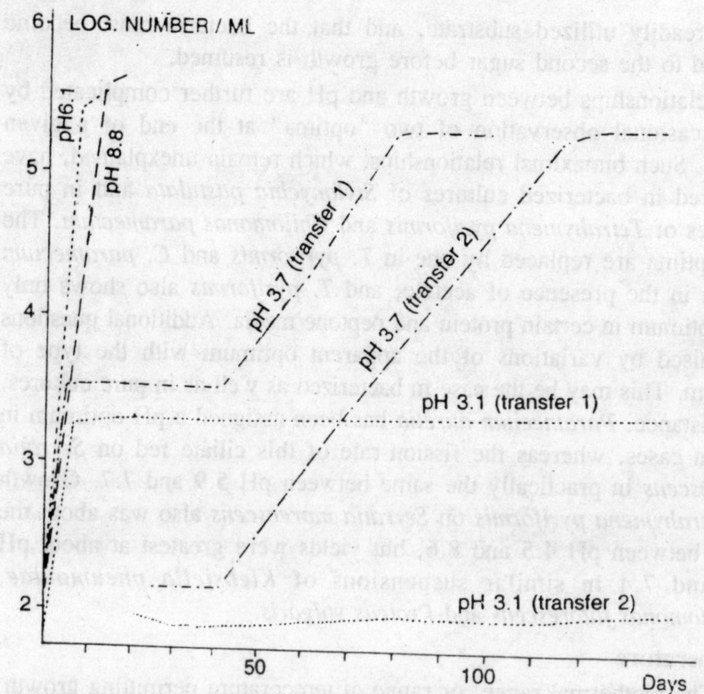

Fig. 5.14. Growth of Astasia longa (strain J) in relation to pH of the medium.

resumption of growth apparently cannot be attributed to the slight rise in pH (0.1) during incubation. Only the first phase of growth is observed in a medium at pH 3.1 and second transfers in medium at the same pH show no significant growth after four months. A delayed growth phase seems to be limited to distinctly acid media since it has not appeared within the pH range, 6.0-9.6. A particularly interesting feature of these populations is the early increase in acid media, even at a pH level which inhibits later growth.

The data suggest the possibility that inocula from a healthy culture may contain enough critical reserves to insure a 20- to 25-fold increase in number, in an unfavourable environment. This reserve apparently is exhausted before the flagellates are completely ajdusted to the new environment, and in media which are not too acid, a period of "adaptation" precedes the resumption of growth. Two periods of logarithmic growth separated by an appreciable stationary phase—Monod's phenomenon of "diauxie—have been observe also in bacteria grown on a mixture of two carbohydrates. In such cases, it has been assumed that the first phase of growth ends with exhaustion of the more readily utilized substrate, and that the bacteria must become adapted to the second sugar before growth is resumed.

Relationships between growth and pH are further complicated by the occasional observation of two "optima" at the end of a given period. Such bimaximal relationships, which remain unexplained, have appeared in bacterized cultures of *Stylonychia pustulata* and in pure cultures of *Tetrahymena pyriformis* and *Chilomonas paramecium*. The two optima are replaced by one in *T. pyriformis* and *C. paramecium* grown in the presence of acetate, and *T. pyriformis* also shows only one optimum in certain protein and peptone media. Additional questions are raised by variations of the apparent optimum with the type of medium. This may be the case in bacterized as well as in pure cultures. For instance, *Paramecium aurelia* has been assigned a pH optimum in certain cases, whereas the fission-rate of this ciliate fed on *Serratia marcescens* in practically the same between pH 5.9 and 7.7. Growth of *Tetrahymena pyriformis* on *Serratia marcescens* also was about the same between pH 4.5 and 8.6, but yields were greatest at about pH 5.0 and 7.4 in similar suspensions of *Klebsiella pneumoniae*, *Pseudomonas fluorescens* and *Proteus vulgaris*.

Temperature

The biothermal range, or range of temperature permitting growth, extends from about 54 to approximately 0°C. for Protozoa. Adaptation to the higher temperatures within this range is rare, although certain flagellates (at 54°), shelled rhizopods (at 51°), amoebae (at 50-52°) and ciliates (at 46°) have been reported from hot springs.

Except for the unusual thermophilic species, active stages are killed by temperatures approaching or exceeding 45°. *Euglena gracilis*, at pH 7.0, is killed within eight minutes at 44°; *Entamoeba gingivalis*, within 20 minutes at 45° *Paramecium caudatum*, within nine seconds at 40°; *Spirostomum ambiguum*, at 36°; *Colpoda cucullus*, at 37-45° after exposures of 0.5-10.0 minutes. Termite flagellates are eliminated from their hosts after 24 hours at 36°, and gregarine trophozoites from *Tenebrio* larvae after six days at 37.5° Lethal exposures depend upon time as well as temperature, and the thermal death time at a given temperature also varies with pH of the medium. The resistance of *Euglena gracilis* to high temperatures is greatest at pH 5.0 and is less above pH 7.0 than below. *Paramecium caudatum*, on the other hand, shows greater resistance to 40° above and below pH 7.0 than at the neutral point. Both *E. gracilis* and *P. multimicronucleatum* have shown increasing resistance with increasing density of population. After the maximum is reached, however, resistance decrease gradually in older cultures of the latter.

Cysts are generally more resistant than corresponding active forms. Dried cysts of *Colpoda cucullus*, for example, resist 100° dry heat for three hours, although moistened cysts die within 30 minutes at temperatures of 49-55°. Excystment is retarted by non-lethal exposures to 37-48°. Somewhat higher lethal temperatures, in 5-minute exposures to moist heat, have been reported for intestinal parasites; *Entamoeba coli*, 76°; *E. histolytica*, 68°; *Endolimax nana*, 64°; *Giardia lamblia*, 64°; *Chilomastix mesnili*, 72°; *Iodamoeba büschlii*, 64°. Unsporulated oocysts of *Eimeria miyairii* are quickly killed at 53°.

Short exposure to temperatures below 0° C. is often not lethal to active stages. Cultures of *Leishmania donovani* have remained viable after intermittent exposure to –12° over a period of 10 days and *Entamoeba gingivalis* may live almost 10 hours at 0°. Fission may continue slowly—for example, a fission every two weeks in *Paramecium caudatum* at temperatures just below zero. Cytoplasmic division is more susceptible than nuclear division to extremes of temperature in *Amoeba proteus*, so that binucleate forms are occasionally seen toward the limits of the range. Freezing and prolonged exposure to sub-zero temperatures are fatal to active stages of many species, although cysts of *Colpoda* have survived exposure to liquid air.

Little is known about biothermal ranges of individual species. However, fission occurs in *Amoeba proteus* at 11-30°, in *Astasia longa* at 15-30° in peptone media and at 22-30° in ammonium-N media; and in *Chilomonas paramecium* between 9.5 and 35°. An optimum for fission has been reported in a few species; *Paramecium aurelia*, 24-28.5°, *Chilomonas paramecium*, 26-30.5° *Astasia longa*, 30°; *Tetrahymena geleii*, 28.5° in the range, 7.8-28.5°. *Euglena gracilis*, in peptone medium, has shown an optimum of 10° in darkness. With supplementary acetate, the optimum is shifted to about 23° which is approximately that for growth in light.

Temperature coefficients (Q_{10} values) and thermal increments (μ values) for fission have been calculated in several cases. For *Paramecium aurelia*, Q_{10} = 2.7 at 21.5-31.5° for *P. aurelia*, μ = 23,000 calories at 12-25° for fission of *Amoeba proteus* μ = 16,500 calories at 11-30°, and for cytoplasmic division μ = 20,500 (11-21°) and 7,300 (21.30°). For *Tetrahymena pyriformis*, Q_{10} and μ values vary with the temperature range: at 7.8-12.3°, Q_{10} = 9.7 and μ = 35,800 cal.; at 12.3-20°, Q_{10} = 3.0 and μ = 18,400; at 20-28.5°, Q_{10}

= 1.5 and μ = 7,350. Reported Q_{10} values (22-28°) for *Astasia longa* vary with the medium—2.10 in peptone, 2.17 in acetate and peptone, 1.28 in acetate and ammonium-N, and 8.03 in an inorganic medium.

The use of thermal coefficients and thermal increments in biology has been based upon the assumption that Q_{10} and μ values related to the nature of the reaction, and upon the hope that a study of such data might furnish clues to the fundamental nature of various biological phenomena. The Q_{10} value is the coefficient of increase in the velocity of a reaction for each 10° increase in temperature. Q_{10} values are calculated from the equation,

$$\log Q_{10} = \frac{10(\log k_1 - \log k_2)}{t_1 - t_2},$$

in which k_1 represents the reaction velocity at temperature t_1 and k_2 the velocity at temperature t_2. Log *k* is a linear function of temperature (Centigrade). For a particular reaction, Q_{10} values vary with temperature and usually increase as the temperature decrease. For example. Q_{10} may be 10 or greater for a given reaction at low temperatures, as compared with 2 or less for a higher range.

The thermal increment, described by the law of Arrhenius, is calculated from the equation,

$$\mu = \frac{4.6(\log k_2 - \log k_1)}{\frac{1}{T_1} - \frac{1}{T_2}}$$

in which T_1 and T_2 are absolute temperature values. The μ value represent that heat of activation, or the number of caloris required to transform one gram equivalent of "inactive molecules" of the reacting substance into "active" ones. There is a close relationship between μ and Q_{10} values, and the latter may be derived from the former for short temperature ranges. A Q_{10} value of 2.0 corresponds to a μ of about 13,200; a Q_{10} of 10, to a m of about 44,000 calories. Unfortunately, the biological significance of thermal coefficients and thermal increments in uncertain.

Light and Darkness

A source of light is obviously important for chlorophyll-bearing flagellates, in which the relation to photosynthesis doubtless accounts for various effects on growth. However, Dusi (106) has reported that under constant illumination *Euglena gracilis* grew well in poptone medium but poorly in inorganic medium, whereas *E. klebsii* grew well in inorganic medium under the same conditions. *E. virdis*, on the other hand, failed to grow under constant illumination. Temperature as the significant factor, rather that illumination, apparently was not completely excluded in these cases.

Light and darkness also may influence the effects of other factors on growth. Thus the thermal optimum for *Euglena gracilis* in peptone medium is about 10° in darkness and 25° in light. Accelerating effects of certain organic acids are relatively greater in darkness, while oxalate is slightly stimulatory in light and without effect in darkness. Plant auxins also have accelerated growth of *E. gracilis* in light but not in darkness.

Even less is known about growth of higher Protozoa in relation to light Richards, in analyzing data on growth of several ciliates, noted that the seasonal rhythms reached a peak in July. On this basis, he suggested that temperature is less important than sunlight when both are variables. On the other hand, light of high intensity is lethal to pigmented *Blepharisma undulans*, the effect being attributed to a photooxidation of the pigment with irreversible damage to protoplasmic components. Indirect effects have been reported for *Plasmodium cathemerium*. Exposure of the hosts to artificially prolonged periods of "day" and "night" lengthen the cycle of merogony.

Effects of Certain Toxins and Venoms

Bacterial exotoxins are relatively inactive against Protozoa. Exposure of *Paramecium aurelia*, *P. calkinsi*, and *P. caudatum* to diphtheria toxin has not affected fission rate or death rate although undiluted culture filtrates containing this toxin may be lethal. Tetanus toxin (150 MLD) and botulinus toxin in various concentrations are without action on *P. caudatum*. On the other hand, a thermostable cytolysin produced by *Pseudomonas aeruginosa* is lethal to *Glaucoma scintillans*. Although ricin is inactive, certain snake venoms, in minimal concentrations of 1.4-150 μg/ml, are lethal to *P. caudatum*. Locomotion is inhibited and rupture of the cortex and disintegration of the ciliate occur sooner or later. Susceptibility to *Carotalus atrox* venom varies with the species. *Bursaria truncatella*, *P. aurelia*, and *Stentor coeruleus* are killed within an hour, *Frontonia leucas*, *Oxytricha fallax*, and *Volvox* after longer periods, while certain other species are not harmed. This apparent resistance of certain species may be largely a matter of degree. For example, the MLD (minimum lethal dose) of *Crotalus atrox* venom for *Coleps hirtus* is nine times that for *Oxytricha fallax*. Sensitivity of 14 species to *Cobra* venom shows no apparent correlation with sensitivity to *Crotalus* venom. Adequate doses of antiserum completely protect *Paramecium multimicronucleatum* against lethal concentrations of *Cobra* venom.

Effects of Certain Therapeutic Drugs

In addition to their action on growth of Protozoa, certain drugs have shown specific effects on metabolic activities. From a practical standpoint, such results are of interest because they help to plan attacks against parasites at vulnerable points. As more is learned about food requirements and metabolic activities, the development of specific drugs for particular parasites more closely approaches realization. Another interesting possibility is that the determination of specific effects of chemotherapeutic drugs may reveal additional tools for the analysis of protozoan metabolism.

Specific effects of certain therapeutic drugs have been reported for malarial parasites and trypanosomes. Hydrolysis of proteins by *Plasmodium gallinaceum* is retarded by atebrin and quinine, and the oxidation of carbohydrates also is retarded by these drugs. The antimalarial activity of a series of napthoquinones seems to be related to their effects on succinic dehydrogenase. Oxygen consumption of *Plasmodium cathemerium* is inhibited by sulfanilamide and sulfathizole and that of *P. knowlesi* by sulfanilamide and quinine. Surprisingly, however, sulfanilamide has no effect on respiration of *P. inui*, and little or no therapeutic action, whereas the drug eradicates infections with *P. knowlesi* in the same host. Trivalent arsenicals (halarsol, reduced atoxyl, reduced tryparsamide) are powerful inhibitors of respiration in *Trypanosoma rhodesiense*. Triose-phosphate dehydrogenases of *T. hippicium* are sensitive to oxophenarsine and the activity of hexokinase also is inhibited by arsenicals in trypanosomes.

The susceptibility of Protozoa to certain antibiotics varies with the species. *Euglena gracilis* var. *bacillaris* remains viable in concentrations of penicillin at least five times as great as those tolerated by *Tetrahymena geleii* and the difference inresistance to streptomycin is of the same order. Tyrothricin, in chicken, has shown parasiticidal activity against extracellular merozoites of *Plasmodium gallinaceum*. Aureomycin likewise has shown activity against *Entamoeba histolytica*, and comparable effects have been reported more recently for terramycin.

One of the most interesting effects reported so far is the bleaching action of streptomycin on certain green flagellates, first reported in *Euglena gracilis* var. *bacillaris*. The effect, which involves a loss of the ability to synthesize chlorophyll, presumably is the result of specific damage to certain enzyme systems. As mentioned above for pantothenic acid, pteroylglutamic acid and nucleic acid derivatives, certain vitamin analogues retard or inhibit growth and oxygen consumption of several Protozoa. In addition to the general interest of such findings and their bearing on the determination of vitamin requirements, the therapeutic value of certain pantothenate analogues in chickens infected with *plasmodium gallinaceum* indicates that results of practical value may be expected in further exploration of this field.

Effects of Carcinogenic Hydrocarbons

Stimulation of fission by several carcinogenic hydrocarbons has been reported for *paramecium*, but these findings have not been confirmed. In the only investigation on ciliates in pure culture. Tittler has obtained no evidence that 3,4-benzpyrene, methylcholanthrene, or 1,2,5,6- dibenzanthrene significantly influences growth of *Tetrahymena pyriformis*.

Effects of Irradiation

Irradiation is a tool of potential value in the study of various problems. One of the least explored is the possibility of inducing biochemical mutations in Protozoa, and the prospects grow more intriguing as protozoan food requirements become better known. Effects on rates of fission, as well as the immediate and pathological effects of irradiation, have interested a number of workers, so that such information is available for a few species.

Beyond the violet end of the visible spectrum extend the overlapping *ultraviolet*, *X-ray*, and *gamma-ray* spectra. The ultraviolet spectrum includes radiation from wave-lengths of about 390 mμ (3900A., or Angstrom units) to 1.5 mμ or less. Rays of 390-200 mμ are transmitted through quartz and are sometimes termed the "quartz spectrum." Below 200 mμ lies the Schumann-Lyman-Millikan region in which the rays are absorbed by water, air, and most other materials. In the Schumann range (approximately 200-125 mμ), fluorite is used for transmission.

The reported effects of ultraviolet irradiation vary with the wave length, the dosage, the species, and physiological condition of the organisms. In the quartz spectrum radiation is relatively harmless at the longer wave lengths. Heavy dosage at 313 mμ is not lethal to *Paramecium multimicronucleatum* and there is almost no effect on *Euglena* at 313 and 365 mμ. Excystment of *Colpoda duodenaria* is slightly retarded at 313 mμ in a dosage of not less than 30,000 ergs/ mm^2, but tripled dosage at 366 mμ is without effect *Peranema trichophorum* is killed at 253 mμ but not at longer wave lengths. Radiation at 302 mμ in a dosage of 28,000 ergs/mm^2, and also the shorter wavelengths in lighter dosage, are lethal to *P. multimicronucleatum*. Selective effects of particular wave lengths have been noted. Motor responses of *P. trichophorum* are most rapid at 302 mμ. Fission of *Paramecium caudatum* is retarded more markedly at 280.4 than by equivalent dosage at 265.4 mμ, although recovery from exposure to the longer wave lengths is more rapid. This difference in rate of recovery is attributed to greater absorption by nucleoproteins at 265.4. Since absorption is essentially the reverse for cytoplasmic proteins, fission is delayed to a greater extent at the longer wavelength. At a particular wavelength, the specific effects may vary with the dosage. At 280.4 mμ immobilization of *P. caudatum* required about 11,800 ergs/mm^2, while fission is retarded by 2,000-3,000.

Effects of ultraviolet vary also with the species. *Fabrea salina* is about six times as resistant as *Tetrahymena pyriformis* and the latter is twice as resistant as *Blepharisma undulans* and *Spirostomum ambiguum* to radiation at 253.7 mμ. Differences also have been noted within the genus *Paramecium* and among several strains of *P. multimicronucleatum*. Physiological condition of the organisms also influences susceptibility. *Tetrahymena pyriformis* in old cultures is much less resistant than in young populations, and starved specimens of *Paramecium* are more susceptible than well-fed ciliates. Sensitivity of *P. caudatum* seems to be greatest in early stages of fission. Sensitivity of *Paramecium* also increase with rising temperature within the range, 0-30°, and preliminary exposure to ultraviolet increases susceptibility to high temperatures.

Effects on fission also have been noted. Fission of *P. caudatum* may be accelerated by light dosages and in the absence of serious injury, recovery from heavier dosage may be accompanied by accelerated fission. Still heavier dosages retard or inhibit fission, and successive exposures may interrupt fission of *P. caudatum* to form chains of several individuals. The ultraviolet action spectrum of *Paramecium* has been described by Giese.

Certain morphological effects are to be expected. Liquefaction of the cortex in *Amoeba dubia* and *A. proteus* is followed by temporary liquefaction of the endoplasm and then, after heavier dosage, by gelation of the endoplasm. Comparable changes occur in *Spirostomum ambiguum*. After short exposures the endoplasm becomes fluid and cyclosis is accelerated, while prolonged exposure causes gradual increase in viscosity, and finally coagulation and vacuolation. Locomotion is accelerated at first, but ceases as lethal dosage is approached. Fragmentation of the macronuclear chain is common. The cortex eventually ruptures after lethal exposure. Vacuolation and cytolysis, following immobilization, also have been described in 11 other species of ciliates. Cytolysis has not been observed in *Peranema trichophorum*, although immobilization, distortion of the body and coagulation of the endoplasm are characteristic effects. Euglena gracilis disintegrates after heavy dosage with quartz ultraviolet, but green flagellates are less sensitive than colourless strains of the species.

Beyond the ultraviolet, the X-ray spectrum extends into the region of gamma-rays emitted by radium, decreasing wave length being correlated with increasing power of penetration ("hardness"). Of the radium emanations (alpha, beta, and gamma rays), alpha-rays are softest and gamma-rays hardest. X-rays produced at 1,000 kv or more extend into the gamma-ray region.

The morphological effects of X-rays and radium are similar to those of ultraviolet. Movement of *Paramecium* and *Colpidium* is first accelerated and then retarded in lethal exposure to X-rays, and amoeboid movement is similarly affected by radium. Vacuolation of the cytoplasm, upon exposure to radium, has been noted in *Amoeba diploidea*, *A. vahlkampfia*, *Entamoeba histolytica*, and *Spirostomum ambiguum*, and the effect of X-rays on *Euplotes taylori* is similar. Cytolysis of *Paramecium*, although not of *Peranema trichophorum*, is caused by lethal dosage with X-rays, and lethal exposure to radium induces cytolysis of *Spirostomum ambiguum*. Heavy dosage with alpha rays (polonium source) causes immobilization and cytolysis of *Polytoma uvella*. Lighter dosage may be followed by fission, but the daughter flagellates undergo cytolysis. Sublethal exposure of *Eudorina elegans* to radium induces deformed daughter colonies containing less than the normal number of flagellates. Unusually large organism, resulting from continued growth but retarded fission, occur in *Colpidium colpoda* after exposure to X-rays, and in *Bodo caudatus* and *Entamoeba histolytica* after exposure to radium. The sensitivity of *Paramecium* to X-rays is increased by preliminary treatment with, vital dyes and other reagents, and is less at 15° than at lower or higher temperatures. *Paramecium bursaria* seems to be less sensitive to X-rays than its symbiotic algae, which are sometimes eliminated at certain dosages. Lethal effects of X-rays on *Tetrahymena pyriformis* in pure culture have been attributed to the production of H_2O_2 in culture media, which become toxic whether irradiated directly or prepared from irradiated distilled or tap water.

Effects of radium and X-rays on growth of populations have been described for several species. Growth of *Entamoeba histolytica*, exposed to gamma-rays primarily or to unscreened radium for 24-48 hours, reaches a maximum one to several days sooner than in the controls. Exposure of *P. caudatum* and *P. multimicronucleatum* to X-rays for 10 minutes to four hours has retarded fission for 2-5 days. Longer exposures, or exposures repeated at intervals of several days, may increase the fission rate. *Bodo caudatus*, exposed continuously to gamma-rays in serial transfers, shows retarded fission and no acclimatization. Such effects may persist for several weeks after removal of the radium, although recovery is complete after three months.

In a given transfer, the lag phase is prolonged almost three hours in irradiated cultures and the period of greatest sensitivity occun about 2.0-2.5 hours before the first fission in a new culture. Although the generation time is essentially normal thereafter, irradiated populations cannot catch up with the controls before the end of the incubation period. Slower growth in the young irradiated population is correlated with larger individual size. Irradiation for part of the incubation period, so as to allow 8-11 subsequent hours of growth, is followed by acceleration of growth to produce populations exceeding 90 per cent of the normal density. The production of ammonia (per culture and per flagellate) by *B. caudatus* is increased after exposures which produce maximal effects on size and fission-rate.

Locomotion

Locomotion in free-living Protozoa is of two basic types: *swimming*, which depends upon the activity of flagella, cilia, or their derivatives; and *creeping*, which is dependent upon direct contact with a substratum. Creeping in Amoebida and similar organisms usually involves pseudopodial activity and is termed *amoeboid movement*.

Amoeboid Movement

Several explanations have been proposed for amoebaid movement. According to one view, locomotion in *Amoeba proteus* is a "walking" process in which extended pseudopodia become attached to the substratum and then contract to pull the body forward (91). A rolling movement has been attributed to *Amoeba verrucosa*. A given point on the surface passes forward on the upper surface, downward at the anterior end, remains on the lower surface for a time as the body rolls forward, and then passes upward at the posterior end to repeat the cycle.

Locomotion in *Amoeba limax* has been interpreted as "fountain streaming", in which there is a forward streaming of endoplasm through

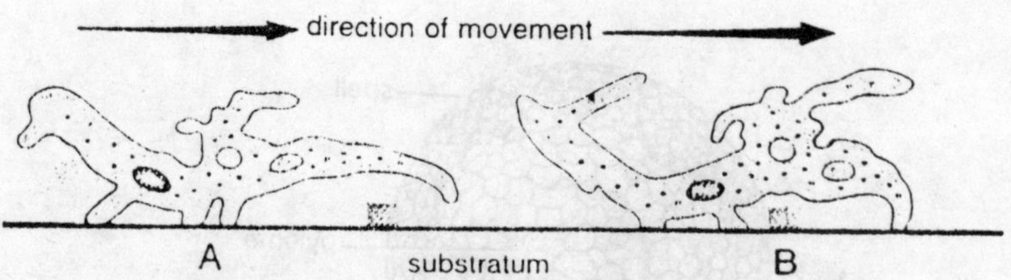

Fig. 5.15. Amoeboid movement.

a tubular layer of ectoplasm. During movement, endoplasm is continually converted into ectoplasm at the anterior end, and ectoplasm into endoplasm posteriorly. According to this interpretation, the flow on the upper surface is backward, instead of forward as in rolling movement. Mast has resolved the ectoplasm of *A. proteus* into a thin elastic *plasmalemma*, or surface layer, and a thicker *plasmagel*. Between the two there is usually a hyaline fluid, except where the plasmalemma is attached to the substratum. During locomotion the plasmalemma flows forward, as in rolling movement. The pasmagel remains a tube which is converted into endoplasm (*plasmasol*) posteriorly and is formed from plasmasol anteriorly as a pseudopodium grows.

Locomotion is attributed to several processes: (1) The plasmalemma becomes attached to the substratum. (2) There is a local, partial liquifaction of the plasmagel. (3) The rest of the plasmagel, which is under tension, forces the plasmasol against this weakened area to produce a bulge, the beginning of a pseudopodium. (4) Posteriorly, the inner surface of the contracting plasmagel is converted into plasmasol. (5) Anteriorly, the plasmagel tube is continuously regenerated by gelation of the plasmasol as the pseudopodium grows. The major factor is thus assumed to be a contraction of the ectoplasm, or plasmagel of Mast. The nature of this contraction remains uncertain, although it has been suggested that contraction represents the elastic recoil of a plasmagel under continues tension that syneresis of the plasmagel causes contraction, and that the process of gelation involves or causes a contraction. In locomotion of shelled rhizopods, such as *Arcella* and *Difflugia*, a developing pseudopodium extends from the mouth of the shell and swings about freely until it makes contact with the substratum and adheres to it.

Contraction of the pseudopodium then pulls the body forward. If movement is to continue, a new pseudopodium is extended to repeat the process. Locomotion of creeping Foraminiferida, by means of myxopodia, is similar. The myopodia are extended, become attached

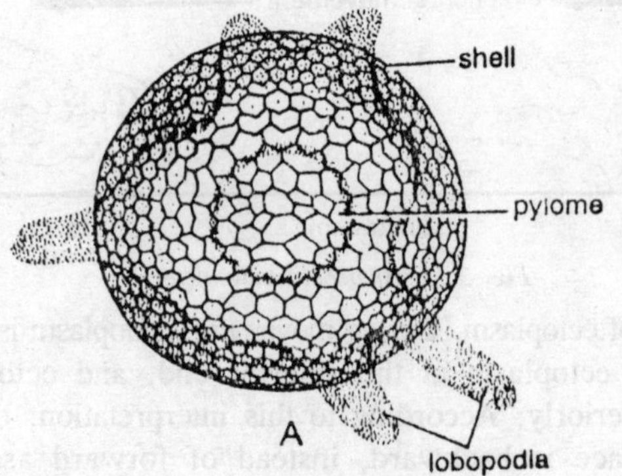

Fig. 5.16. Arcella.

to the substratum, and then contract to pull the organism toward the point of attachment. Axopodia also may function to a limited extent in movement along substratum. The mechanism in *Acanthocystis* apparently involves terminal adhesion of axopodia, followed by a contraction which rolls the body toward the point to attachment.

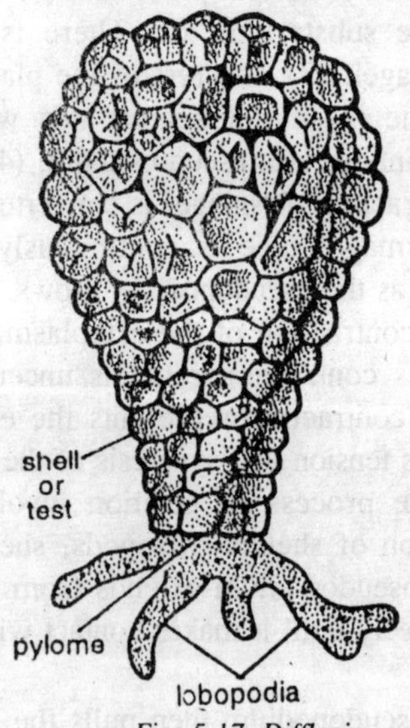

Fig. 5.17. Difflugia.

Flagellar Locomotion

The mechanical aspects of flagellar activity have been disputed and various explanations have been suggested for locomotion in

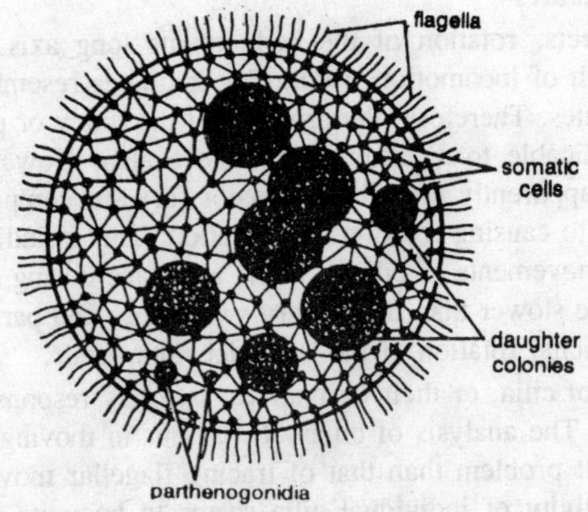

Fig. 5.18. Volvox.

flagellates. Perhaps the most plausible mechanism is that suggested by Lowndes whose data indicate that the basic function of the flagellum, at least in uniflagellate species, is to produce rotation of the organism on its major axis as well as gyration about an axis which marks the general direction of locomotion.

Inflagellar activity, waves pass spirally along the flagellum with increasing amplitude from base to tip, producing two distinct components of force. The resultant of these two components, acting on the anterior end of the flagellate, cause both rotation and gyration which, in an elongated organism, supply the force for propulsion, the principle being that of the screw or propeller. An additional forward component may be supplied by the flagellum itself if it is swung backward as in *Euglena viridis*, but not if it is merely swung outward more or less at a right angle as in *Rhabdomonas incurvum*.

In such colonies as *Volvox*, the flagella are believed to act as propellers, drawing water toward the points of attachment and thus creating forward components of force. The stroke of the flagellum is so directed that the *Volvox* colony usually rotates in swimming, although rapid swimming without rotation also may occur. The ingenious experiments of Brown produced data which agree with the interpretations of Lowndes, and indicate further that gyration of a flagellum alone also may produce a fairly effective locomotor force. This possibility may explain gliding in *Peranema trichophorum*, which Lowndes apparently could not reconcile with his observation on other flagellates.

Swimming in Ciliates

In two respects, rotation of the body on its long axis and the usually spiral path of locomotion, swimming inciliates resembles that in various flagellates. Therefore, the principle of the screw or propeller would seem applicable to swimming in ciliates also. However, the cilia themselves apparently contribute a major forward component of force in addition to causing rotation and gyration. This is indicated in the "browsing" movements of ciliates along a surface during feeding. Movement may be slower than in ordinary swimming, and particularly in various hypotrichs, rotation of the body does not occur.

The activity of cilia, or their derivatives, is solely responsbile for such movements. The analysis of ciliary behaviour in moving ciliates is a more difficult problem than that of tracing flagellar movements. However, the activity of individual cilia seems to be quite variable and may even include spiral, flagellum-like undulations. Such a range of activity is presumably correlated with the variety of manoeuvers to be observed in ciliates. The spiral path followed in swimming, as traced by Bullington in 164 species, shows a width, length, and direction rather characteristic of each species. Both rotation and gyration are attributed to the combined action of all the body cilia rather than a particular group.

In ciliates normally tracing left spirals, the cilia beat obliquely backward to the right for forward movement. When the same ciliate swims backward, the cilia beat obliquely forward to the left. Although a given type of spiral is more or less characteristic of a species, five species of *Paramecium* (40) and four of *Frontonia* may follow either right or left spirals, although swimming is always more rapid in one direction than in the other. Certain other ciliates swim either in right or in the left spirals, but not in both. A right spiral is characteristic of backward swimming in both *Paramecium* and *Frontonia*, and is independent of the spiral followed in forward locomotion.

Responses to Stimuli

Reactions of Protozoa of different stimuli vary with the species as well as with the nature and intensity of the stimulus. Some species may show no reaction to a stimulus which evokes marked reactions in others. The responses studied most extensively are motor reactions which usually tend to move a sensitive organism toward or away from the source of stimulation with some regularity. The response typically involves the organism as a whole, and the morphological nature of the response depends upon and is limited by the structure of the organism.

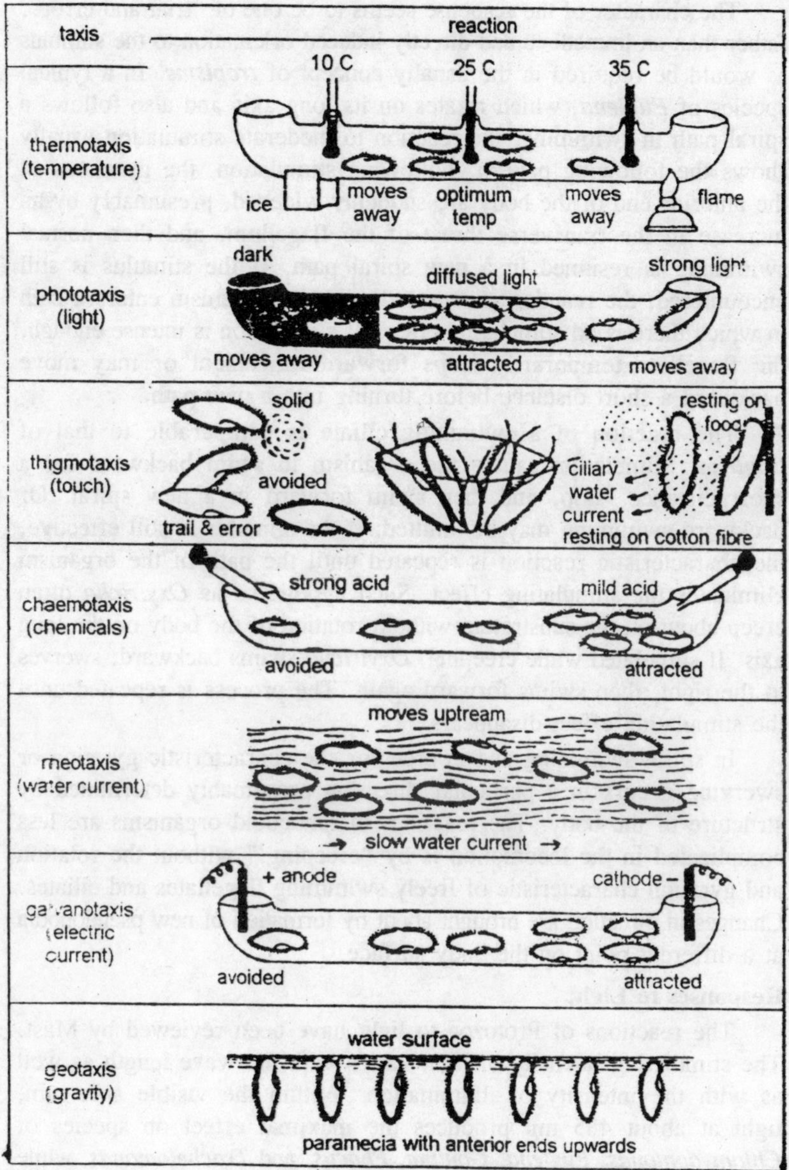

Fig. 5.19. Responsiveness.

In other words, the response is a stereotyped reaction with depends primarily upon structural features of the species rather than upon the nature of the stimulus.

The character of the response seems to be one of "trial and error", rather than an immediate and directly induced orientation to the stimulus as would be required in the usually concept of *tropisms*. In a typical species of *Euglena*, which rotates on its long axis and also follows a spiral path in swimming, the reaction to moderate stimulation usually shows the following pattern. Following stimulation, the gyrations of the anterior end of the body are suddenly widened, presumably by an increase in the transverse thrust of the flagellum, and then normal swimming is resumed in a new spiral path. If the stimulus is still encountered, the reaction is repeated until the organism enters a path in which there is no stimulating effect. If stimulation is intense enough, the flagellate temporarily stops forward movement or may move backward a short distance before turning into a new path.

The reaction of a swimming ciliate is comparable to that of *Euglena*. Stimulation causes the organism to swim backward for a short distance, stop, and then swim forward in a new spiral. Or backward swimming may be omitted. If the stimulus is still effective, the characteristic reaction is repeated until the path of the organism eliminates the stimulating effect. Such hypotrichs as *Oxytricha* often creep about on the substratum without rotation of the body on the long axis. If stimulated while creeping, *Oxytricha* swims backward, swerves to the right, then swims forward again. The process is repeated until the stimulating effect disappears.

In spiral swimming and in creeping the characteristic gyration or swerving occurs in a particular direction presumably determined by structure of the body. The reactions of amoeboid organisms are less complicated in the locomotion is by "creeping," without the rotation and gyration characteristic of freely swimming flagellates and ciliates. Changes in direction are brought about by formation of new pseudopodia at a different point on the body surface.

Responses to Light

The reactions of Protozoa to light have been reviewed by Mast. The stimulating intensity of light varies with the wave length as well as with the intensity of illumination. Within the visible spectrum, light at about 485 mμ produces the maximal effect on species of *Chlamydomonas*, *Euglena*, *Gonium*, *Phacus*, and *Trachelomonas*, while light at 535 mμ is most effective for *Eudorina*, *Pandorina*, and *Spondylomorum*. The stimulatory spectrum of *Volvox* is similar to that for *Euglena*. Many flagellates—species of *Euglena*, *Chlamydomonas*, *Cryptomonas*, and *Gonium*, among others—react so that the path of locomotion is definitely oriented to the source of light. Others, such as *Peranema*, may show merely a shock reaction which is not followed by definite orientation.

Species of *Euglena* respond to a sudden change in the intensity of illumination by their characteristic motor reaction, and the response is repeated until the stigma is equally illuminated at each point in the spiral path of locomotion. As a result, photopositive specimens swim toward the source, and photonegative specimens away from the source of light. Illumination of *Amoeba proteus*, which is photonegative in strong light, causes an increase in thickness of the plasmagel by inducing gelation of the adjacent plasmasol in the stimulated region. This increase in elastic strength causes a contraction of the plasmagel in the stimulated area. Therefore, the formation of pseudopodia in this region is inhibited and new pseudopodia will tend to develop at the opposite end of the body.

A small increase in illumination may do not nothing more than retard temporarily the growth of a pseudopodium. The result of the first type of reaction is a photonegative response, while the second type produces only a delay in locomotion. The photonegative *Stentor coeruleus*, one of the few ciliates known to react definitely to light, shows a typical motor reaction to increased illumination, and the response is repeated until the organism is equally illuminated throughout its spiral course and is moving away from the source of light.

Reactions to Electric Current

Although reactions to the electric current can scarcely be considered part of the adjustment to natural environments, many Protozoa show rather specific responses. In the genus *Amoeba*, reactions vary with the species. *Amoeba proteus* shows a well defined orientation in direct current and moves toward the cathode, whereas *A. dofleini* shows no response. The reaction of *A. proteus* depends upon an induced solation at the cathodal surface, resulting in a decreased elastic strength of the plasmagel in this area. The response of the organism depends upon its orientation when stimulated. Amoebae moving toward the anode show reversal of protoplasmic flow at the cathodal end, followed by cessation of flow at the anodal end.

If the current is too strong and the medium is not acid, disintegration of the organism begins at the anodal surface, whether the amoeba is moving toward or away from the cathode. With weaker currents, the direction of locomotion is reversed. Ciliates usually react to a direct current by reversal of the ciliary stroke on the cathodal surface. As a result, the body is turned so that the organism swims toward the cathode. In a strong but sub-lethal current, ciliary reversal may be so extensive that the ciliate swims backward toward the anode.

Responses to Temperature

Reactions to unfavorable temperatures, as described for various ciliates, involve typical motor responses similar to those noted under stimulation of light in certain species. The response is repeated until the path of locomotion takes the organism into a region with a more favourable temperature.

6

PROTOZOANS IN ENVIRONMENT

With regard to their habitats, protozoa may be divided into free-living forms and those living on or in other organisms as parasitic protozoa.

FREE-LIVING PROTOZOA

The vegetative or trophic stages of free-living protozoa occur in every type of fresh and salt water, soil or decaying organic matter. Even in the circumpolar regions or at extremely high altitudes, certain protozoa are found at times in fairly large numbers. The factors which influence their distribution and population in a given body of water are temperature, light, chemical composition, acidity, kind and amount of food present in the water and degree of adaptability of the individual protozoa to various environmental changes. Their early appearance as living organisms, their adaptability to various habitats, and their capacity to remain viable in the encysted conditions, probably account for the wide distribution of the protozoa throughout the world.

Temperature

The majority of protozoa are able to live only within a certain range of temperature variation, although in the encysted condition they can withstand a far greater temperature fluctuation. The lower limit of the temperature is marked by the freezing of the protoplasm, and the upper limit by the destructive chemical change within the body protoplasm. The temperature toleration seems to vary among different species of protozoa; and even in the same species under different conditions. For example, *Glaser* and *Coria* (1933) cultivated *Paramecium caudatum* on dead yeast free from living organisms at 20-25°C. (optimum 25°C.) and noted that at 30°C. the organisms were killed. *Doudoroff*

(1936), on the other hand, found that in *P. multimicronucleatum* its resistance to raised temperature was low in the presence of food, but rose to a maximum when the food was exhausted, and there was no appreciable difference in the resistance between single and conjugating individuals. The thermal waters of hot springs have been known to contain living organisms including protozoa.

Uyemura (1936, 1937), made a series of studies on protozoa living in various thermal waters of Japan, and reported that many species lived at unexpectedly high temperatures. Some of the protozoa observed and the temperatures of the water in which they were found are as follows: *Amoeba* sp., *Vahlkampfia limax*, *A. rdiosa*, 30-51°C; *Amoeba verrucosa*, *Chilodonella* sp., *Lionotus fasciola*, *Paramecium caudatum*, 36-40°C; *Oxytricha fallax*, 30-58°C. Under experimental conditions, it has been shown repeatedly that many protozoa become accustomed to a very high temperature if the change be made gradually. *Dallinger* (1887) showed a long time ago that *Tetramitus rostratus* and two other species of flagellates became gradually acclimatized up to 70°C in several years.

In nature, however, the thermal death point of most of the free-living protozoa appears to lie between 36°C and 40°C and the optimum temperature, between 16° and 25°C. On the other hand, the low temperature seems to be less detrimental to protozoa than the higher one. Many protozoa have been found to live in water under ice, and several haematochrome-bearing Phytomastigia undergo vigorous multiplication on snow in high altitudes, producing the so-called "red snow". *Kuhne* (1864) observed that *Amoeba* and *Actinophrys* suffered on ill effects when kept at 0°C for several hours as long as the culture medium did not freeze, but were killed when the latter froze.

According to *Greeley* (1902), when *Stentor coeruleus* was slowly subjected to low temperatures, the cilia kept on beating at 0°C for one to three hours, then cilia and gullet were absorbed, the ectoplasm was thrown off, and the body became spherical. When the temperature was raised, this spherical body is said to have undergone reverse process and resumed its normal activity. If the lowering of temperature is rapid and the medium becomes solidly frozen, *Stentor* perishes. The sporozoites of *Plasmodium vivax*, *P. falciparum*, and *P. ovole* are said to have survived at -70°C for 375, 183 and 70 days respectively (*Jeffery* and *Rendtorff*, 1955).

Jeffery (1957) later reported that the blood parasites and sporozoites of all four species of human *Plasmodium* survived low temperature preservation for at least two years with little if any loss in vitality. *Fulton* and *Smith* (1953) found *Emtamoeba histolytica* in cultures that contained a single species of bacteria and 5 per cent glycerol, survived -79°C for 65 days. Seemingly, during the complete cessation of metabolic activity in these cases, life continues.

Light

In the Phytomastigia which include chromatophore bearing flagellates, the sun light is essential to photosynthesis. Therefore, sun light further plays an important role in those protozoa which are dependent upon chromatophore- possessing organisms as chief source of food supply. Hence the light is another factor concerned with the distribution of free-living protozoa.

Chemical Composition of Water

The chemical nature of the water is another important factor that influences the very existence of protozoa in a given body of water. Protozoa differ from one another in morphological as well as physiological characteristics. Individual protozoan species requires a certain chemical composition of the water in which it can be cultivated under experimental conditions, although this may be more or less variable among different forms. *Kolkwitz* and *Marsson* (1908, 1909) distinguished four types of habitats for a few animals which were based upon the kind and amount of inorganic and organic matter and amount of oxygen present in the water: namely, katharobic, oligosaprobic, mesosaprobic, and polysaprobic.

Katharobic protozoa are those which live in mountain springs, brooks, or ponds, the water of which is rich in oxygen, but comparatively free from organic matter. *Oligosaprobic* forms are those that inhabit waters which are rich in mineral matter, but in which no purification processes are taking place. Many Phytomastigia, various testaceans and ciliates, such as *Frontonia*, *Lacrymaria*, *Oxytricha*, *Stylonychia*, *Vorticella*, etc. inhabit such waters. *Mesosaprobic* protozoa live in water in which active oxidation and decomposition of organic matter are taking place.

The majority of fresh water protozoa belong to this group: namely, numerous Phytomastigia, Zoomastigia, Heliozoida, and all orders of Ciliata. Finally polysaprobic forms are capable of living in water which because of dominance of reduction and cleavage processes of organic matter, contain at most a very small amount of oxygen and are rich in carbonic acid, gas and nitrogenous decomposition products.

The black bottom slime contains usually an abundance of ferrous sulphide and other sulphurous substances. *Lauterborn* (1901) called this *sapropelic*. Examples of polysaprobic protozoa are *Pelomyxa palustris*, *Euglypha alveolata*, *Pamphagus armatus*, *mastigamoeba*, *Treponomas agilis* etc.

Certain free-living protozoa which inhabit waters rich in decomposing organic matter are frequently found in the faecal matter of various animals. Their cysts either pass through the alimentary canal of the animal unharmed or are introduced after the faeces are voided, and undergo development and multiplication in the faecal infusion. Such forms are collectively called *coprozoic* protozoa. Some of the cop-orozoic protozoa are: *Scytomonas pusilla*, *Rhynchomonas nasuta*, *Cercomonas longicauda*, *C. crassicauda*, *Trepomonas agilis*, *Naegleria gruberi*, *Acanthamoeba hyalina*, *Sappinia diploidea*, *Chalmydophrys stercorea*, *Tillina magna*, etc. As a rule, the presence of sodium chloride in the sea water prevents the occurrence of numerous species of fresh- water inhabitants.

Certain species, however, have been known to live in both fresh brackish water or salt water. The following species have been reported to occur in both fresh and salt waters: Mastigophora: *Amphidinium lacustre*, *Ceratium hirundinella*; Sarcodina: *Lieberkuhnia wagneri*; Ciliata: *Mesodinium pulex*, *Prorodon discolor*, *Lacrymaria olor*, *Amphileptus claparedei*, *Lionotus fasciola*, *Nassula aurea*, *Trochiliaides recta*, *Chilodonella cucullulus*, *Trimyema compressum*, *Paramecium calkinsi*, *Colpidium camphylum*, *Platynematum sociale*, *Cinetochilum margaritaceum*, *Pleuronema coronatum*, *Caenomorpha medusula*, *Spirostomum minus*, *S. teres*, *Climacostomum virens*, and *Thuricola folliculata*; Suctoria: *Metacineta mystacina*, *Endosphaera engelmanni*.

It seems probable that many other protozoa are able to live in both fresh and salt water, judging from the observations such as that made by *Finley* who subjected some species of freshwater protozoa to various concentrations of sea water, either by direct transfer or by gradual addition of the sea water. He found that *Bodo uncinatus*, *Uronema marinum*, *Pleuronema jaculans* and *Colpoda aspera* are able to live and reproduce even when directly transferred to sea water, that *Amoba verrucosa*, *Euglena*, *Paramecium*, *Stylonychia*, etc., tolerate only a low salinity when directly transferred, but, if the salinity is gradually increased, they live in 100% sea water, and that *Arcella*, *Colpoda cucullus*, *Halteria* etc. could not tolerate 10% sea water even when the change was gradual.

Finley noted no morphological changes in the experimental protozoa which might be attributed to the presence of the salt in water, except *Amoeba verrucosa*, in which certain structural and physiological changes were observed. When the salinity increased, the pulsation of the contractile vacuole became slower. The body activity continued up to 44 per cent sea water and the vacuole pulsated only once in forty minutes, and after systole, it did not reappear for ten to fifteen minutes. The organism became less active above this concentration.

Soil

Some protozoa inhibit soils of various types and localities. Under ordinary circumstances they occur near the surface, the maximum abundance being found at the depth of about 10-12 cm. It is said that a very few protozoa occur in the subsoil. The majority of Testacea inhibit moist soil in abundance.

Food

The kind and amount of food available in a given body of water also controls the distribution of protozoa. The food is ordinarily one of the deciding factors of the number of protozoa in a natural habitat. Species of *Paramecium* and many other holozoic protozoa cannot live in waters in which bacteria or minute protozoa do not occur. If other conditions are favourable, then the greater the number of food bacteria, the greater the number of protozoa. *Noland* (1925) studied more than sixty-five species of fresh water ciliates with respect to various factors and came to the conclusion that the nature and amount of available food has more to do with the distribution of these organisms than any other one factor. *Didinium nasutum* feeds almost exclusively on paramecia; therefore, it cannot live in the absence of the later ciliate.

As a rule, *euryphagous* protozoa which feed on a variety of food organisms are widely distributed, while *stenophagous* forms that feed on a few species of food organisms are limited in their distribution. The adaptability of protozoa to varied environmental conditions influences their distribution. The degree of adaptability varies a great deal, not only among different species, but also among the individuals of the same species. *Stentor coreuleus* which grows ordinarily under nearly anaerobic conditions, is obviously not influenced by alkalinity, pH, temperature of free carbon dioxide in the water (*Sprugel*, 1951).

Parasitic Protozoa

Some protozoa belonging to all groups live on or in other organisms. The sporozoa are exclusively parasites. The relationships between the hosts and the protozoa differ in various ways, which make the basis for distinguishing the associations into three types as follows: *commensalism*, *symbiosis*, and *parasitism*.

Commensalism

Commensalism is an association in which an organism, the *commensal*, is benefited, while the *host* is neither harmed nor benefited. Depending upon the location of the commensal in the host body, the term *ectocommensalism* or *endocommensalism* is used. *Ectocommensalism* is often represented by protozoa which may attach themselves to any aquatic animals that in- habit the same body of water, as shown by various species of chonotrichs, peritrichs, and Suctoria.

In *endocommensalism*, there is a definite relationship between the commensal and the host. For example, *Kerona polyporum* is found on various species of *Hydra*, and many thigmotrichs are inseparably associated with certain species of mussels. Endocommensalism is often difficult to distinguish from endoparasitism, since the effect of the presence of a commensal upon the host cannot be easily understood. On the whole, the protozoa which live in the lumen of the alimentary canal may be looked upon as endocommensals. These protozoa undoubtedly use part of the fuod material which could be used by the host, but they do not invade the host tissue. As examples of endocommensals may be mentioned: *Endamoeba blattae*, *Lophomonas blattarum*, *L. striata*, *Nyctotherus ovalis*, etc. of the cockroach; *Entamoeba coli*, *Iodamoeba butschlii*, *Endolimax nana*, *Chilomastix mesnili*, etc., of the human intestine; numerous opalinids in the colon of frogs and toads, etc. the term *parasitic Protozoa*, in its broad sense, included the commensals also.

Symbiosis

Symbiosis, on the other hand, is an association of two species of organisms, which is of mutual benefit. *Zooxanthellae* containing yellow or brown chromatophores, which live in Foraminiferida and Radiolaria, and certain algae belonging to *Chlorella* (zoochlorellae) containing green chromatophores, which occur in some fresh-water protozoa, such as *Raramecium bursaria*, *Stentor amethystinus*, etc. are looked upon as holding symbiotic relationship with the respective protozoan host.

Many species of the highly interesting Hypermastigida, which are present commonly and abundantly in various species of termites and the woodroach *Cryptocercus*, have been demonstrated to digest the cellulose material which makes up the bulk of woodchips the host insects take in and to transform it into glycogenous substances that are used partly by the host insects. If deprived of these flagellates by being subjected to oxygen under pressure or to a high temperature, the termites die, even though the intestine is filled with wood-chips. If removed from the gut of the termite, the flagellates perish, *Cleveland* (1949-1957) found that the molting hormone produced by *Cryptocercus* induces sexual reproduction in several flagellates inhabiting its hind-gut. Thus, the association here may be said to be an absolute symbiosis.

Parasitism

Parasitism is an association in which one organism called *parasite* lives at the expense of the other called *host*. Here also *ectoparasitism* and *endoparasitism* occur, although the former is not commonly found. *Hydramoeba hydorxena* feeds on the body cells of *Hydra* which die on an average in 6.8 days as a result of the infection and the amoebae disappear in from four to ten days, if removed from a host *Hydra*. The parasites absorb by osmosis the vital body fluid, feed on the host cells or cell-fragments by pseudopodia or cytostome, or enter the host tissues or cells themselves, living on the cytoplasm or in some cases on the nucleus of host cells. Consequently they bring about abnormal or pathological conditions upon the host which often succumbs to the infection.

Endoparasitic protozoa of man are *Entamoeba histolytica*, *Balanitidium coli*, species of *Plasmodium*, *Leishmania*, and *Trypanosoma*, etc. The Sporozoa and Cnidosporidia as was stated before, are without exception *coelozoic*, *histozoic* or *cytozoic* parasites. Because of their modes of living, the endoparasitic protozoa cause certain morphological changes in the cells, tissues, or organs of the host. The active growth of *Entamoeba histolytica* in the gland of the colon of the victim, produces first slightly raised nodules which develop into abscesses and the ulcers formed by the rupture of abscesses, may reach 2 cm or more in diameter, completely destroying the tissue of the colon wall. Similar pathological changes may also occur in the case of infection by *Balantidium coli*.

In *Leishmania donovani*, the victim shows an increase in number of the large macrophages and mononucleates and also enlargement of the spleen. *Trypanosoma cruzi* brings about the degeneration of the infected host cells and an abundance of leucocytes in the infected tissues, followed by an increase of fibrous tissue. *T. gambiense*, the causative organism of African sleeping sickness, causes enlargement of lymphatic gland and spleen, followed by changes in meninges and an increase of cerebro-spinal fluid. Its most charastistics changes are the thickening of the arterial coat and the round-celled infiltration around the blood vessels of the central nervous system. Malarial infection is invariably accompanied by an enlargement of the spleen; the blood becomes watery; the erythrocytes decrease in number; the leucocytes, subnormal; but mononuclear cells increase in number; pigment granules which are set free in the blood plasma at the time of merozoite-liberation are engulfed by leucocytes; and enlarged spleen contains large amount of pigments.

In *Plasmodium falciparum*, the blood capillaries of brain, spleen and other viscera may completely be blocked by infection erythrocytes. Practically all Microsporida are cytozoic, and the infected cells become hypertrophied enormously, producing in one genus the so-called Glugea cysts. In many cases, the hypetrophy of the nucleus of the infected cells is far more conspicuous than that of the cytoplasm. The microsporidan, *Nosema apis*, attacks solely the gut epithelium of the honeybee, but the ovary of an infected queen bee degenerates to varying degrees. For the great majority of parasitic protozoa, there exists a definite host-parasite relationship and animals other than the specific hosts possess a natural immunity against an infection by a particular parasitic protozoa.

Immunity involved in diseases caused by protozoa has been most intensively studied in haemozoic forms, expecially *Plasmodium* and *Trypanosoma*, since they are the causative organisms of important disease. Development of these organism in hosts depends on various factors such as the species and strains of the parasites, the species and strains of vectors, and immunity of the host. *Boyd* and co-workers showed that reinoculation of persons who have recovered from an infection with *Plasmodium vivax* or *P. falciparum* with the same strain of the parasites, will not result in a second clinical attack, because of the development of homologous immunity, but with a different strain of the same species or different species, a definite clinical attack occurs, thus these being no heterologous tolerance.

The homologous immunity was found to continue for at least three years and in one case for about seven years in *P. vivax*, and for at least four months in *P. faciparum* after apparent eradication of the infection. As to the mechanism of immunity, the destruction of the parasites by phagocytosis of the endothelial cells of the spleen, bone marrow and liver and continued regenerative process to replace the destroyed blood cells, are the two important phases in the cellular defense mechanism. Besides, there are indication that humoral defense mechanism through the production of antibodies is in active operation in infections by *Plasmodium knowlesi* and trypanosomes.

Origin of Parasitism

With regard to the origin of parasitic protozoa, it is generally agreed among biologists that the parasite in general evolved from the free-living form. The protozoan association with other organisms was begun when various protozoa which lived attached to, or by crawling on, submerged objects happened to transfer themselves to various invertebrates which live in the same water. These protozoa benefit by change in location as the host animal moves about, and thus enlarging the opportunity to obtain a continued supply of food material. Such ectocommensals are found abundantly; for example, the peritrichous ciliates attached to the body and appendages of various aquatic animals such as larval insects and microcrustaceans.

Ectocommensalism may next lead to ectoparasitism as in the case of *Costia* or *Hydramoeba* and then again instead of confining themselves to the body surface, the protozoa may bore into the body wall from outside and actually acquire the habit of feeding on tissue cells of the attached animals as in the case of *Ichthyophthirius*. The next step in the evolution of parasitism must have been reached when protozoa, accidentally or passively, were taken into the digestive system of the metazoa. Such a sudden change in environment appears to be fatal to most protozoa. But certain others possess extraordinary capacity to adapt themselves to an entirely different environment. For example, *Dobell* (1918) observed in the tadpole gut, a typical free-living limax amoeba, with characteristic nucleus, contractile vacuoles, etc., which was found in numbers in the water containing the faecal matter of the tadpole.

Hyperparasitism

Certain parasitic protozoa have been found to parasitize other protozoan or metazoan parasites. This association is names hyperparasitism. The microsporidan *Nosema notabilis* is an exclusive parasite of the myxosporidan *Sphaerospora polymorpha*, which is a very common inhabitant of the urinary bladder of the toad fish. A heavy infection of the microsporidan results in the degeneration and death of the host myxosporidan trophozoite. Thus *nosema notabilis* is a hyperparasite.

7

MOVEMENT

Locomotion means the displacement or progression through the medium. The purpose of this process is to obtain food from different places and different sources. It includes a study of the locomotory organelles and the mechanism by which movement is achieved.

LOCOMOTORY ORGANELLES

Protozoa possess a variety of locomotory organelles, and usually each type is characteristic of a class. They include:

1. Pseudopodia
2. Flagella
3. Cilia
4. Myonemes.

Pseudopodia

These are temporary cytoplasmic extensions of the body surface of naked protozoa which do not possess a definite pellicle. They can be with drawn quickly and formed fresh rapidly depending on the activity of the animal. Four main types of pseudopodia are distinguished as follows:

Axopodia

These are also called *actinopodia*. The axopodia are long, stiff, semitransparent extension of cytoplasm with pointed ends. They radiate out from the circular body in all directions. Each axopodium is strengthened by an internal axial filament consisting of a number of fibrils. The axial filament usually originate from a central granule in the middle of the protozoa body, sometimes within the nucleus.

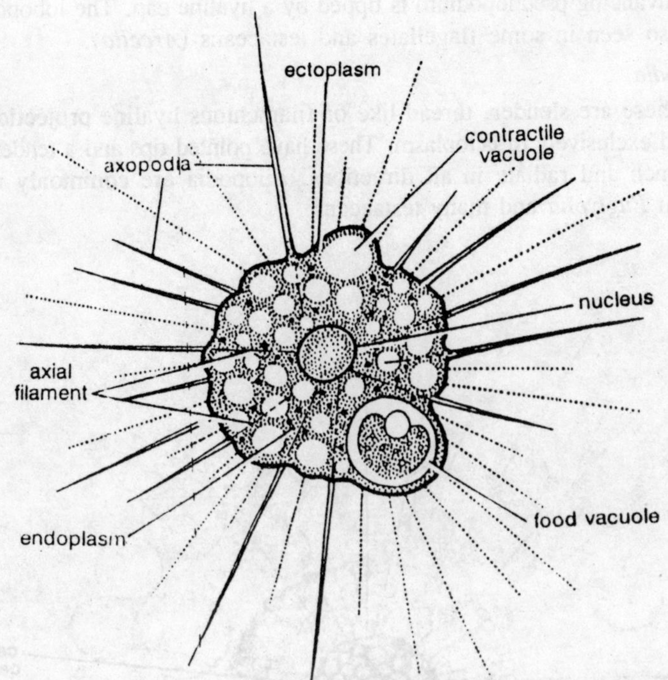

Fig. 7.1. Actinophrys.

Lobopodia

These are characteristic of amoeboid protozoa. These are broad, finger-like or lobe-like projections of both the ectoplasm and endoplasm.

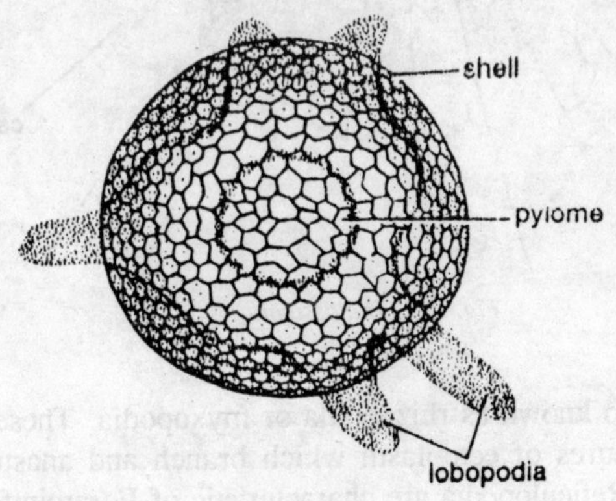

Fig. 7.2. Arcella.

The advancing pseudopodium is tipped by a hyaline cap. The lobopodia are also seen in some flagellates and testaceans (*Arcella*).

Filopodia

These are slender, thread-like or filamentous hyaline projections, formed exclusively of ectoplasm. These have pointed tips and a tendency to branch and radiate in all directions. Filopodia are commonly met with in *Euglypha* and many testaceans.

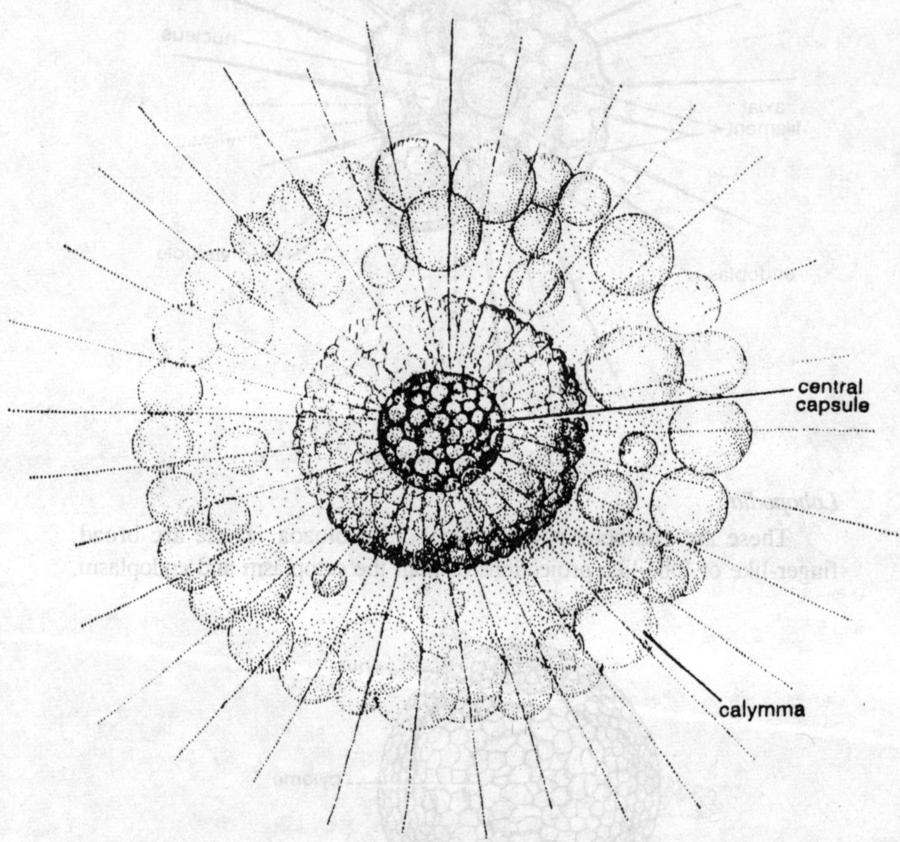

Fig. 7.3. Thalassicola.

Reticulopodia

These are also known as rhizopodia or myxopodia. These are also filamentous structures of ectoplasm which branch and anastomose to form a network. Reticulopodia are characteristic of Foraminifera, such as *Elphidium*, *Chamydophrys* etc.

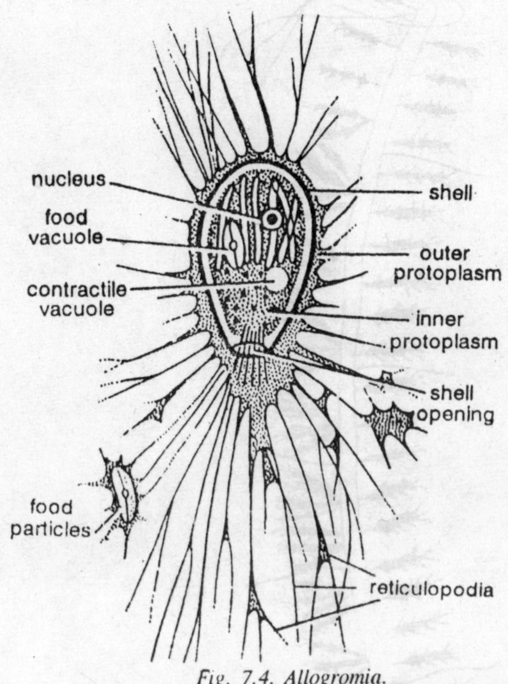

Fig. 7.4. Allogromia.

Flagella

Flagella are extremely fine, delicate, highly vibratile extensions of the protoplasm characteristic of all Masugophora and the flagellate stages of some sarcodme protozoa. *Pitelka* (1949) studied the ultra structure of flagellum of *Euglena gracilis*. A flagellum consists of a central stiff but elastic shaft, the axoneme surrounded by a protoplasmic sheath. The axoneme surrounded by a protoplasmic sheath. The central fibres are closed in a membraneous inner sheath. Each peripheral fibre is made up of two fibres and double row of short arms. In many flagella lateral filaments, arising out of one, two or more of the peripheral fibrils, may be found. These are called *mastigonemes*. On the basis of arrangement of mastigonemes the following five types of flagella are distinguished:

1. *Acronematic*. The lateral mastigonemes are absent but the distal end bears a terminal filament and hence assumes a whip-like appearance that is why the flagellum is known as *lash* or *whip flagellum*. It occurs in *Polytoma*, *Chlamydomonas*, *Cercomonas* etc.

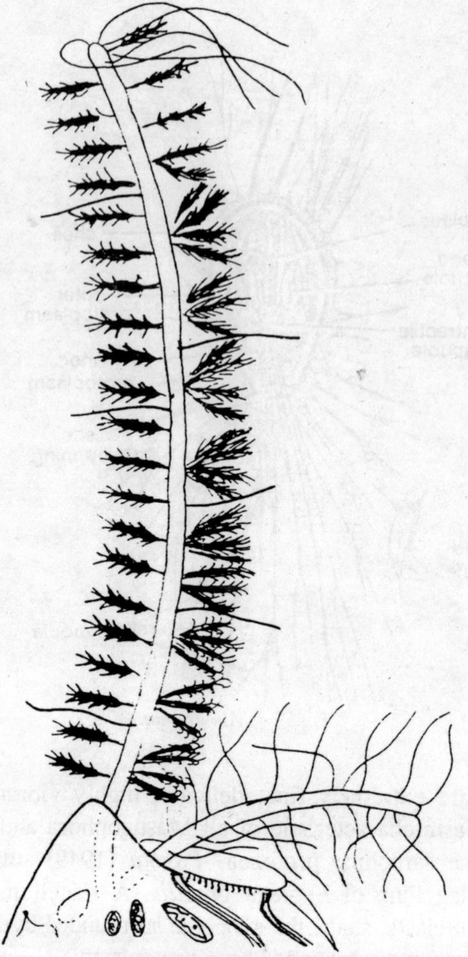

Fig. 7.5. Ochromon.

2. *Pentachronematic*. It bears a terminal filament or mastigoneme as well as one or two rows of mastigonemes e.g. *Paranema*.
3. *Stichonematic*. It bears a single row of lateral mastigonemes, as in *Euglena*.
4. *Simple* or *Animatic*. Flagellum without mastigonemes and terminal filament e.g. *Noctiluca*.
5. *Pantonematic*. The mastigonemes are arranged in two or more rows as found in *monas* etc.

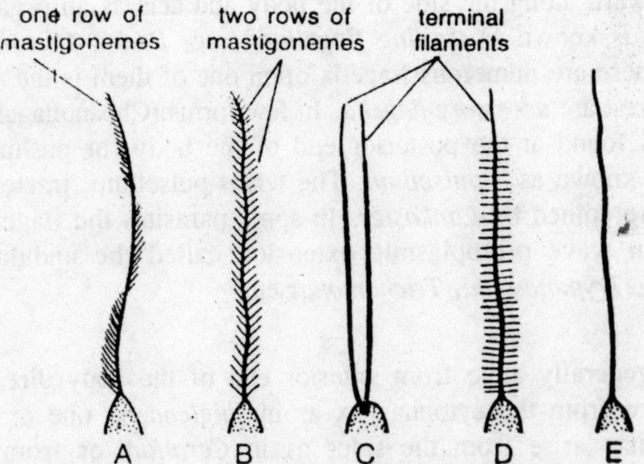

Fig. 7.6. Arrangement of mastigonema on flagellum.

Number of Flagella

The number of flagella varies from species to species. In most cases only one or two flagella are present, but some parasitic forms have more e.g., *Giardia limblia* has 8 flagella and *Trichonympha* has numerous. When 2 or more flagella are present they may be equal in length and similar in function as in *Volvox* or one dragging the body and is directed forward and known as *tractellum* while the other is

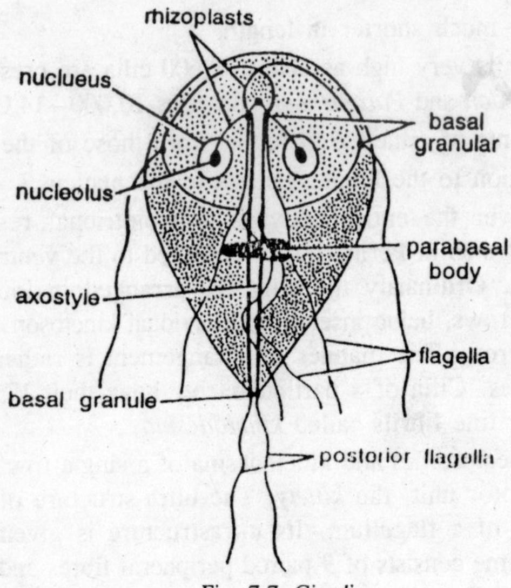

Fig. 7.7. Giardia.

directed backward along the side of the body and acts as an organ of propulsion. It is known as *trailing flagellum* (e.g. *Bodo*, *Anesonema* etc.). When there are numerous flagella often one of them is the *main flagellum* and rest are *accessory flagella*. In few forms (Choanoflagellata) a flagellum is found at the posterior end of the body for pushing it forward. It is known as a *pulsellum*. The terms pulsellum, tractellum and trailing are coined by *Lankaster*. In some parasites the flagellum borders a thin wave protoplasmic extension called the undulating membrane e.g. *Trypanosoma*, *Trichomonas* etc.

Origin

Flagella generally arise from anterior end of the body directly, e.g. *Volvox*, or from the cytopharynx as in *Euglena* by one or two roots. They may arise from the sides as in *Ceratium* or from the posterior end as in *Trypanosoma*. A flagellum arise from *basal granule* or *blepharoplast* or *kinetosome*. The blepharoplast may lies at the surface or deep in the cytoplasm, on the nuclear membrane or within the nucleus. The blepharoplast is connected with organelles like centrioles, parabasal body etc., when these are present.

Cilia

The ciliary locomotion is characteristic of Ciliata. The cilia are extremely fine thin extensions of the ectoplasm. These are fine vibratile processes, which resembles the flagell in structure but differ from later in following respects:

1. The cilia are much shorter in length.
2. Their number is very high as about 11,600 cilia are present on the body of *Protodon* and *Paramecium* possesses 10,000—14,000 cilia.
3. The movements of cilia are different from those of the flagella.
4. Lack of relation to the blepharoplasts to the nucleus.

The cilia cover the entire body as in Holotricha; restricted to within a few circles as in Peritricha; or confined to the ventral surface as in Hypotricha. Ordinarily the cilia are arranged in longitudinal, diagonal or spiral rows, being inserted to individual kinetosome disposed on a ridge or furrow. The manner of arrangement is rather constant for a given species. Cilia of a particular row have their kinetosomes interconnected by fine fibrils called *kinetodesma*.

The cilia, kinectosomes and kinetodesma of a single row constitute a function locomotor unit, the *kinety*. The ultra-structure of a cilium is similar to that of a flagellum. Its ultrastructure is given by *Satir* (1968). The axoneme consists of 9 paired peripheral fibres and 2 central fibres. The central fibres are enclosed in the protoplasmic sheath. The

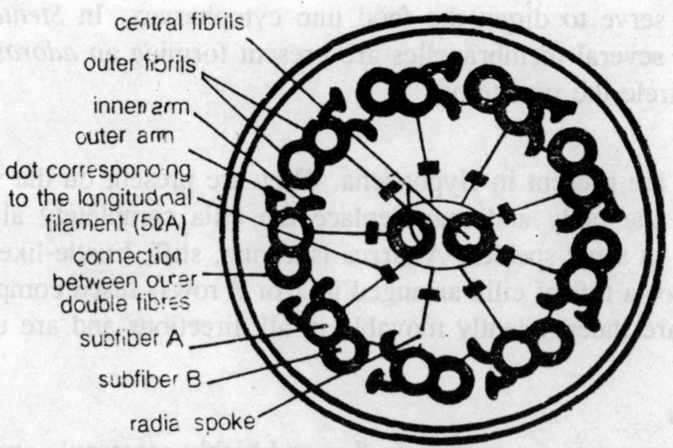

Fig. 7.8. T.S. of a cilium.

peripheral fibres made up of two fibrils each (doublet). One of the fibril double row of short arms (dynein arms). In some cases the cilia becomes fused variously forming compound organelles such as *undulating membranes*, *membranelles* and *cirri*.

Undulating membrane

It is a delicate membrane formed by the union of long row of cilia. Their wave-like rippling movements drive food particles down to cytopharynx e.g. *Cyclidium*, *Pleuronema* etc.

Membranelles

These are rectangular plates formed by the fusion of several short rows of cilia in the region of the peristome. They typically occur in

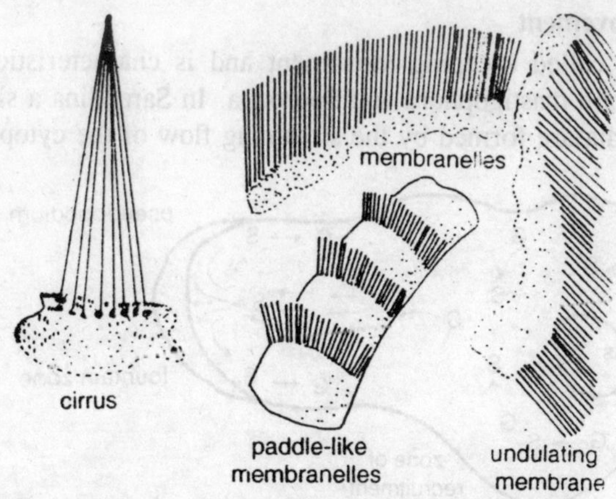

Fig. 7.9. Specialized organelles formed by cilia.

series and serve to direct the food into cytopharynx. In *Stentor* and *Vorticella*, several membranelles are present forming an *adoral zone*, which encircle the peristome.

Cirri

These are present in Hypotricha. They are present on the ventral surface of the body and may replace the cilia completely although both occur in same species. A cirrus is a long, stiff, bristle-like organ composed of a tuft of cilia arranged in 2 or 3 rows, fused completely. The cirri are independently movable in all directions and are used in crawling.

Myonemes

The myonemes are extremely fine and highly contractile structures running in all the directions in the pellicle or ectoplasm of flagellates, ciliates, sporzoa and infusorians. The myonemes are best developed in ciliates. In certain stalked ciliates (*Vorticella*) the longitudinal myonemes of the body proper converge basally into the stalks forming the *spasmonemes*.

Modes of Locomotion

There are four modes of locomotion as there are 4 lomotory organelles. These are:

1. Pseudopodial or Amoeboid movement
2. Flagellar movement
3. Ciliary movement
4. Metaboly or gliding.

Amoeboid Movement

It is also called creeping movement and is characteristic of all Sacodina, certain mastigophora and Sporozoa. In Sarcodina a single or few pseudopodia are formed by the streaming flow of the cytoplasm in

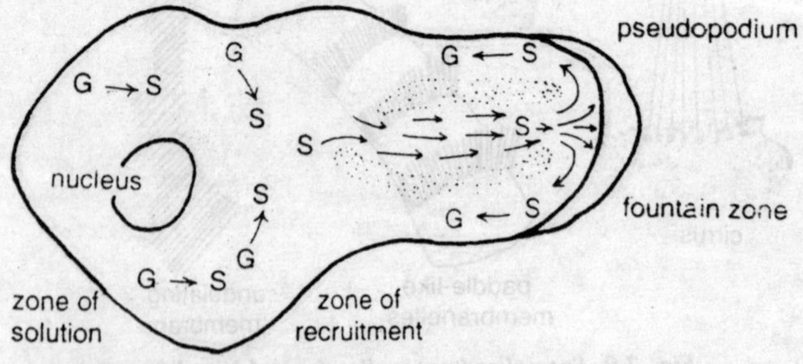

Fig. 7.10. Schematic diagram of pseudopodial formation.

the direction of movement. A number of theories have been proposed to explain the formation of the pseudopodia. These are:

1. Surface Tension Theory of *Rhumbler* (1989).
2. Contraction Theory of *Dellinger* (1906).
3. Change of viscocity of Sol-gel Theory of *Hyman* (1917) supported by *Mast* (1923-25) and *Pentin* (1923).
4. Front Contraction Theory or Fountain zone Theory of *Allen* (1961) and *Goldacre* (1961-64).
5. Rolling Movement of *Jennings* (1904).

Of the various theories advanced about the formation of pseudopodia, the Sol-gel. Theory of *Hyman* (1971) enjoys a wide acceptance. All the theories are described in the chapter of *Amoeba*.

Flagellar Movement

The flagella are so extremely fine structures so that their movement is difficult to observe under light microscope. The mechanism of flagellar movement is not properly understood. A number of theories have been proposed some of them are as follows:

Sidewise lashing movement

According to *Ulhela* and *Krijsman* (1925) a flagellum ordinarily moves in a sidewise lash during rapid locomotion consisting of an *effective stroke* and a *recovery stroke*. In the effective stroke the flagellum is held out rigidly with slight concavity in the direction of stroke, while in recovery stroke the flagellum is relaxed, strongly curved and is brought forward again. This draws the body forwards.

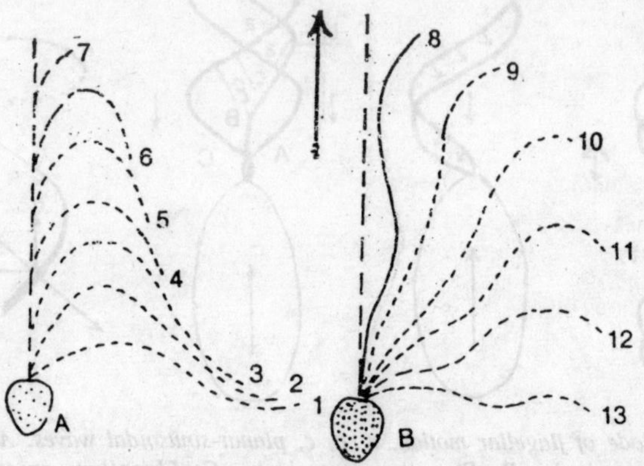

Fig. 7.11. Action of flagellum. A—Recovery stroke, B—Effective stroke.

Usually the flagellum beats obliquely so that during forward movement, the animal also rotates on its longitudinal axis.

Screw propeller theory

According to *Butschlii* a flagellum undergoes a series of lateral movement during which pressure is exerted at right angles to its surface. This can be resolved into (a) a force parallel to the body and (b) a force at right angles to the main axis. The first drives the animal forward and the second rotates the animal on its axis. The animal is propelled like a screw.

Circular beat theory

Metzner (1920) suggested that flagellum beats in a circle tracing a cone producing sufficient suction to drive the animal forward.

Undulating movement

This is the latest theory *Verworn* observed it in *Paranema*. According to this theory the flagellum undergoes spiral undulations with the wave on it usually passing from base to tip. The flagellum is usually held backward of the body during swimming. The lateral

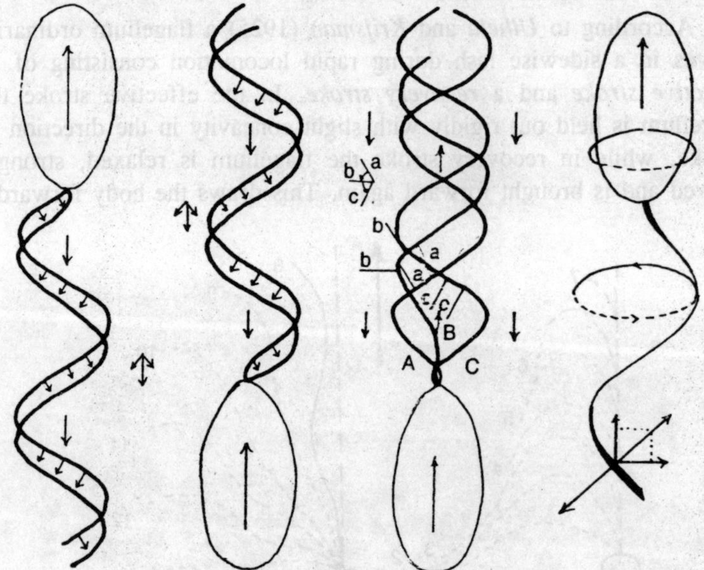

Fig. 7.12. Mode of flagellar motion, a, b, c, planar-sinusoidal waves. A—Flagellum posteriorly, B—Flagellum anteriorly, C—Flagellum anteriorly, with mastigonemes, A, B, C,—Three different phases of a sine wave progressive proximal-distal, D—Flagellum posteriorly, helicoidal wave.

undulations are *sinusoidal* and the spiral undulations are *helicoidal*. Helicoidal motion adds a gyrational (rotating on its axis) component to movement and the animal is propelled forward in water like a screw (*Grell*, 1973). The locomotive force produced by a given flagellar wave form depends on the amplitude, wavelength and frequency of the movement of waves along the flagellar length. Mastigonemes increase the surface area of a flagellum and consequently its effectiveness also. A backward movement can be caused if the undulations progress from base to tip. The flagellates can cover distance of 15-300μ per second.

Ciliary Movement

Ciliary movement is brought about by the cilia. The cilia work like oars. Each beating of cilia consists of an effective stroke and a recovery stroke. In the effective stroke the cilia beat back as straight rigid rods. In the recovery stroke the cilia return to their original position in a limp flexes state. The effective stroke push the water back and propel the body forwards. During the effective stroke water strikes the cilia at right angle to its length, the animal moving in the direction opposite to it. Cilia are usually disposed in linear series. A cilium in a particular row beats slightly in advance of the cilium behind it and slightly later than the cilium just ahead of it. Thus the cilia of a longitudinal row beat metachronously, while those of a transverse row beat synchronously. The bending of cilia is caused by contraction of their peripheral fibrils.

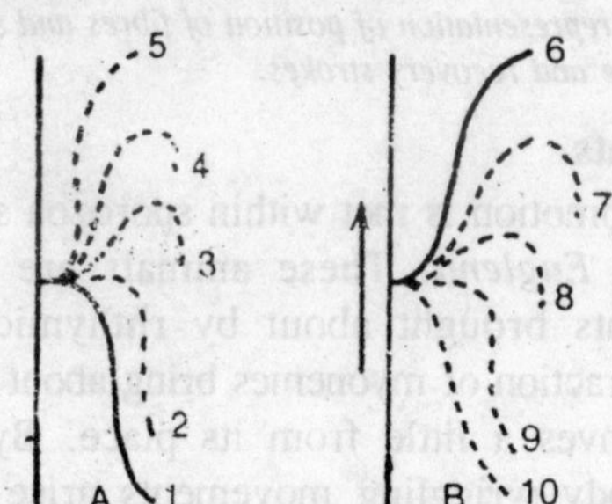

Fig. 7.13. Cilia. A—Effective stroke, B—Recovery stroke.

According to *Giese* (1979) Ciliary movement is caused by the sliding of the members of the doublet of peripheral fibres in relationship to one another by a ratchet like mechanism. The sliding of peripheral doublets is resisted by the radial spoke which leads to the bending of the cilium.

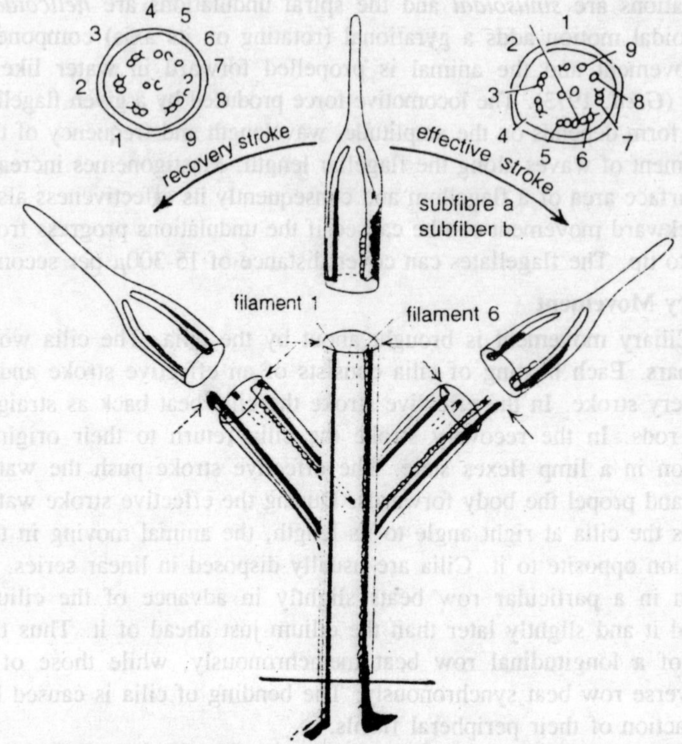

Fig. 7.14. Diagrammatic representation of position of fibres and subfibres in the cilium during effective and recovery strokes.

Metabolic Movements

This type of locomotion is met within sporozoa such as gregarians and flagellates like *Euglena*. These animals are capable of slow wriggling movements brought about by rhthymic contractions of myonemes. The contraction of myonemes bring about a change in shape of animal which moves a little from its place. By contraction and expansion of the body wriggling movements arise in *Euglena* as a result of which it moves forward. This type of locomotion is called *gregarine* or *Euglenoid* or *metabolic* movements.

8

EXOSKELETON

In some protozoa, the body is naked and not covered or enclosed in any covering, in others, the body is covered by a firm layer produced as differentiation or secretion from the most superficial part of the body. A beginning of such as differentiation is seen in *Amoeba verrucosa* in which a firm pellicle covers the body surface but is elastic to allow amoeboid movement. Advance is seen in some flagellates where the thin cuticle of the body allows euglenoid movement. Still in others a firms covering gives shape to the body which varies from a thin flexible cuticle to the rigid inflexible *lorica*. The lorica takes the form of rigid, cellulose covering in Phytomastigina.

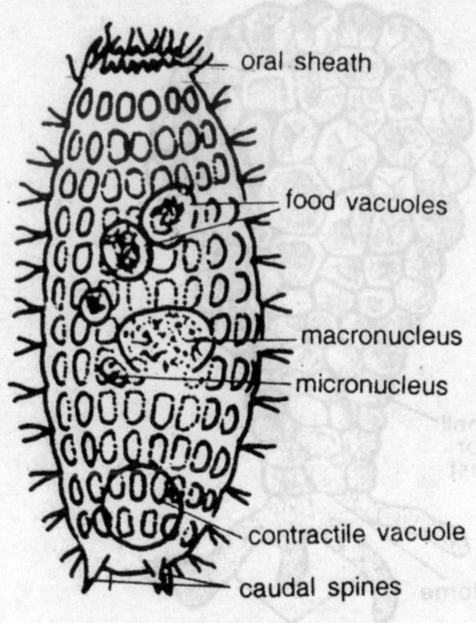

Fig. 8.1. Coleps.

In some of the dinoflagellates it consists of two pieces, the upper *epitheca* and lower *hypotheca*. In other dinoflagellates the cover may be more complex, formed of a number of plates: *cingular*, *precingular*, *post-cingular*, *apical* and *end-apical plates*. In Ciliophora such as Peritricha and Suctoria, the cover takes the form of continuous lorica of various forms. A close fitting cuticle or lorica which is a part of the body is different from the structure built externally or internally as shell, test etc. Such structures exhibit infinite variety but may be classified under two headings: (i) External structures, (ii) Internal structures.

External Structures

Such a structure lies on the surface of the body and may be composed of materials secreted by the organism (antophya) or from foreign particles taken from the surroundings (xenophya). The skeleton elements secreted by the organism may be organic or inorganic in nature. In the later case it may be gelatinous or cellulose or tectin. In the later case it may be calcarious or silicious.

The calcarious skeleton as formed of calcium carbonate and the silicious of amorphous silica. In xenophya the skeleton has a base of gelatinous or chitinous layer on which flowing particles are cemented. Some forms like *Difflugia* shows no special projections for any particular

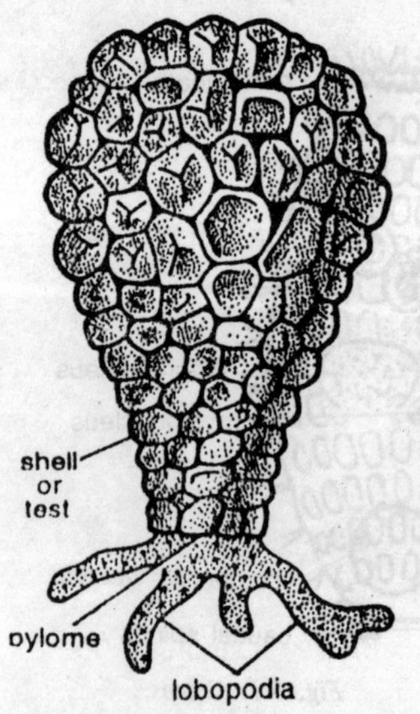

Fig. 8.2. Difflugia.

materials. Many species, however, forms test of particular materials. In *Haliphysma* the test is formed of sponge spicules and in *Technitella* of calcarious plates from echinoderms. The simplest architecturae type of shell is a single chambered or *monothalamus* chitinous shell of *Sqammulina*. In majority of foraminifera the simplest form gets complicated. The protoplasm of the original chamber over-flows through

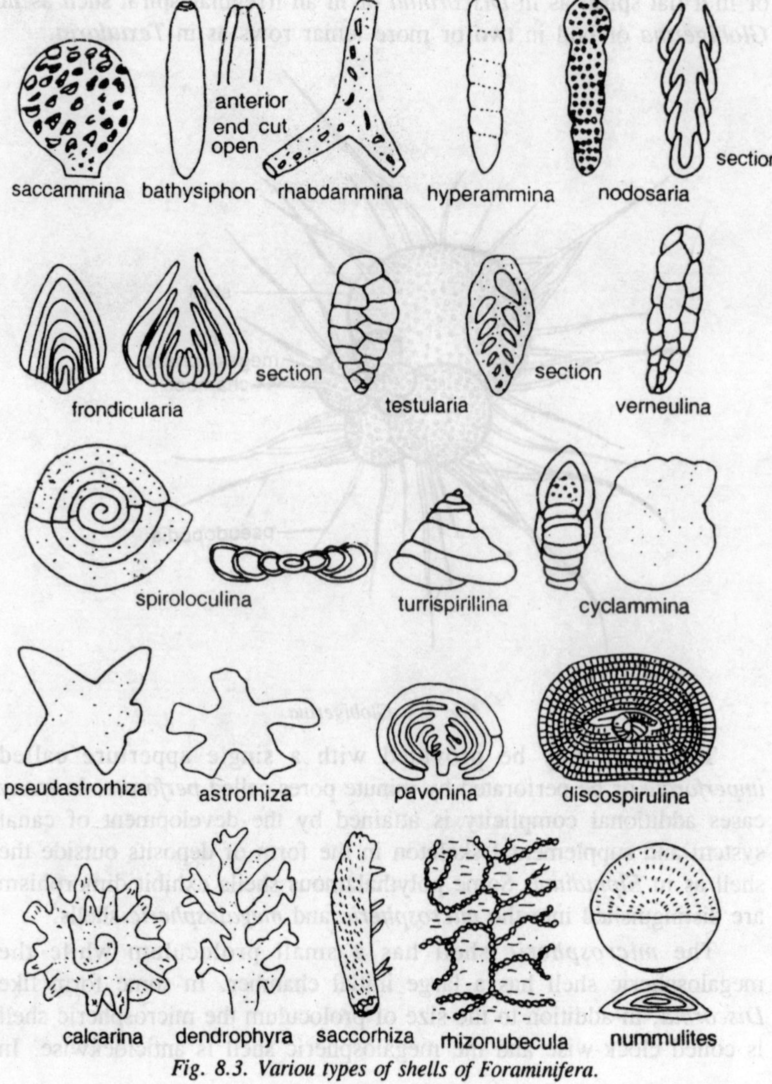

Fig. 8.3. Variou types of shells of Foraminifera.

its aperture and the extruded protoplasm surrounding itself secretes a shell in connection with the original chamber and communicating with it through its aperture. In this way a two-chambered shell is produced and by repetition of this process many chambered or *polythalamous* shell is formed. The new chambers may be joint in a leniar rows as in *Nodosaria* or alternatively as the opposite side of the original chamber as in *Spiroloculina* or each chamber the predecessor as in *Frondicularia* or in a flat spiral as in *Discorbina* or in an irregular spiral such as in *Globigerina* or laid in two or more leniar rows as in *Textularia*.

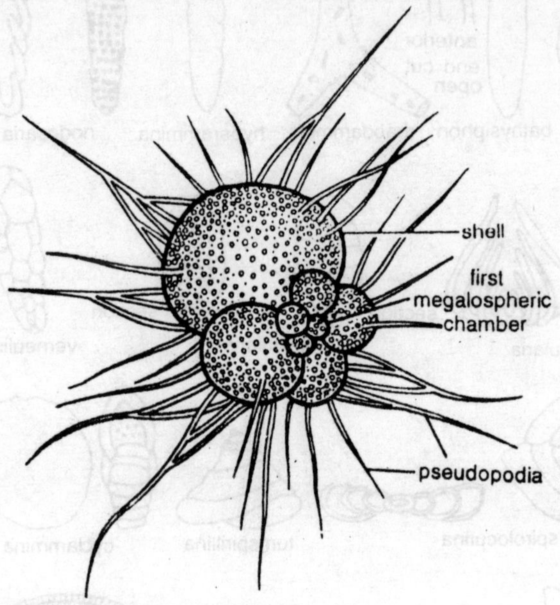

Fig. 8.4. Globigerina.

The shell may be provided with a single apperture called *imperforate* or be perforated by minute pores called *perforate*. In many cases additional complicity is attained by the development of canal system and supplemental skeleton in the form of deposits outside the shell as in *Elphidium*. Some polythalamous shells exhibit dimorphism are distinguished into the *microspheric* and *megalospheric* shells.

The *microspheric* shell has a small proloculum while the megalospheric shell has a large initial chamber. In some form like *Discorbia*, in addition to the size of proloculum the microspheric shell is coiled clock-wise and the megalospheric shell is anticlockwise. In some Heliozoa like *Nuclearia* the skeleton is in the form of gelatinous investment. In other Heliozoa like *Rapidiophrya* the body is surrounded by a capsule of loosely woven fibres.

Internal Structure

In many cases the body form is maintained by the internal supporting structure. In some forms like *Allogromia* the protoplasm of the body flows round it, so that the shell becomes an internal structure. Such a internal structure is, however, different from one which arise inside the body. The true internal structure in many Heliozoa are of temporary in nature like the axial rods in pseudopodia. These rods stiffens the pseudopodia and are absorbed when pseudopodia are retracted. In other cases the structures are of permanent nature.

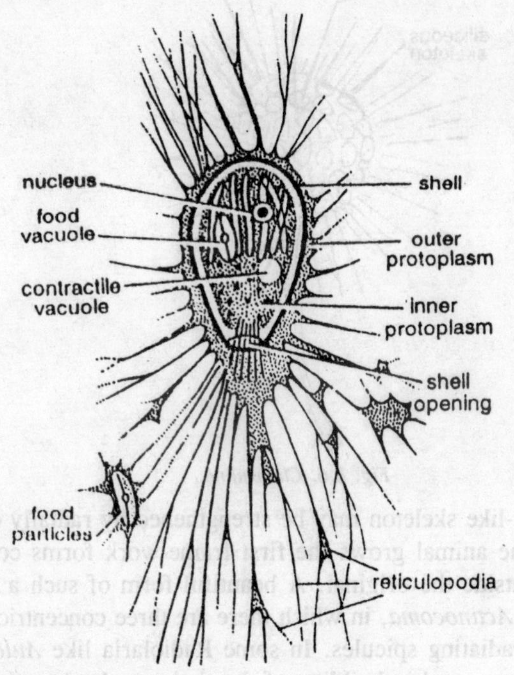

Fig. 8.5. Allogromia.

From such structure lacks the axostyle and axial rods of pseudopodia, it is easy to derive in which the organic bases in indurated with mineral substances or spicules composed entirely of mineral substances. In either case the spicule grows by deposition of fresh inorganic substances upon that already laid. If such a deposition takes place at one end of the rod-shaped spicules the opposite extremity pushed outwards and ultimately projects beyond the limits of the body. As regards the material, the spicules are calarecous, silicious or of strontium sulphate. In the simplest case the spicules are rod-shaped or needle like disposed radially or tangentially.

A simple type spicular skeleton is seen in *Acanthocytis*, in which the silicious spicules are arranged like rods and project some distance beyond the surface. In others like *Sphaerozoeum* the spicules are arranged tangentially. From this simple type more complicated types evolved by fusion of the spicules to form a frame work. The commonest type is a net-work formed by the fusion of tangentially disposed spicules as in *Clathrulina*.

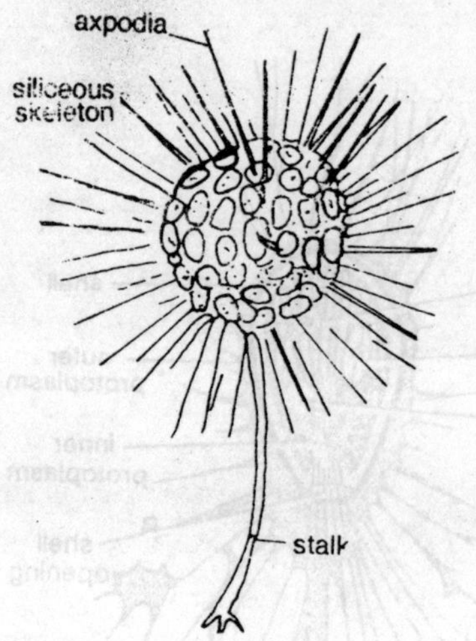

Fig. 8.6. Clathrulina.

The lattice-like skeleton may be strengthened by radially disposed branches, as the animal grows the first frame-work forms concentric frame work outside the original. A beautiful form of such a skeleton is exhibited by *Actinocoma*, in which there are three concentric spheres connected by radiating spicules. In some Radiolaria like *Aulocanthus* foreign bodies are used in building of the skeleton. In *Acantharia* it is formed of strontium sulphate laid in the form of 20 rods radiating from the centre. These rods emerge out of the animal in fine circles comparable to the equatorial tropical and circumpolar circle of the globe. The arrangement is known as Mullar's law. The central capsule of Radiolaria may also be included in the internal skeleton. It is composed of chitinous, pseudochitinous or tectinuous materials and has perforation in it one or groups which permit cytoplasmic continuity between the extracapsular and intera-capsular part of the cytosome.

Functions of Skeleton

The skeleton performs several functions. These are:

1. It gives a definite shape to the organism.
2. It provides surface for the attachment of myonemes.
3. It protects animal from its enemies.
4. It protects the animalcule from mechanical changes taking place around it.

9

PARASITIC PROTOZOANS

Protozoa of the Human Mouth

The Human Mouth is in some respects a fairly rigorous environment for Protozoa. Foods and drinks vary widely in temperature and chemical nature, and the disturbances involved in the practice of dental hygiene add further complications. Nevertheless, a flagellate (*Trichomonas tenax*) and an amoeba (*Entamoeba gingivalis*) manage to infect a significant proportion of the population. The life-cycles of these two parasites do not include cysts, so that infections are spread by the transfer of trophozoites. Under experimental conditions, a trace of moisture has kept *E. gingivalis* alive long enough for droplet transfer, and for transfer by way of contaminated cups and other utensils. Transfer by direct oral contact entails much less risk for the parasite.

Trichomonas tenax

This flagellate probably was first described as *Cercaria tenax* by O.F. Müller in 1774. Many years later, the organism was found again and described as *Tetratrichomonas buccalis* Goodey. *T. tenax* shows a size range of about 5-21 x 3.8-7.6μ. The undulating membrane is usually rather short and the membrane-flagellum may not extend beyond the membrane. Autotomy occasionally produces forms with projecting axostyles and disproportionately long membranes.

Mitosis has been described by Hinshaw, and *T. tenax* has been compared with other trichomonads of man by Wenrich. Although rarely present in the healthy mouth, infection with *T. tenax* may approach an incidence of 90 per cent in cases of advanced pyorrhea. However, a casual relationship to pyorrhea has not been established. In addition to

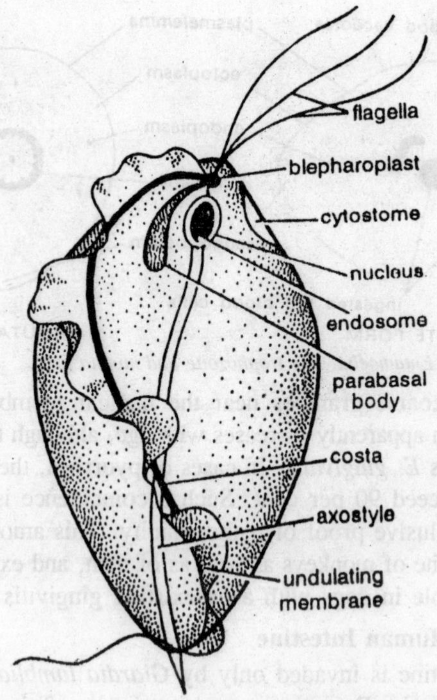

Fig. 9.1. Trichomonas.

their occurrence in the mouth, the flagellates have been found occasionally in pus from infected tonsils and, rarely, in material from the lungs.

Entamoeba gingivalis

This species, described as *Amoeba gingivalis* by Gros in 1849, evidently was the first amoeba reported from man. The specific name, *Endamoeba buccalis*, was proposed later by Prowazek who had overlooked the paper by Gros. Several detailed descriptions have been published more recently and mitosis have been described by Stabler and Noble. Literature on the species has been reviewed by Kofoid. The amoeba measures 6-60μ in length, usually shows clear pseudopodia, and may contain a number of food vacuoles containing leucocytes, or less commonly, bacteria. The amoebae ingest living leucocytes and consequently are not mere scavengers.

In cultures, both leucocytes and red corpuscles are ingested. The nucleus, 2-6μ in diameter, often shows a central clump of granules, as

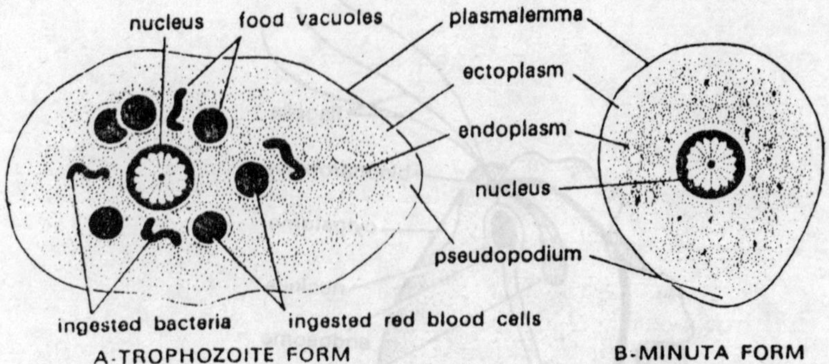

Fig. 9.2. Entamoeba. The trophozoite and minuta form.

well as a zone of coarse granules near the nuclear membrane. The incidence of infection apparently increases with age, although the healthy mouth rarely harbors *E. gingivalis*. In cases of pyorrhea, the incidence is high and may exceed 90 per cent. Such a coincidence is tempting but there is no conclusive proof of pathogenicity. This amoeba seems to be a natural parasite of monkeys as well as of man, and experimental infections are possible in dogs with a preexisting gingivitis.

Flagellates of the Human Intestine

The small intestine is invaded only by *Giardia lamblia*, whereas the colon may contain *Retortomonas intestinalis*, *Tricercomonas intestinalis*, *Chilomastix mesnili* and *Pentatrichomonas hominis*.

Retortomonas intestinalis (Wenyon and O'Connor) Wenrich

This flagellate, often known as *Embadomonas intestinalis* (Wenyon and O'Connor) Chalmers and Pekkola, has been reassigned to *Retortomonas* by Wenrich on the basis that *Embadomonas* Mackinnon is a synonym of *Retortomonas Grassi*. Objections of this conclusion have been presented by Bishop. *R. intestinalis* is a small (4-9 x 3-4μ) organism with two unequal flagella. The longer flagellum extends anteriorly, the other anterolaterally from a pit (the "oral pouch" or "cytostome"). Shape varies somewhat although the anterior end is usually rounded. The cytoplasm often contains food vacuoles. Binary fission has been described by Bishop. The cyst which measures 4.5-7.0 x 3.0-4.5μ, is ovoid to pear-shaped and contains one nucleus; nuclear division apparently does not occur.

Tricercomonas intestinalis (Wenyon and O'Connor)

This species may or may not be identical with *Enteromonas hominis* Fonseca. This question has been discussed by Dobell and O'Connor and by Wenyon, and the latter has pointed out that *Enteromonas hominis* was described with only three flagella. Flagellated stages have three anterior flagella and a fourth which may seem to lie within the cytoplasm and emerge at or near the posterior end of the body. The size range is 4-10 x 3-6μ. The cysts measure 6-8 x 3-4μ and contain 1-4 nuclei. A flagellate apparently identical with *T. intestinalis* has been found in monkeys.

Chilomastix mesnili (Wenyon) Alexeieff

This flagellate seems to be specifically identical with one in apes and monkeys. The active stage, 6-20μ in length, has three anterior flagella and a shorter fourth which usually lies in the cytostomal groove, a depression extending obliquely from near the anterior end to about the middle of the body. A second groove often arises near the left anterior margin of the cytostomal cleft and extends posteriorly in one or two spiral turns. Solid food is ingested through a cytostome at the posterior end of the cytostomal groove.

Just beneath the surface, a cytostomal fibril of uncertain significance extends along the cytostomal groove. Fission has been described by Geiman and by Boeck and Tanabe. Encysted stages measure 7-10 x 4.5-6.0μ, contain one or two nuclei, and often granules and fibrils representing the blepharoplasts, cytostomal fibrils and possibly flagellar axonemes. Mitosis has been described in encysted stages.

Pentatrichomonas hominis (Davaine) Kirby

Two flagellates from the human colon are described in the literature as *Trichomonas hominis* (Davaine) Leuckart, with four anterior flagella, and *Hexamitus ardindelteili*, later transferred to the genus *Pentatrichomonas* on the basis of its fifth free flagellum. In confirming observations of Wenrich. Kirby concluded that *T. hominis* normally has fifth free flagellum and that the two supposedly distinct flagellates should be recognized as *Pentatrichomonas hominis*.

Trichomonads apparently identical with *P. hominis* have been found in monkeys, cats, dogs, and rats. The flagellate measures 8-15 x 3.5μ. The undulating membrane is about as long as the body and there is usually a free posterior portion of the membrane-flagellum. The fifth flagellum, which may be trailed posteriorly, beats in a rhythm different from that of the other four anterior flagella. A costa extends beneath the base of the undulating membrance and the axostyle usually projects beyond the posterior end of the body. Encysted stages are unknown.

Giardia Lamblia (Stiles)

This flagellate, described by Lambl in 1859 as *Cercomonas intestinalis*, probably was first seen by Leeuwenhoek in 1681. Although the organism is often referred to as *Giardia intestinalis*, Lambl's name had already been used for a parasite of Amphibia and the correct specific name is *Giardia lamblia*. Nuclear division and fission have been described. The flagellate measures 9-21 x 5-11μ. The body is flattened dorso-ventrally with a rather convex dorsal surface and a more flattened ventral surface, the anterior part of which forms a concave "sucker". Two flagella emerge from the posterior pole, while three other pairs extend from the lateral and antero-lateral surfaces.

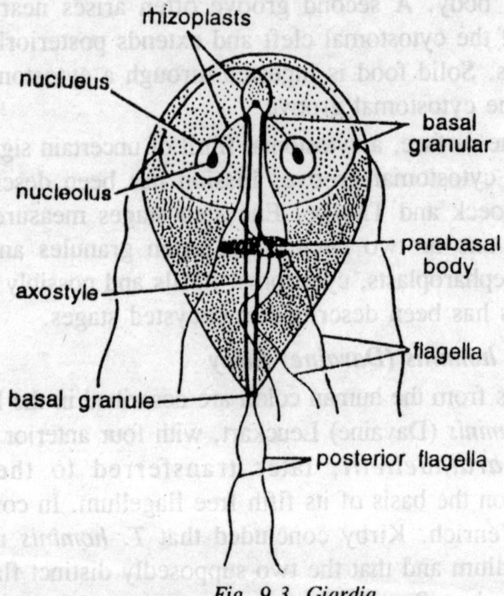

Fig. 9.3. Giardia.

The paired axostyles, which may appear fused in stained preparations, extend to the posterior end of the body. Two parabasal bodies, sometimes fused together, lie near the axostyles in the posterior third of the body. The cysts measure 8-14 x 6-10μ and may contain 2-16 nuclei, axostyles, parabasal bodies, and fibrils which are possibly flagellar axonemes.

The cyst is the stage most commonly found in stool examinations, since the flagellated forms are discharged primarily during attacks of diarrhea. Although it was suspected, at one time, that rodents may serve as reservoirs, this suspicion has not been confirmed. Experimental infection of rats has been reported, but such infection has been temporary and has not led to production of cysts.

Flagellosis

The effects of flagellate infections have been evaluated primarily by correlating clinical observations with incidence of infection. On such a basis, there is no evidence that *Retortomonas intestinalis* and *Tricercomonas intestinalis* are harmful. *Chilomastix mesnili* has been associated with abnormal stool frequency often enough to arouse suspicion, but there is no reason for considering this species a serious parasite. *Pentatrichomonas hominis* has been found in diarrheic patients and *Pentatrichomonas ardin-delteili* was first observed in patients with dysentery and later in cases of dysentery and chronic diarrhea. Therefore, it is often assumed that *P. hominis* is occasionally a caustive or contributary factor in digestive disturbances. Invasion of tissues—agonal and possibly postmortem—has been reported, but probably does not occur in the usual infection.

Infection with *Giardia lamblia* is frequently correlated with digestive disturbances, although there is no invasion of tissues. Heavy infections might interfere with normal absorption, since the flagellates adhere to the mucosa. Symptoms include chronic diarrhea, attacks of diarrhea alternating with constipation, chronic stomache, occasional cramping and colic, nausea, abdominal tenderness, loss of appetite, chronic headaches, and irritability.

Chemotherapy

The treatment of flagellate infections is a less pressing problem and has attracted less attention than the treatment of amoebiasis. Some of the drugs used for *Entamoeba histolytica* has been tried also in flagellate infections, but the results are not always directly comparable. Atebrin has been used effectively for elimination of *Giardia lamblia*, although occasional infections are not cured. Giardiasis in children also has been treated with bismuth-salicylate, followed by treparsol, and with acranil. Diodoquin is said to be active against *P. hominis*, and good results with gentian violet in combination with argyrol enemas also have been reported.

Amoebae of the Intestinal Lumen

The human colon may be invaded by *Endolimax nana*, *Dientamoeba fragilis*, *Entamoeba coli*, and *Iodamoeba bütschlii*. In addition, natural infection with *Entamoeba polecki*, a parasite of monkeys, has been observed.

Endolimax nana

First described as *Entamoeba nana*, this species was later transferred to the genus *Endolimax* by Brug in 1918. An apparently identical amoeba has been reported from monkeys. The amoeboid stage, usually observed only in loose stools, is small (6-15μ) sluggish form with clear pseudopodia and food vacuoles containing bacteria. The stained nucleus often shows no peripheral granules, although such can be demonstrated after adequate fixation. The endosome is large, usually irregular but sometimes ovoid or spherical, and may be central or eccentric.

Precystic stages have been reported as rounded forms without food vacuoles. Mitosis has been described. The mature 5-12 x 4-6μ, contain four, or rarely eight, nuclei. The shape is usually ovoid, and one surface is often more convex than the opposite side. Stored glycogen may be present in young cysts but disappears gradually as the cysts mature. Occasionally, small filaments have been reported as possible chromatoid bodies. The nuclei of the mature cyst are appreciably smaller than those of the trophozoite, and the endosome is often eccentric.

10

MULTIPLICATION

In many Protozoa, reproduction occurs at frequent intervals, with relatively short periods of growth intervening under favourable conditions. In other cases, growth may extend over a period of several to many days, so that reproduction occurs at comparatively long intervals. Depending upon the species, reproduction may or may not be preceded regularly by sexual phenomena. Of the species which do show sexual activity, some normally undergo syngamy as a prelude to a reproductive phase while others show sporadic sexual activity.

METHODS OF REPRODUCTION

The less complex Protozoa reproduce either by binary fission or by simple budding. In either case, the nucleus undergoes mitosis; or mitosis of the micronucleus and "amitosis" of the macronucleus occur in Ciliophora. Cytoplasmic division is approximately equal in fission, unequal in budding. Although reproduction in uninucleate species is comparable in some respects to cell division in higher organisms, the structural specialization of many Protozoa introduces complications. The new organisms must be equipped with various organelles, the nature of which varies with the species.

Parental organelles such as flagella are often inherited equally or unequally by the daughter organisms which later produce enough new structures to complete their equipment. Even the paraglycogen reserves of *Stentor coeruleus*, normally stored posteriorly, are shifted to the middle of the body and then shared between the daughter organism in transverse fission. Blepharoplasts, basal granules, kinetoplasts, and sometimes chromatophores and pyrenoids, are self-reproducing. Their duplication during fission fills the needs of the daughter organisms.

Other structures, including the cirri of certain ciliates and the parabasal apparatus of certain flagellatles, undergo resorption so that each daughter organism must develop a set of its own—in the case of cirri, apparently by outgrowth from inherited basal granules. The resorption of parental structures is sometimes extensive. Reproduction may thus involve dedifferentiation of the old boy as well as the differentiation of new structures in the developing daughter organisms. The beginning of differentiation, in two new centres of organization within the parental body, possibly supplies the stimulus for subsequent dedifferentiation.

Reproduction of multinucleate Protozoa, or of multinucleate stages in the life-cycle, may involve either budding or fission. In many Sporozoa a young uninucleate stage grows, with repeated mitoses, into a multinucleate plasmodium which then reproduces by *schizogony*.

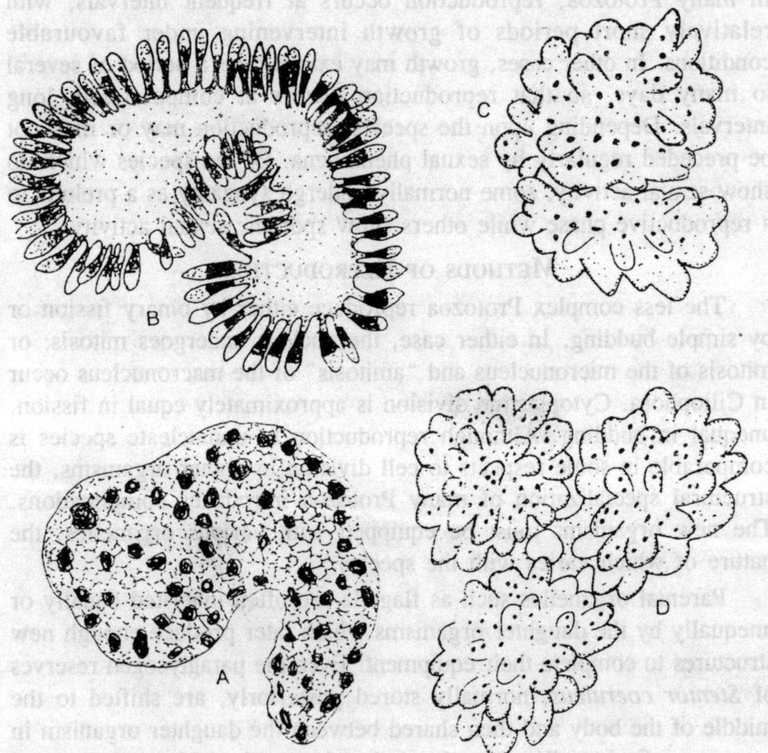

Fig. 10.1. A. Multinucleate stage (schizont) of Ovivora thalassemae; superficial section; B. Schizogony in O. thalassemae; C, D. Plasmotomy in Pelomyxa carolinensis; division into two and three individuals.

Essentially, schizogony is multiple budding in which separation of uninucleate buds from a residual mass of protoplasm is completed within a short time. Certain multinucleate Protozoa normally divide into several organisms, each of which receives some of the parental nuclei. This process, not necessarily synchronized with nuclear division, is known as *plasmotomy*. Both schizogony and plasmotomy have been described in *Coelosporidium*, while plasmotomy is characteristic of *Pelomyxa*. In the latter, plasmotomy produces 2-6 smaller organisms, among which the parental nuclei are distributed at random. Less variation is characteristic of *Coronympha octonaria*, in which all eight nuclei usually undergo mitosis and the daughter nuclei separate in groups of eight before plasmotomy occurs.

Binary Fission

In Mastigophora, fission may occur in the active stage, within a cyst, or in non-flagellated palmella stages. The plane of fission is most frequently longitudinal and the division-furrow usually appears first at the anterior end. Among the dinoflagellates, however, fission is often oblique and may be almost transverse in late stages. *Spirotrichonympha bispira* divides transversely, although related species undergo longitudinal fission. Mitosis in Trichomonadida is commonly followed by migration of the karyomastigonts to opposite sides of the body and fission is then completed by cytoplasmic constriction. Cytoplasmic structures may undergo division, resorption followed by origin *de novo*, or partial resorption followed by growth and differentiation. Duplication of blepharoplasts, apparently by division, is characteristic of fission in the flagellate stage.

The behaviour of blepharoplasts in non-flagellated stages of Phytomastigophorea is mostly unknown, although they persist as division-centres in *Eudorina illinoisiensis*. The fate of other cytoplasmic structures in fission seems to be variable. The stigma divides in *Chlamydomonas nasuta*, whereas the old stigma passes to one daughter flagellate in *Platydorina caudata*. Division of the chromatophores has been reported in certain Euglenida. Division of pyrenoids has been described in *Eudorina illinoisiensis*, resorption of the old pyrenoid and differentiation of new ones occur in *Chlamydomonas nasuta*. Flagella probably do not split in fission and the few reports of such a process are based upon inadequate evidence. Retention of the old flagella has been described most frequently.

In biflagellate and multiflagellate species, each daughter organism may receive one or more of the original flagella and develop the necessary new ones, as in *Heteronema* and *Trichonympha*. However, flagellar resorption occurs in *Monas* and in Phytomonadida which divide within a parental theca. The old flagella and associated structures also degenerate in *Lophomonas* and related genera. The axostyle of trichomonad flagellates, the pharyngeal-rod apparatus of *Heteronema*, the siphon of *Entosiphon*, and the cresta of devescovinid flagellates undergo resorption, whereas the costa of trichomonads apparently is retained by one of the daughter flagellates.

The kinetoplast of Trypanosomidae divides but parabasal bodies of other flagellates usually do not. One of the exceptions is *Chilomonas paramecium* in which each daughter receives part of the old parabasal apparatus. Parabasal bodies are sometimes retained intact, as in *Barbulanympha laurabuda*; or partial or complete resorption may occur. Although complete resorption of the parabasal body sometimes occur in *Trichomonas termopsidis* and various devescovinid flagellates, a portion often remains attached to its blepharoplast. In these cases, the parabasal of one daughter in regenerated from the persisting fragment while that of the other is differentiated *de novo*. The rigid theca of *Ceratium* and related dinoflagellates is divided in fission and the mission portions are regenerated.

On the other hand, such testate flagellates as *Trachelomonas volvocina* usually undergo fission within the test, one daughter emerging to produce a new test. The simpler Sarcodina often show little of cytological interest aside from division of the nucleus. However, the cytoplasmic changes in *Amoeba proteus* indicate that the physical aspects of fission are not particularly simple. The presence of a shell complicates reproduction of many Sarcodina.

In primitive genera (*Cochliopodium*, *Pseudodifflugia*), the simple test is divided in fission. *Euglypha alveolata* secretes reserve shell-plates and stores them until the next fission, when they are passed to one of the daughter organisms. The other retains the old test. In typical for aminiferida, schizogony has replaced binary fission. Fission in ciliates is typically transverse and, in at least certain species, there seems to be a definite division-plane which is not displaced by amputations just before fission. However, the plane of fission in Peritrichida passes from the oral to the aboral end and is morphologically longitudinal.

The plane of fission in Opalinidae also is oblique or almost longitudinal *Cyathodinium piriforme* is unusual, in that the plane of fission passes through the originally longitudinal axis of the body but separates the posterior ends of the daughter ciliates in late fission.

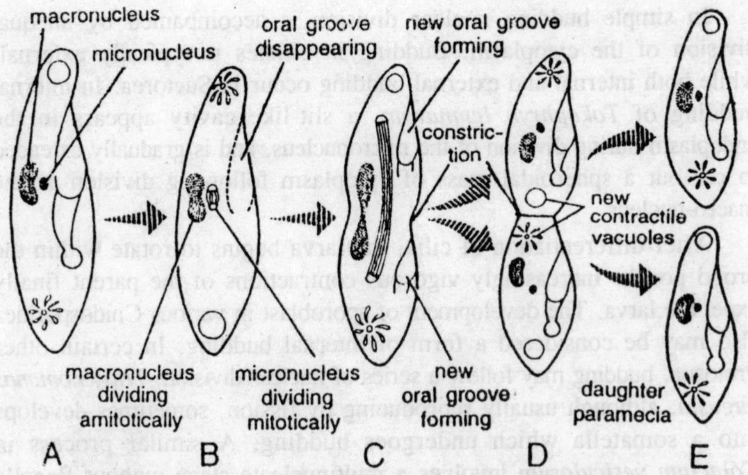

Fig. 10.2. Binary fission in Paramecium.

Reorganization in ciliates is often striking, and may involve macronuclei as well as cytoplasmic structures. The old cirri are resorbed in *Uronychia*, dedifferentiation of the peristomial area occurs in *Bursaria*, and resorption of the peristomial membranelles in *Fabrea*.

In *Chilodonella uncinatus*, the old pharyngeal basket, cytostome, and many body cilia are resorbed. On the other hand, *Euplotes*, *Colpidium*, *Glaucoma*, and *Stentor* retain the peristomial organelles. Division of the parental cytostome and peristome occurs in *Cyclochaeta astropectinis* and possibly in other peritrichs. The infraciliature shows genetic continuity through multiplication of basal granules, as traced in *Chilodonella*. Foettingeriidae, *Opalina*, and *Ichthyophthirius*, among others. In *Tetrahymena* and similar ciliates the development of a new mouth for the posterior daughter involves the multiplication of basal granules at a particular level in the stomatogenous row. These basal granules later give rise to membranelles of the new peristomial area.

The continuity of basal granules is especially striking in *Podophrya fixa*, which shows the usual ciliated larva and non-ciliated adult of the Suctorea. Basal granules persists in the adult, and during reproduction, those in the cortex of the bud multiply and form rows from which the cilia of the larva arise. All the cilia, and apparently their basal granules also, are resorbed in *Cyathodinium*. New infraciliatures appear as endoplasmic units which migrate to opposite surfaces of the body, where cilia then arise from the new basal granules. This process resembles the formation of new mastigonts in *Lophomonas*.

Budding and Schizogony

In simple budding nuclear division is accompanied by unequal division of the cytoplasm. Budding in ciliates is typically external, while both internal and external budding occur in Suctorea. In internal budding of *Tokophrya lemnarum*, a slit-like cavity appears in the endoplasm during division of the micronucleus, and is gradually extended to cut out a spheroidal mass of cytoplasm following division of the macro-nucleus.

After differentiation of cilia, the larva begins to rotate within the brood pouch. Increasingly vigorous contractions of the parent finally expel the larva. The development of sporoblast in various Cnidosporidea also may be considered a form of internal budding. In certain other Protozoa, budding may follow a series of nuclear division. *Tritrichomonas augusta*, although usually reproducing by fission, sometimes develops into a somatella which undergoes budding. A similar process in *Colacium vesiculosum* involves a multinucleate stage without flagella or reservoirs. These structures appear in each bud before it is separated from the parental somatella.

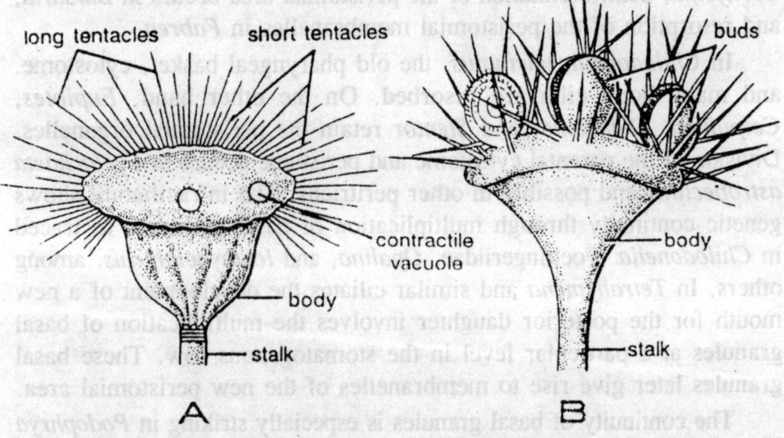

Fig. 10.3. Budding in Ephelota.

Schizogony, involving the production of several to many buds more or less simultaneously, is characteristic of certain Protozoa. This process is especially efficient in many sporozoa in which the plasmodium often contains many nuclei before Schizogony. A schizont of *Eimeria bovis*, for example, may produce as many as 170,000 merozoites.

Nuclear Division

Although mitosis has been reported in most species which have been studied carefully, the small size of many nuclei has made it difficult to interpret the structure of chromosomes in early mitosis and in the interphase. The interphase chromatin of Cryptomonadida and *Zelleriella elliptica* has been described as fine granules dispersed on a network; that of *Pelomyxa carolinensis*, as Feulgen-positive granules and short filaments. The Feulgen-positive interphase chromatin of Euglenida, according to different reports, ranges from perindosomal granules to a continuous spireme which in optical section simulates separate granules. Actually, it has been impossible to find stages suggesting an chromatic network containing chromatin granules in some of the Euglenida and Dinoflagellida. Instead, beaded chromosomes seem to persist through vegetative stages.

In general, however, chromosomes of the later prophases seem to develop from some sort of a "reticulum" and, in such favourable material as *Pamphagus hyalinus*, the process has been traced in living material. In certain species of *Entamoeba* and in *Naegleria gruberi*, the chromosomes develop from a finely granular or reticular zone of Feulgen-positive material around the endosome. The persisting "peripheral chromatin" granules, apparently adherent to the nuclear membrane in *Entamoeba*, may be give rise to chromosome-like bodies perhaps analogous to the nucleoli of *Zelleriella*.

Interpretations are even more difficult in *Endamoeba blattae* because the interphase nucleus is Feuglen-negative, although Feulgen-positive chromosomes appear in mitosis. The origin of chromosomes from an endosome or a karyosome, in nuclei supposedly containing no interphase "*chromatin* granules," has been reported in certain Protozoa. In some of these, such as *Endolimax nana*, periendosomal material has been demonstrated in more recent investigations. Furthermore, Noble reports a functional separation of endosomal granules and periendosomal chromatin in *Entamoeba gingivalis*, in spite of their intermingling during early prophases. However, the interphase precursors of the chromosomes have not yet been identified with certainty in many Protozoa and much remains to be learned about the earliest stages of mitosis in most species. The observations of Cleveland on *Holomastigotoides* have shown that the chromosomes persist as such throughout the mitotic cycle.

The diagrammatically clear behaviour of the chromosomes in this genus supplies a logical pattern for interpreting mitosis in a smaller nuclei. Each chromosome consists of a coiled *chromonema* embedded in a matrix. Only the matrix is distinguished in heavily stained preparations, but both components can be detected with phase-contrast microscopy and also by ordinary microscopy in suitably stained preparations. The disappearance of major coiling and the apparent lengthening of each chromonema, as the chromosomal matrix disappears late in mitosis, result in long twisted filaments. This stage, in small nuclei containing a number of chromosomes, would suggest the interphase "reticulum" described in various species. If the uncoiled chromosomes are very slender, optical sections of a small nucleus might suggest a granular organization of the chromatin.

Origin of chromosomes from such a "granular" or "reticular" interphase, in the light of chromosomal behaviour in *Holomastigotoides*, may involve only a condensation of preexisting chromosomes. Each chromonema becomes more and more tightly coiled (in the "major coils" of Cleveland) as a new matrix is developed. The result is the more or less compact chromosome of the later prophases. According to Cleveland, duplication of each chromonema occurs before the development of the new matrix.

Eumitosis and Paramitosis

Differences in chromosomal behaviour and the structure of the achromatic division-figure have formed the bases for various classifications of protozoan mitoses. Among these systems, that of Belar has the advantage of simplicity in recognizing two general types, *eumitosis* and *paramitosis*. Characteristic features of *eumitosis* are longitudinal splitting of the chromosomes, the development of compact prophase chromosomes, and the apperance of an equatorial belt of chromosomes within the spindle. Many protozoan mitoses can be fitted into such a scheme.

Nuclear division in *Dimorpha mutans* is representative. Mitosis is initiated by division of the centrosome, and the subsequent development of an amphiaster is accompanied by the formation of short chromosomes from the interphase chromatin. The nucleus moves into the spindle, the nuclear membrane is said to disappear, and the chromosomes form an equatorial plate. Additional examples are found in *Actinophrys*, *Pelomyxa* and *Zelleriella*. In paramitosis, condensation of the prophase chromosomes is less marked and a typical equatorial plate is not developed.

In *Aggregata eberthi*, only one end of each chromosome extends into the equatorial zone of the spindle. Since the daugther chromosomes

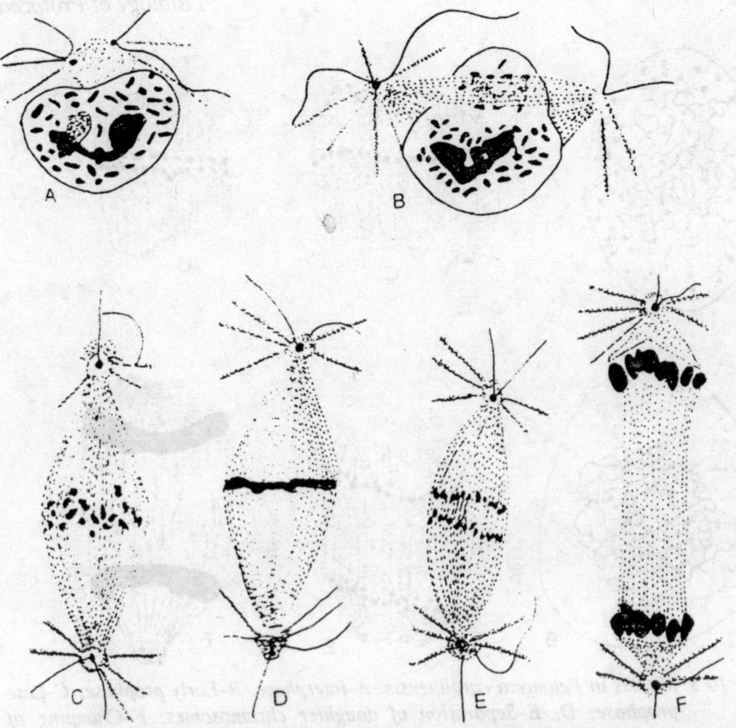

Fig. 10.4. Mitosis in Dimorpha mutans; eumitotic type, basal portions of flagella and a new axopodia are indicated at the poles of the division.

separate before they are shortened, later stages of mitosis suggest transverse division of long chromosomes. Long chromosomes persist also in certain Radiolarida, Dinoflagellida, Euglenida, and in *Teratonympha*. The picture presented during separation of the daughter chromosomes depends upon the position of the centromeres and the length of the chromosomes.

Terminal centromeres, which have been demonstrated in *Holomastigotoides* probably occur in *Aggregata* and in Euglendai and Dinoflagellida. The appearance of V-shaped daughter chromosomes during anaphases, as described in *Pleurotricha lanceolata*, would suggest median instead of terminal centromeres. Persistence of the nuclear membrane throughout mitosis is characteristic of many Protozoa. However, a large portion of the old membrane is discarded after division of the nucleus in *Holomastigotoides* and disappearance of the membrane in the prophase has been described in *Dimorpha mutans*.

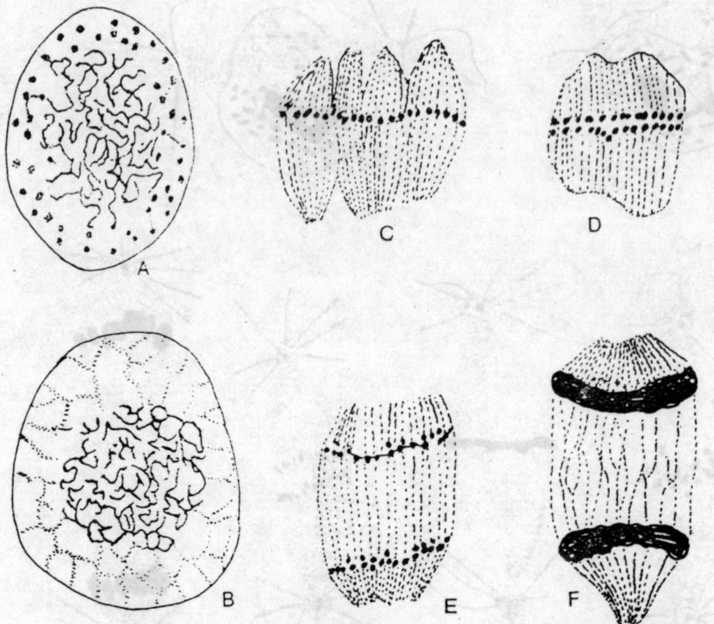

Fig. 10.5. Mitosis in Pelomyxa carolinensis. A–Interphase; B–Early prophase; C–Late prophase; D, E–Separation of daughter chromosomes; F–Clumping of chromosomes in the anaphase.

Micronucleus of Ciliates

The small size of the micronucleus increases the difficulty of interpreting chromosomal behaviour. Longitudinal splitting of the chromosomes has been reported in some species and transverse division in others, but decisions are difficult for the almost spherical chromosomes found in certain ciliates. Longitudinal splitting has been described in *Pleurotricha*, *Stylonychia*, *Conchophthirius*, and in pregamic division in *Euplotes*. In the last three cases, members of each pair of daughter chromosomes slip past each other toward the poles of the spindle.

Achromatic Figure

Both extranuclear and intranuclear achromatic figures have been described in Protozoa. The extranuclear figure is sometimes represented merely by the centrosomes and a *paradesmose* which ranges from a delicate fibril to a bundle of fibrils in different species. The fibrillar paradesmose, as seen in *Gigantomonas*, differs mainly in degree from the extranuclear spindle of *Pseudotrichonympha* and similar types.

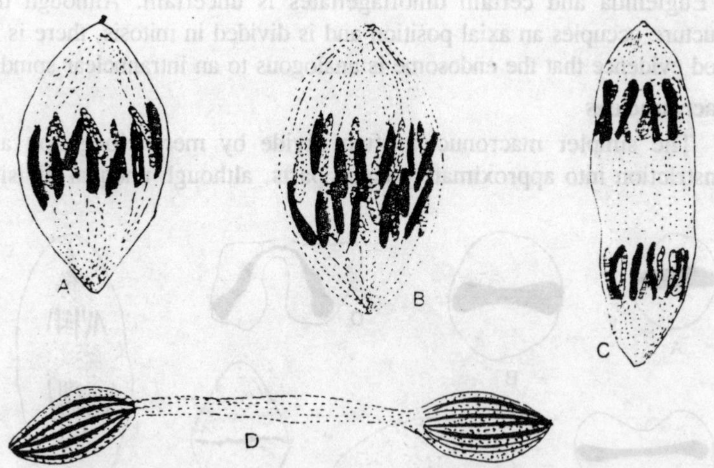

Fig. 10.6. Mitosis in the ciliate, Conchophthirius anadonatae; A–Longitudinal splitting ofthe chromosomes. B, C–Separation of daughter chromosomes. D–Nuclear division nearly completed.

Comparable extranuclear spindles occur in certain dinoglagellates and in *Aggregata*. Since the nuclear membrane persists in such forms as *Pseudotrichonympha*, some of the astral rays, during development of the spindle, make connections with the centromeres at the nuclear membrane. Each chromosome in *Holomastigotoides*, for example, ends in a terminal centromere which remains anchored to the nuclear membrane.

Duplication of the centromere parallels that of the chromonema. In some of the Hypermastigida, development of the spindle and its connections with the chromosomes has been followed in living flagellates. Pulls exerted on the achromatic figure cause corresponding movements of the chromosomes; when the tension is released, the fibrils and chromosomes snap back into place. Intranuclear figures have been described in the micronuclei of ciliates, in *Actinophrys*, *Monocystis* and *Euglypha*, among others. In some of these cases, the spindle ends in centrosomes which seem to be embedded in the nuclear membrane or else adherent to it. An intranuclear spindle is typical of the dividing micronucleus although little is known about division-centres in ciliates.

The spindle sometimes extends into achromatic masses ("polar caps") which may or may not contain "centrioles." Only a granule has been described at each pole in certain ciliates, and even the granules seem to be missing in others. The significance of the persisting endosome in Euglenida and certain dinoflagellates is uncertain. Although this structure occupies an axial position and is divided in mitosis, there is no good evidence that the endosome is analogous to an intranuclear spindle.

Macronucleus

The simpler macronuclei often divide by mere elongation and constriction into approximately equal parts, although unequal division

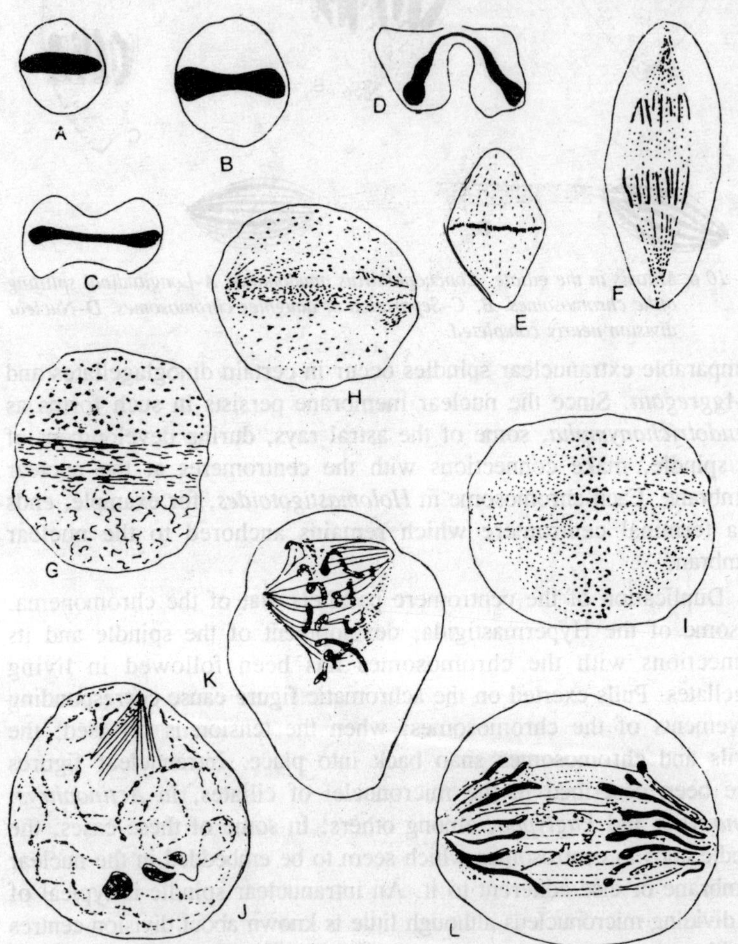

Fig. 10.7. Behaviour of the endosome during mitosis in Heteronema acus; A, B, D, E Intranuclear spindle, micronucleus of Stentor coeruleus. F. Intranuclear spindle, micronucleus of Stylonychia pustulata. G-I. Intranuclear spindle in Oxymonas grandis. J-L. Intranuclear spindle in Pyrsonympha; early stage in development.

occurs occasionally. Division of the compact macronucleus is not always simple, however. A regular elimination of material from the macronucleus during division, or from the daughter afterward, has been described in such genera as *Ancistruma*, *Colpoda*, *Tillina*, *Chilodonella*, *Colpidium*, *Glaucoma*, and *Urocentrum*. The significance of this process is unknown. Ciliates with more than one macronucleus and those with long beaded or band-like macronuclei may show more complicated nuclear changes. The C-shaped macronucleus of *Euplotes* is shortened and thickened, and undergoes changes in staining reactions which suggest progressive internal changes. The two macronuclei of *Stylonychia pustulata* fuse into a single body which then divides.

The macronuclear chains of *Spirostomum*, *Stentor*, and *Blepharisma* also undergo extensive condensation. In *Spirostomum ambiguum* and *Stentor coeruleus* the macronuclear nodes gradually fuse into a compact

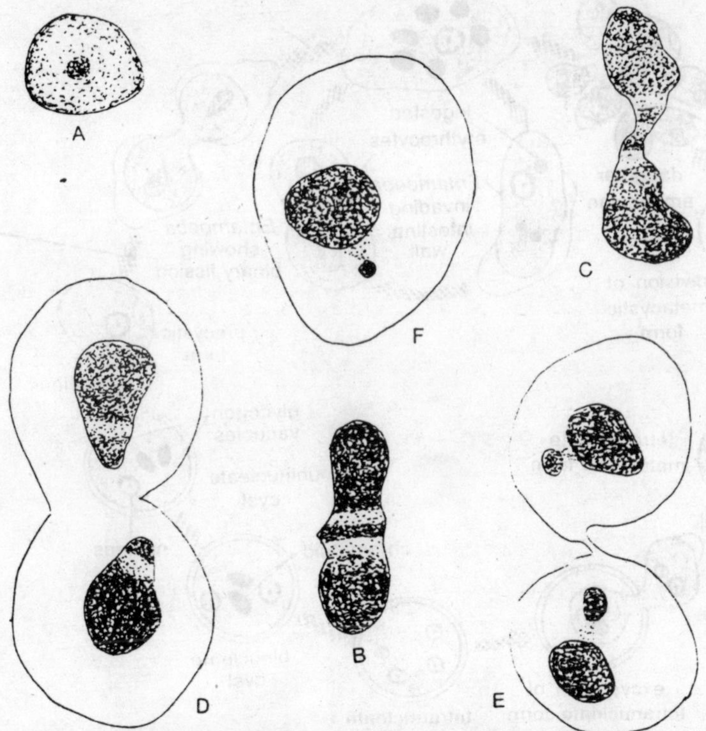

Fig. 10.8. Elimination of chromatin during macronuclear division in Colpidium colpoda. A–Central chromatin mass evident just before division. B–E–Stages in division. F–Separation of discarded mass from a daughter macronucleus.

central body, which then undegoes moderate elongation and a final central body, which then undergoes moderate elongation and a final constriction. In *Blepharisma undulans*, the anterior and posterior macronuclear nodes fuse into two masses, while the middle nodes gradually disappear. The anterior and posterior masses then fuse into one body which elongates and undergoes division.

LIFE-CYCLES

The simple life-cycles of many species include only an active phase and a cyst. With the cyst apparently eliminated, the "cycle" reaches the limit of simplicity in such types as *Entamoeba gingivalis* and *Pentatrichomonas hominis*. Other modifications of this basic pattern include: (a) the development of two or more stages in the active phase;

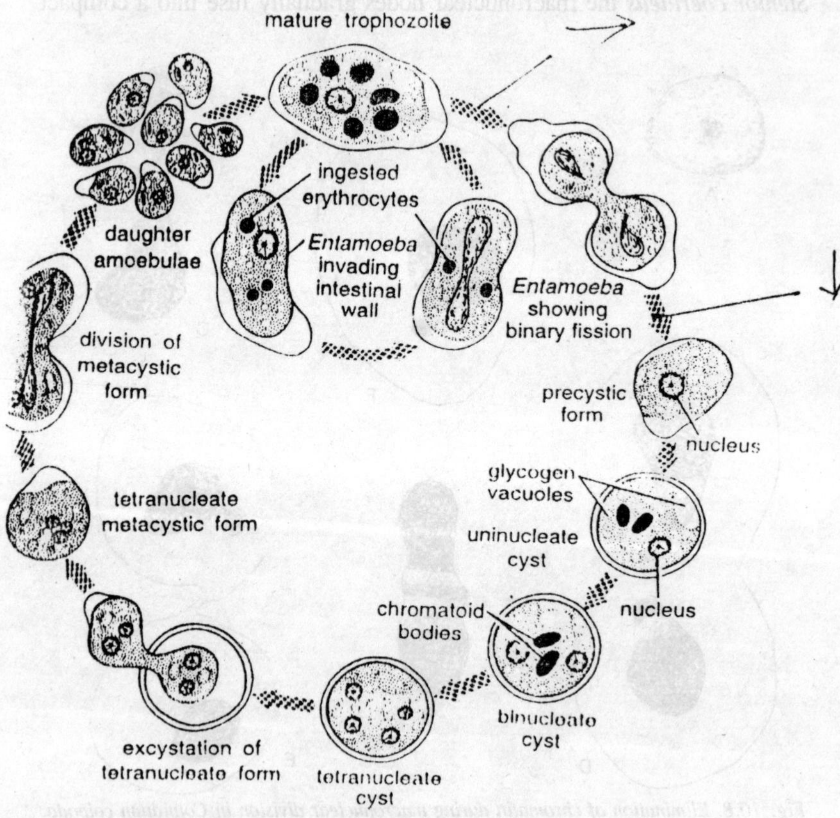

Fig. 10.9. Life cycle of Entamoeba.

(b) the introduction of sexual phenomena, which may appear in a sexual phase alternating with an asexual phase in the cycle. Two or more active stages occur in the life-cycles of many Protozoa. In addition, immature and adult forms of a single organism may be quite different in appearance and behaviour. Examples include the ciliated larva and non-ciliated adult of Suctorea and the stalkless telotroch and the stalked adult of vorticellid ciliates. Dimorphism sometimes involves the alternation of amoeboid and flagellate stages.

The flagellate stage may be temporary, as in *Naegleria*; or it may be the dominant stage, as in *Tetramitus* and certain Chrysomonadida. The flagellate, *Gigantomonas herculea*, shows amoeboid-flagellate dimorphism in which reproduction is limited to the amoeboid phase. Reproductive stages in *Haematococcus* and related genera also are typically non-flagellated. The dominant phase in *Colacium* is a non-flagellated form which occasionally produces flagellate buds. Dimorphism also may involve the alternation of a gamete-producing stage and one which undergoes asexual reproduction, as in foraminiferid.

Life-cycles characterized by more than two active stages are found in certain Trypanosomidae, in many Sporozoa, and in some of the Ciliophora. Protozoan life-cycles may be considered adaptive in that they represent responses to changes in the environment, and perhaps favour, or insure survival when such changes occur. Occurrence of a cycle as such probably parasitic species which must reach a susceptible host in order to complete the cycle, or in many instances even to survive for more than a short time. Within a suitable host, there is often reasonable security during completion of a life-cycle, but establishment in a host does not necessarily insure independence of external conditions. For example, the development of *Plasmodium vivax* in the mosquito may be retarded or prevented by unfavorable temperatures.

A modification of environmental conditions may induce a marked change in the cycle, in parasitic as well as free-living species. Maintenance of *Plasmodium gallinaceum* in chick-tissue cultures has caused a normal stock to lose its ability to produce pigmented erythrocytic stages. Chicks inoculated from such cultures always died from exoerythrocytic infections, always without showing normally pigmented erythrocytic stages, and often without any erythrocytic parasites at all. In some cases it has been possible to eliminate cyclic changes by strict control of invironmental conditions, as in the prevention of conjugation and encystment inciliates by Woodruff, Beers, and others. Such elimination of cyclic changes does not necessarily mean that the particular life-cycles have no significance. Since a given cycle presumably adapts a species to changing environments it may normally encounter, a perfectly uniform environment may fail to evoke the cycle.

Cysts

Encysted stages, in which the organism is enclosed within a cyst membrane, are a common feature of protozoan life-cycles. On the basis of apparent functions, *protective* and *reproductive* cysts have been distinguished. Protective cysts may be developed directly from active stages, from zygotes in *Volvox* and Gregarinida, or from sporoblasts (division-products of the zygote) in Coccidia. Such cysts usually possess rather firm walls, the composition of which varies from group to group. The cyst membranes of many ciliates are probably composed largely of proteins, although the inner membrane (endocyst) may be carbohydrate in nature. In Endamoebidae and *Giardia*, the properties of the cyst wall resemble those of keratins. Siliceous cyst walls are characteristic of Chrysomonadida, and walls composed largely of sand grains are produced in *Diffugia*. Many of the thick-walled cysts shows spines, ridges, or other surface markings.

A compound cyst wall, composed of two or more membranes, is not uncommon. In such cases one of the membranes—the ectocyst of *Bursaria*, the mesocyst of *Didinium*, the outer membrane of *Volvox*—is often thicker and more rigid than the others. This heavy membrane may be continuous like the others, or it may as in *Bursaria*, contain an "emergence-pore" closed by a thin membrane. The two- layered cyst of *Naegleria* contains several analogous pores. Double or multiple resting cysts are sometimes produced in Colpodidae. The double cyst of *Tillina magna* shows only one ectocyst, but each of the contained ciliates has its own mesocyst and endocyst. The protective qualities of cysts vary with the species.

Dried cysts of *Colpoda cucullus* have remained viable for more than five years. Cysts of *Naegleria gruberi* also withstand drying. Drying at room temperature prolongs the life of protective cysts of *Stylonethes sterkii* but kills those of *Euplotes taylori*. Cysts of *Didinium nasutum* do not survive desiccation although they have remained viable for ten years in sealed containers of hay infusion. Cysts of Endamoebidae also do not survive drying. However, cysts of *Entamoeba histolytica*, kept moist under refrigeration, have remained viable for 46 days. *Woodruffia*

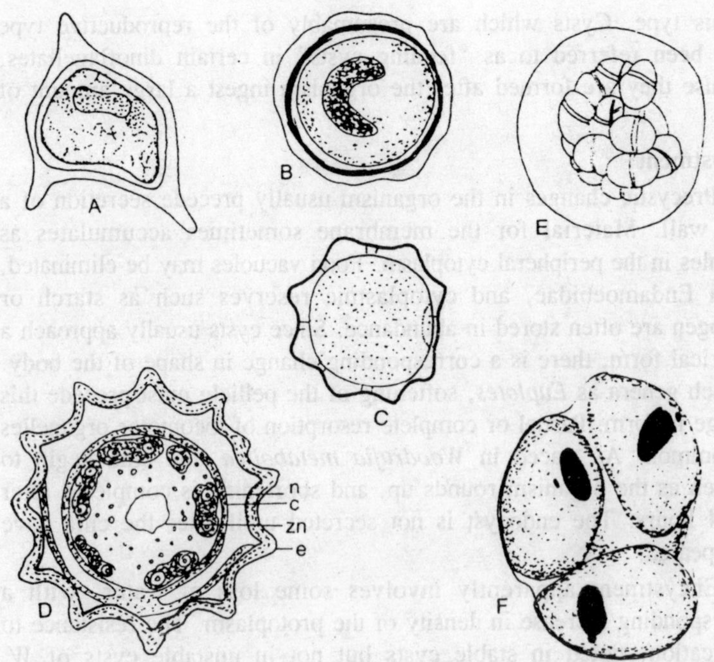

Fig. 10.10. A–Cyst of Ceratium hirundinella. B–Protective cyst of Didinium nasutum; outer (ectocyst) and inner (mesocyst) membranes evident. C–Protective cyst of Bursaria truncatella. D–Encysted zygote of Volvox globator; diagrammatic. E–Reproductive cyst in Gyrodinium sp. F–Reproductive cyst in Colpoda cucullus.

metabolica produces two types of resting cysts, a stable one which resists desiccation, and an unstable type which does not.

Resistance of protective cysts to unfavorable temperatures is sometimes striking. Thoroughly dried cysts of *Colpoda* have resisted exposure to dry heat at 100º for three hours, and immersion in liquid air for 13.5 hours. Reproductive cysts are those in which fission, budding, and sometimes gametogenesis and syngamy occur indifferent species. However, reproductive activities are not limited entirely to reproductive cysts, sicnce mitosis occurs in the protective cysts of *Giaridia* and various Endamoebidae. The wall of the reproductive cyst, although sometimes compound, is usually thin and had relatively little protective value. Such cysts are known invarious dinoflagellates and in certain free-living and parasitic ciliates. Fission within a cyst is characteristic of *Colpoda cucullus* and related species. A similar cyst serves also for attachment of *Ichthyophthirius multifilis* to the substratum. The gametocyst of gregarines probably should be included in this type. Cysts which are presumably of the reproductive type have been referred to as "feeding cysts" in certain dinoflagellates, because they are formed after the organism ingest a large amount of food.

Encystment

Precystic changes in the organism usually precede secretion of a cyst wall. Material for the membrane sometimes accumulates as globules in the peripheral cytoplasm. Food vacuoles may be eliminated, as in Endamoebidae, and cytoplasmic reserves such as starch or glycogen are often stored in abundance. Since cysts usually approach a spherical form, there is a corresponding change in shape of the body. In such genera as *Euplotes*, softening of the pellicle must precede this change in form. Partial or complete resorption of locomotor organelles is common. As traced in *Woodrufia metabolica*, the cilia begin to shorten as the organism rounds up, and shortening is completed after 22-44 hours. The endocyst is not secreted until after the cilia have disappeared.

Encystment apparently involves some loss of water, with a corresponding increase in density of the protoplasm. The resistance to desiccation, noted in stable cysts but not in unstable cysts of *W. metabolica*, is attributed to a lower water content of the former. The occurrence of encystment has been correlated with various environmental changes. Encystment of *Euplotes taylori* seems to be related to evaporation of the culture medium, while *Bursaria truncatella* encysts when transferred singly or in groups to food-free spring water. Encystment of *Didinium nasutum* is induced by crowding, either with or without a food supply. *Colpoda* (*duodenaria*) *steinii* encysts when starved, and the percentage of cysts increase with the number of organisms present. Encystment of this ciliate has been attributed to the inactivation of essential enzyme systems by metabolic products.

The lack of materials for synthesis of such enzymes should produce the same effect, and encystment of *C. steinii* in pure culture has followed elimination of thiamine, pyridoxine, nicotinamide, or pantothenic acid from the standard medium, or the omission of foods known to contain several B-vitamins. An abundance of food has been considered essential for encystment of some species, but such a food supply would favour rapid multiplication with subsequent crowding.

Encystment of ciliates also has been related to an unusually low or high pH of the medium although *Didinium nasutum* encysts most frequently within the range favourable to growth. The varied data on encystment obviously hinder selection of any one factor as they key to this process. However, such a theory as that of Taylor and Strickland lends itself to possible correlation with several environmental changes. The inactivation of a critical enzyme system might result from accumulation of metabolic poisons—induction by waste products and by crowding. Inactivation might be accelerated by a deficiency of materials for synthesizing such enzymes—induction by starvation and crowding. Also, the inactivation of an enzyme system might occur more rapidly at one pH than at another.

Excystment

Excystment often includes the regeneration of peripheral organelles as well as a certain amount of internal reorganisation. Rupture of the cyst membranes may involve two different mechanisms. The more important seems to be the absorption of water by the protoplasm early in excystment. The resulting increase in volume farcibly ruptures rigid membranes. In ciliates, the absorbed water may accumulate in a large excystment-vacuole apparently identical with the contractile vacuole of *Euplotes taylori* and *Didinium nasutum*, or in a number of vacuoles as in *Tillina magna*.

The second mechanism involves the secretion of enzymes which digest the endocyst and perhaps other flexible membranes. This enzymatic action, first described in *Colpoda cucullus*, probably occurs also in *tillina* and *Didinium*. The first signs of excystment in *Didinium nasutum* are the beginning of cyclosis and the appearance of a small posterior vacuole. When the vacuole grows to about half the volume of the body, a bluge appears at the opposite pole of the cyst. A little late, the mesocyst and ectocyst are ruptured and the organism slips out, still within the endocyst. The ciliate soon becomes very active within the endocyst, which gradually increases in diameter. The membrane becomes thinner and thinner, and finally seems to dissolve in the medium.

Excystment is completed within four hours. At emergence, the meridionaly arranged cilia extend from the anterior ciliary girdle about halfway to the posterior end of the body. Later on, the posterior cilia of the longitudinal rows develop into a posterior girdle, while the intermediate cilia disappear. The primitive ciliary pattern of Holotrichida is thus recapitulated to some extent during excystment of *D. nasutum*. Excystment of *Bursaria truncatella* is strikingly different. Cyclosis begins early, and a hyaline area of cytoplasm just beneath the emergence-pore becomes more apparent. After a time, the opercular

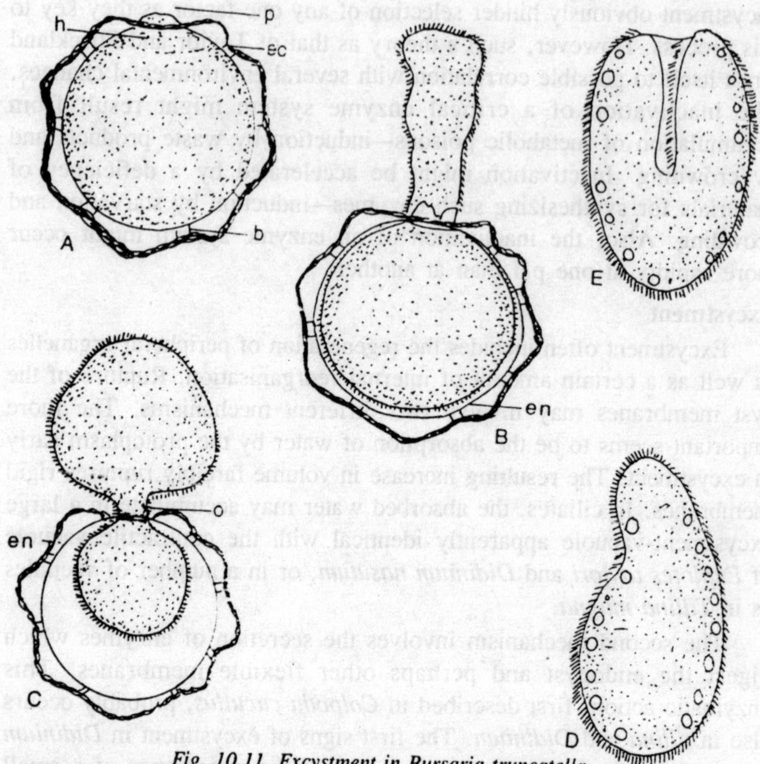

Fig. 10.11. Excystment in Bursaria truncatella.

membrane bulges outward, and then breaks suddenly as a column of cytoplasm erupts through the pore. The endoplasm streams into the protruded part of the body and ciliar activity, which now increases, tends to move the body through the pore in repeated thrusts. Emergence is completed, posterior end first, and the immature organism swims away. During the next hour the peristomial groove and membranelles are differentiated, and the adult form is gradually assumed. The physiological aspects of excystment are probably no less complicated than the morphological changes.

Excystment of *Colpoda steinii* involves several stages. In an initial phase, the length of which is influenced by temperature but not by oxygen tension, essential organic substances are absorbed from the medium. The activities of three later periods, distinguishable by varying susceptibility of the cysts to X-rays, are influenced by oxygen tension but not by organic components of the medium. Weyer suggested that excystment of *Gastrostyla steinii* is induced solely by organic substances elaborated by bacteria in the medium. Excystment of *Didinium nasutum*, in various media, depends upon the presence of living bacteria. Previously bacterized culture fluids are inactive after being heated or filtered to remove the bacteria, but only at a low oxidation-reduction potential. Barker and Taylor apparently were the first to show that excystment can be induced specifically by adding certain animal or plant extracts to basal media.

Some substance or group of substances was active for *Colpoda steinii* in dilutions as high as 1: 100,000,000. In attempts to isolate these factors, two concentrates from the extractor were found to be active separately, and also to show complementary effects in combinations. The activity of hay extracts was next related to salts of organic acids (acetic, citric, fumaric, malic, and tartaric), the effectiveness of which was quadrupled by a co-factor prepared from hay and replaceable by certain sugars in dilute solutions. Two crystalline substances, prepared from corn leaves, proved active at concentrations of 2.0-4.0 $\times 10^{-8}$ gm/ml in the presence of suitable co-factors. The co-factors, prepared from corn extract and essentially inactive themselves, could be replace in part by a sugar solution and certain combinations of thiamine, nicotinic acid, nicotinamide, adenylic acid, citrate, and malate.

A later report indicates that potassium ions, not replaceable by sodium ions, are essential to excystment of *C. steinii*. Several vitamins proglutamate, malate, and propionate) and adenosine triphosphate showed some activity. Requirements for excystment are less complex in certain other ciliates. Distilled water induces excystment of *tillina*, *magna* and *Colpoda cucullus*, and dilution of the original medium is effective for *Euplotes taylori*.

Sexual Phenomena

Varieties of Sexual Phenomena

Although sexual processes are not necessarily a prerequisite to reproduction as they so commonly are in Metazoa, and although many Protozoa undergo such activity at irregular intervals, the life-cycles of certain species cannot be completed without syngamy. For example, the mosquito phase of the life-cycle in *Plasmodium* must be initiated by gametogenesis and syngamy. The same thing is true for the formation of spores (protective cysts)) in *Eimeria* and related genera. Various kinds of sexual phenomena have been described in Protozoa. *Syngamy*, in which two gametes fuse completely to form a zygote, may involve gametes which are similar in appearance (*isogamy*), or are of two types (*anisogamy*). *Pedogamy* appears to be an unusual type of syngamy in which the two gametes are not more than one or two cell-generations removed from a single gametocyte. *Autogamy* involves the formation of two gametic nuclei, and their subsequent fusion to form a synkaryon (zygotic nucleus) within a single organism. *Parthenogenesis*, or the development of a gamete without syngamy, has been reported but its status in Protozoa is uncertain. Typical *conjugation* involves the exchange of haploid pronuclei (gametic nuclei) between two paired organism, the formation of a synkaryon in each, and then nuclear reorganization.

Meiosis in Relation to the Life-cycle

A reduction of the chromosomes to the haploid number may occur in gametogenesis (*gametic meiosis*), in an early division of the zygote (*zygotic meiosis*), or in one of the pregamic divisions in conjugation (*conjugant meiosis*). The type of meiosis varies indifferent Protozoa. Available data indicate that the Heliozoida, Foraminiferida, Cnidosporidia, and Ciliophora are diploid throughout most of the life-cycle.

Among the Mycetozoida, some of the Plasmodiophorina are said to be predominantly haploid. Nuclear fusion, supposedly occurring at the end of the vegetative phase, may be followed immediately by meiosis. In such cases, meiosis might be considered zygotic, although the uninucleate haploid products promptly encyst, becoming 'spores". Some of the Eumycetozoina are believed to undergo syngamy just before development of the plasmodium begins, and presumably are diploid throughout the vegetative phase. In such cases, meiosis apparently precedes the formation of 'spores," which give rise to the gametes after excystment. The Coccidia and a number of the Gregarinida are haploid organisms, although a few of the gregarines seem to be diploid. Among the flagellates, gametic meiosis has been reported in two species, and zygotic meiosis in a number of others.

Syngamy

In addition to many established cases of syngamy in Protozoa, a number of descriptions need confirmation. The lack of critical evidence does not in itself justify dismissal of such reports. Syngamy in zoomastigophorea was described occasionally in the older literature but most protozoologists remained unconvinced. The investigations of Cleveland have supplied cytological evidence that was previously lacking. Certain descriptions of syngamy in trypanosomes do not approach the cytological standards set by Cleveland. However, the trypanosomes are not particularly favourable material for studying chromosomal behaviour and the accumulation of adequate evidence will be correspondingly difficult. The status of sexual phenomena in Phytomastigophore other than the Phytomonadida remains uncertain. A fairly recent description of syngamy in *Euglena* has not been confirmed, and the often cited case of '*Copromonas subtilis*' is questionable.

In "*C. subtilis*" the so-called reduction- divisions involved the extrusion of small granules ('polar bodies") from the nucleus, whereas meiosis, as demonstrated in many Protozoa, is a genuine nuclear division. Other reports of syngamy in Euglenida also offer inadequate evidence. Among the Dinoglagellida, syngamy has been reported in *Ceratium hirundinella*, *Coccodinium mesnili* and *Nocitluca milaris*. Syngamy and formation of zygotes have been described in *Glenodinium lubiniensiforme*, a heterothallic species which apparently undergoes zygotic meiosis. These accounts receive additional support from a brief account of meiosis in *paradinium poucheti*.

Among the Sarcodina, descriptions of syngamy have been published for several Testacida and Amoebida. Careful studies of chromosomal behaviour have not been reported. Some supposed instances of syngamy in Amoebida have appeared in peculiar life-cycle which seem to be eliminated by the use of pure-line cultures and the occurrence of sexual phenomena in this order is still unproven. Although isogamy has been reported in some Sarcodina, gregarines and Phytomonadida, certain of these examples involve gametes which are similar in size and form but are distinguishable by vital staining or other means. Mühl noted that members of each pair, in syzygy of certain gregarines, show different staining reactions with neutral red.

These observations have since been confirmed and extended. Such differences in staining reactions have been related to differences in oxidation reduction potentials of the two gametocytes, which may differ also in the quantity and distribution of cytoplasmiic inclusions. Physiological differentiation of similar gemetes also has been reported in Chalmydomonadidae. Species of *Chalamydomonas* may be homothallic (synoecious) or heterothallic (heteroecious). In homothallic species a single culture will develop gametes of both "sexes." Every motile flagellate in such a culture is a potential gametes capable of uniting with a flagellate of the opposite sex in the same culture. As observed in the laboratory, syngamy occurs in heterothallic species only when two cultures containing gametes of opposite sexes are mixed under favorable conditions. Moewus has attributed such differentiation to specific substances produced by *Chlamydomonas*.

The original assumptions were based upon certain effects produced by fluids from cultures. A *motility factor* in culture fluid stimulates rapid formation of flagella upon addition to a culture containing palmella stages. *Termones* determine the sex of the gametes derived from palmella stages in heterothallic races. Ganotermones cause production of female gametes; androtermones, the production of male gametes. *Gamones*, concerned mainly with mutual attraction of the gametes, modify sexually inactive flagellates so that they can undergo syngamy. Androgamones from male cultures cause agglutination of female gametes under facorable conditions; gynogamones from female cultures have comparable effect on male gametes.

Spectroscopic analysis of active substances, concentrated from large volumes of culture filtrates, indicated that they were carotenoid derivatives. In subsequent tests, effects of culture filtrates were more or less duplicated by certain derivatives of protocrocin. Accordingly, it was assumed that protocrocin, synthesized by the flagellates, is broken down in the presence of light in a series of reactions, each controlled by a particular gene. The products include picrocrocin, which in turn yields safranal and glucose, and crocin, which is decomposed into gentiobiose and *cis-* and *trans-* dimethylcrocetin esters. Crocin (or a related glycoside of crocetin) seems to be the motility factor, active for *C. eugametos* in dilutions as high as $4'10^{-15}$.

The action of gynotermones was duplicated by picrocrocin; that of androtermones, by safranal. The gamones were believed to be mixtures of the *cis-* and *trans-*crocetin esters and the intensity of "maleness" or "femaleness" exhibited by gametes was attributed to the *cis-/trans-* ratio in a given mixture. The behaviour of certain American stocks of *Chlamydomonas* differs to some extent from that reported by Moewus. Strains of *C. reinhardi*, *C. minutissima*, and *C. intermedia*, for example, become motile and develop sexual activity in darkness as well as in light. However, light seems to be required for cluming and pairing of *C. moewusi*.

Prior to the work of Moewus, Schreiber had described + and— strains in *Gonium* and *Pandorina*, mixtures of different clones producing zygotes in some combinations but not in others. Tests with lines started from division-products of zygotes indicated that differentiation occurred in the first or second postzygotic fission. Aside from such biochemical differentiation of similar gametes, the development of minor structural differences apparently preceded the evolution of marked gametic dimorphism. Among the gregarines, for example, primitive anisogamy may involve differences in size of the nuclei, differences in shape, and slight differences in size of the gametes. This trend culminated in the development of small microgametes, resembling spermatozoa in their low cytoplasmic content, and relatively large macro-gametes containing appreciable amounts of stored food. Such extreme differentiation is characteristic of certain Sporozoa (Coccidia, Haemosporidia) and *Volvox*.

Pedogamy

In this process, attributed to *Actinophrys* and *Actinosphaerium*, a single organism encysts and then divides into two or more 'gametocytes." After meiosis occurs, the resulting gametes undergo syngamy. In repeating earlier observations on *Actinophrys sol*, Belar described a reductional division in each gametocyte, followed by degeneration of one of the two haploid nuclei. Fusion of the uninucleate gametes was then followed by encystment of the zygote. The occurrence of syngamy in Heliozoida seems unquestionable but the validity of 'pedogamy' may be less certain. It has been suggested that, as incertain Foraminiferida two associated "gametocytes" secrete a common cyst membrane. However, such as interpretation is not supported by Belar's data.

Autogamy

The older literature contains numerous descriptions of autogamy. In a typical account, the nucleus of an encysted amoeba divides and each daughter nucleus undergoes meiosis. The haploid nuclei then fuse in pairs. Or, fusion may be preceded by degeneration of all except two haploid nuclei, so that only one synkaryon is produced. Believing that such cases are open to more plausible explanations, protozoologists generally had considered autogamy a highly dubious process. The question was reopened by Diller's report of autogamy in *Paramecium aurelia*. Autogamy followed by meiosis of the synkaryon, was reported shortly afterward in *Phacus pyrum*, although this account has not been confirmed. Diller's observations on *P. aurelia* have been followed by descriptions of autogamy in *P. bursaria* and *P. trichium*. Cases of autogamy in which ciliates form a conjugant pair but fail to exchange pronuclei have been referred to as *cytogamy* in *P. caudatum*. In addition, certain genetic data agree with the cytological evidence for autogamy in *Paramecium*. Up to a certain point, nuclear behviour in autogamy parallels that in conjugation.

Maturation divisions are normal and pronuclei are formed. Instead of reciprocal transfer, however, fusion of two pronuclei occurs within the same ciliate. It is uncertain whether autogamy is a normal process in its own right or merely abortive conjugation. Chen has found that, in conjugating trios of *P. bursaria*, a small area of cytoplasmic contact will initiate autogamy in the odd member which is left out of the normal pairing.

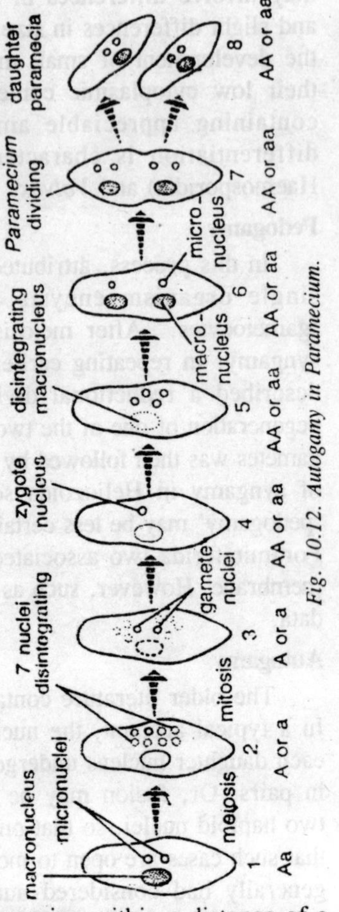

Fig. 10.12. Autogamy in Paramecium.

Conjugation

The onset of conjugation in mass cultures of certain ciliates is indicated by a tendency for the organisms to adhere on contact, sometimes forming clumps containing many individuals. The nature of this mating reaction is uncertain, although such a process suggests that the ciliates develop sticky surfaces. This initial reaction in *Paramecium bursaria* seems to involve chance contact which leads to clumping. In general, such a preliminary reaction seems to be independent of later pairing and may be insignificant, or may not occur at all, in certain clones of *P. bursaria* and in various other ciliates. The stalked conjugant of *Vorticella microstoma* seems to exert some sort of attraction for motile microconjugants passing within a distance of a millimeter. Clumping in *P. bursaria* is followed, after a half hour or so, by gradual breaking up of the aggregates.

At the end of several hours, only pairs and single ciliates remain as a rule. Groups of three or four persist occasionally, but only two members of each group are properly paired for conjugation. Pairing seems to depend upon favorable conditions and may be influenced by temperature and intensity of light. The positions assumed by the paired conjugants and the extent of cytoplasmic fusion vary with the species. Contact commonly involves the peristomial areas of the two conjugants. However, fusion at the posterior ends occurs in *Ancistrocoma myae*, and fusion of oral to aboral surface in *Kidderia mytili*. Among the

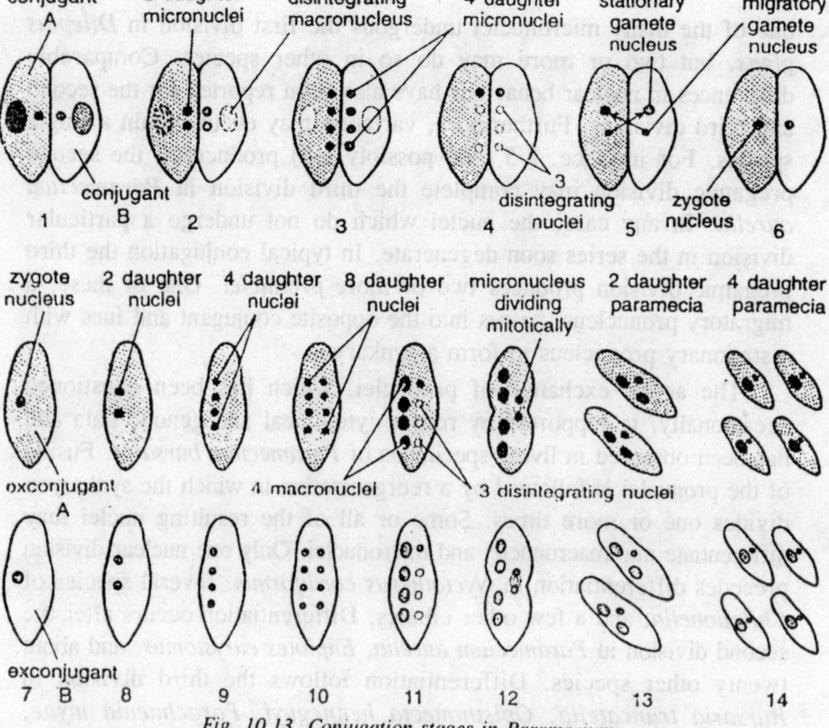

Fig. 10.13. Conjugation stages in Paramecium.

Peritrichida, the microconjugant becomes attached near the abroal end of the body in *Opisthonecta* and *Vorticella*, but near the oral end in *Scyphidia*.

In certain Apostomina, conjugants in lateral contact undergo repeated fission to produce chains and conjugation then proceeds between corresponding members of the chains. The extent of fusion in conjugation apparently is influenced by the nature of the body wall. In ciliates with a firm cuticle, fusion, or sometimes merely adhesion, may involve a limited area of the body such as the left margin of the peristome in *Euplotes*. As a rule, the micronucleus undergoes three pregamic divisions. More commonly the second, but sometimes the first of these, is reductional. However, exceptions to the usual pattern have been noted. The third pregamic division is sometimes omitted in *Paramecium trichium*, and micronuclei may even be exchanged just after the first division.

When several or many micronuclei are present, the number participating in the pregamic divisions varies with the species. Only one of the many micronuclei undergoes the first division in *Dileptus gigas*, but two or more may do so in other species. Comparable differences in nuclear behaviour have also been reported for the second and third divisions. Furthermore, variation may occur within a single species. For instance, 2-5 (and possibly 1-5) products of the second pregamic division may complete the third division in *Paramecium aurelia*. In any case, the nuclei which do not undergo a particular division in the series soon degenerate. In typical conjugation the third pregamic division produces two or more *pronuclei*. One of these, a migratory pronucleus, passes into the opposite conjugant and fues with a stationary pronucleus to form a synkaryon.

The actual exchange of pronuclei, which has been questioned occasionally, is supported by recent cytological and genetic data and has been observed in living specimens of *Paramecium bursaria*. Fusion of the pronuclei is followed by a reorganization in which the synkaryon divides one or more times. Some or all of the resulting nuclei may differentiate into macronuclei and micronuclei. Only one nuclear division precedes differentiation in *Nyctotherus cordiformis*, several species of *Chilodonella*, and a few other ciliates. Differentiation occurs after the second division in *Paramecium aurelia*, *Euplotes eurystomus*, and about twenty other species. Differentiation follows the third division in *Bursaria truncatella*, *Opisthonecta henneguyi*, *Parachaenia myae*, *Vorticella microstoma*, *Paramecium bursaria*, *P. trichium*, *P. caudatum* and a number of other species. Differentiation after a fourth postzygotic division has been reported in *Kidderia mytili* and *Paramecium multimicronucleatum*.

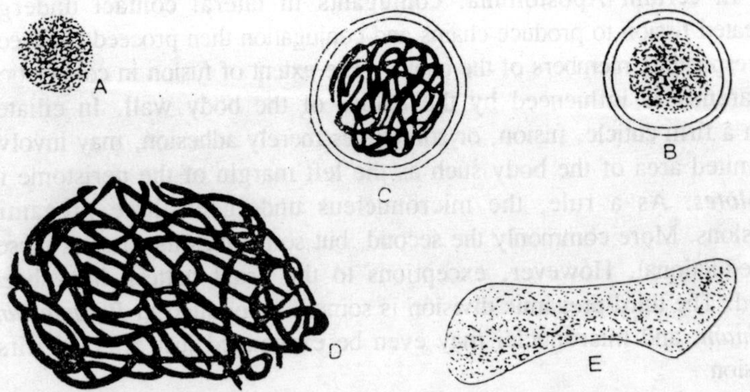

Fig. 10.14. Development of a new macronucleus following conjugation in Nyctotherus cordiformis.

Behaviour of the nuclei inciliates showing two or more postzygotic divisions differs from species to species. All of the nuclei may remain functional, or some of them may degenerate. Variations may occur also in individual species, as in *P. caudatum* and *P. trichium*. Development of the micronucleus usually involves a decrease in size, whereas a differentiating macronucleus grows and often undergoes extensive changes inform as well as internal organization. The young macronucleus of *Nyctotherus cordiformis* soom becomes finely granular and stains more intensely.

Later, the granules give rise to threads during growth of the nucleus and then, as differentiation nears completion, the threads are replaced by the granules characteristic of the mature macronucleus. The early stages of differentiation are similar in *Euplotes eurystomus*. After the threads are replaced by granules the developing macronucleus elongates, extends posteriorly, and makes contact with a remnant of the old macronucleus. Fusion results in a complete macronucleus. Depending upon the species, postconjugant fissions may or may not be necessary to restore the normal nuclear situation. Therefore, the final result of typical conjugation is the formation of 2-8 reorganized ciliates from a pair of exconjugants.

In *Metopus sigmoides*, the pronucleus of one conjugant (the "donor") is accompanied by a large amount of cytoplasm during migration. After separation of the conjugants, the donor eventually dies. Conjugation in *Opisthonecta*, *Urceolaria*, and *Vorticella*, also produces only one functional exconjugant. One conjugant is a microconjugant, produced by budding and the other is a macroconjugant. In *Vorticella microstoma*, a microconjugant becomes attached near the aboral end of a macroconjugant. Fusion then occurs and the endoplasm of the microconjugant gradually flows into the macroconjugant, leaving the pellicle behind. Pregamic divisions and formation of a synkaryon then occur much as in other ciliates.

Conjugation is often considered an orderly process which, once started, goes through a fixed series of nuclear activities. This is not always the case and variations are striking inseveral species. Furthermore, conjugation between particular strains of a species may be abnormal. For instance, in conjugation of certain Russian strains with several American strains of *P. bursaria*, the first pregamic division is usually not completed and all conjugants die before or after separation. The lethal effect is produced after cytoplasmic fusion, but before the exchange of pronuclei. Mixtures of certain abnormal strains of *P. bursaria* with normal strains undergo typical pairing, but separation occurs after a few hours. The micronucleus enlarges slightly but does not start the first pregamic division. Polyploidy seems to have arisen frequently in *P. bursaria*, probably through the fusion of more than two pronuclei in conjugation.

Chromosomal variations also are produced by matings between diploid and polypoid strains, as well as between micronucleate and amicronucleate races. In the latter case, each exconjugant contains a single haploid nucleus which undergoes three divisions and probably produces a new nuclear apparatus. Nuclear behaviour varies also in *Paramecium trichium*. Micronuclei are sometimes transferred just after the second or even the first pregamic division. Occasionally only one of the migratory pronuclei actually migrates, so that conjugants sometimes contain one and three pronuclei. There also may be no exchange of pronuclei, with resulting autogamy in each conjugant. After the second pregamic division, three haploid nuclei sometimes degenerate and the fourth, without dividing again, migrates into the other conjugant.

Each exconjugant thus contains a haploid nucleus which undergoes postzygotic divisions. Heteroploidy occurs frequently in *P. trichium* and has been noted also in *P. aurelia* and *P. caudatum*. The exchange of macronuclear fragments has been observed in *P. trichium*—but not in other species of *Paramecium*—and also in several species of *Chilodenella*.

Factors Inducing Conjugation

The possible causes of conjugation have been discussed for many years. Diverse ancestry was one of the prerequisites suggested by Maupas and the more recent discovery of *mating types* has proven that apparently hereditary differentiation of potential conjugants does exist in certain species. However, conjugation has been observed within single clones, and also among the descendants of a single exconjugant after only a few fissions. Some of these matings between closely related conjugants—as reported in *Paramecium*, *Spathidium*, *Uroleptus*, and *Euplotes*—have not yet been correlated with the basic concepts of mating types. Autogamy might bring about differentiation within clones of *Paramecium*, but such an explanation is of uncertain validity for other ciliates in which autogamy is unknown.

Sexual maturity as a requirement for conjugation also was suggested by Maupas, who believed that strains of ciliates are immature when first established in cultures and must complete a certain number of generations before they can conjugate. In contrast to this view, conjugation has occurred at intervals of only a few days in *Paramecium aurelia* and *P. caudatum*. Jennings has suggested that the duration of "immaturity' in *P. bursaria* varies inversely with the food supply. Starvation is the third factor which Maupas considered essential.

More recently, conjugation of *Paramecium multimicronucleatum*, *Spathidium spathula* and *Uroleptus mobilis*, among others, has been found to follow exhaustion of the food supply. On the other hand, conjugation has occurred in *P. aurelia* just as a rich food supply was beginning to decline, and also in *P. caudatum*, shortly before the populations reached the maximum. The nature of the significant changes with accompany or precede starvation is not yet known. However, the physiological condition of individual ciliates seems to be an important factor, since Boell and Woodruff observed successful conjugation of *P. calkinsi* only between ciliates with subnormal respiratory rates.

A mating reaction between a normal ciliate and one with a low respiratory rate sometimes occurred but conjugation was never completed. Ciliates with high respiratory rates failed to show any mating reactions. Various environmental factors also have been correlated with conjugation. Darkness apparently favors and light suppresses conjugation in *P. aurelia*, although light shows no comparable effect on *P. caudatum* or *Euplotes patella*. Temperature also influences conjugation, and different optima have been noted for different varieties of *P. aurelia*. In one variety the frequency of conjugation has ranged from zero at 24.5° to 68 per cent at 17.6°. Conjugation in *Conchophthirius lamellidens*, parasitic on the gills of a fresh-water mussel, has been observed most frequently on the day following the new moon. Dilution of the medium with weak solutions of aluminium and iron chlorides is said to have induced conjugation of *Paramecium caudatum*, but Ball obtained negative results with several clones of *P. aurelia* and *P. caudatum*. One clone of *P. caudatum* did respond to such treatment but distilled water was just as effective as the salt solutions.

Conjugation of *Glaucoma scintillans* has been stimulated by decreasing the salt content of the medium or increasing the concentration of glucose, and also by adding pyruvic acid to the medium. The bacterial flora of cultures also may influence the incidence of conjugation Chatton and Chatton found that *Glaucoma scintillans* conjugated when fed on *Escherichia coli*, *Proteus vulgaris*, *Shigella dysenteriae*, or *Staphylococus aureus*, but not on *Pseudomonas aeruginosa*, *P. fluorescens* or any one of several other bacterial species.

Conjugation of *P. caudatum* was observed in cultures containing only a gram negative bacillus, but not in other cultures containing at least three kinds of bacteria. Accordingly, it was suggested that so called conjugating and non-conjugating races of ciliates may be determined by the bacterial flora. This conclusion was not supported by Sonneborn and Cohen who induced conjugation invariably in a Johns Hopkins strains and never in Woodruff's strains of *P. aurelia* when both strains were maintained on the same bacterial types.

Mating Types in Ciliates

Following the observations of Sonneborn on *Paramecium aurelia*, *P. calkinsi*, and *P. trichium* and those of Jennings on *P. bursaria*, mating types have been demonstrated also in *P. caudatum*, *P. multimicronucleatum* and *Euplotes patella*. The situation in *P. bursaria* may be illustrated as follows. Two strains, A and B, have been established in pure lines. Conjugation does not occur among ciliates of strain A or among those of strain B, although mixtures of the two do show conjugation. Therefore strains A and B seem to belong to different sexes. A third strains, C, tested in the same way with strain A, behaves like strain B, and consequently might be expected to have the same sex. However, conjugation occurs also in mixtures with strains B and C. A fourth strain, D, is found to conjugate with any of the other three. At this point, conjugation in *P. bursaria* begins to strain basic concepts of bisexuality in animals, and confusion in terminology has been avoided by the substitution of "mating type" for "sex". Further investigation has demonstrated additional groups of mating types. A second group, or variety, contains eight mating types which will not conjugate with the four types in variety I. Mating types N, O, P, and Q have been assigned to a third variety, since they will not conjugate with types belonging to varieties I and II. Variety IV contains types R and S, which do not mate with members of varieties I, II or III. Variety V is represented by mating type T, composed of strains obtained from Russia, and will not mate with members of the other varieties.

A more recently recognized variety VI, including strains from Czechoslovakia, England and Ireland, contains mating types U, V, W, and X. In *Paramecium aurelia* seven varieties have been recognized. Six of these contain two mating types, and one type has been assigned to variety. Normal conjugation occurs between the two mating types of each variety, but not between strains belonging to different varieties. Thirteen varieties, each with two mating types, have been identified in *P. caudatum*. At first, it was believed that conjugation never occurred between members of different varieties in *P. aurelia* and *P. bursaria*, but exceptions have been reported more recently. Type R of variety IV occasionally conjugates with four types of variety II in *P. bursaria*, although the participants die during or shortly after conjugation. Similar cases have been observed in *P. aurelia*. Mating type I will conjugate occasionally with type X, and mating type II with types, V, IX, and XIII. Mating reactions in these intervarietal crosses of *P. aurelia* are always less intense than those within the same variety—only 1-40 per cent as many conjugant pairs in different combinations. In *P. caudatum* intervarietal matings have occurred between variety 10 (type XX) and varieties 8 (type XV) and 9 (type XVII), and also between variety 2 (type IV) and variety 8 (types XV).

Table 10.1. Induction of Conjugation in Euplotes Patella by Fluids from Cultures

	Mating types treated ciliates					
Culture fluids	*I*	*II*	*III*	*IV*	*V*	*VI*
I	−	+	+	+	+	+
II	+	+	+	+	+	+
III	+	−	−	+	−	+
IV	−	−	+	−	+	+
V	+	+	+	+	−	+
VI	−	+	+	+	−	−

The situation in *Euplotes patella* resembles that in *P. bursaria*. Six mating types have been recognized in one variety, and there may be additional varieties. The mating reactions of *E. patella* are especially interesting because specific mating-type substances are released into the culture medium. Fluid from cultures of one mating type will induce conjugation among the ciliates of a single mating type in certain cases. The nature of this effect is uncertain. Kimball apparently favors the view that conjugation is induced in animals which are all of the same mating type, rather than that the mating type is changed in some of the treated ciliates and not in others.

A particular mating-type substance induces conjugation only in a type which does not produce that substance, and these effects have been correlated with the inheritance of mating types in *E. patella*. Certain analogous effects of culture fluid have been observed in *Paramecium bursaria*. Fluid from cultures of several Russian strains (type T) induces conjugation within individual mating types of varieties II, III, IV, and VI, although the effect is usually limited to a small percentage of the ciliates in a culture. The recognition of mating types in certain ciliates has shown that conjugating pairs, in these species at least, are composed of physiologically different organism. However, the relation of mating types of the concept of bisexuality in animals remains uncertain in *Paramecium bursaria* and *Euplotes patella*. On the other hand, *P. aurelia* and *P. caudatum* might possible be interpreted as species composed of "bisexual" varieties which interbreed with difficulty or not at all.

Nuclear Phenomena of Uncertain Significance

Endomixis was originally described in *Paramecium aurelia* as a complete nuclear reorganization occurring in individual ciliates. Macronuclear disintegration and two micronuclear divisions occurs without the usual third pregamic division of conjugation. Only two of these eight micronuclear derivatives persist, so that the first fission leaves each ciliate with one functional nucleus. To nuclear divisions occur. Two of the products then differentiate into macronuclei, while the others divide to form four micronuclei. A second fission completes the reorganization. The significance of endomixis in the life-cycle is still unknown. Woodruff believed that meiosis does not occur— although the second pregamic division is reductional in conjugation of *P. aurelia*— and he suggested that endomixis might be analogous to diploid parthenogenesis.

The discovery of autogamy in *P. aurelia* and the accumulation of genetic data have thrown doubt upon the occurrence of endomixis in *P. aurelia*. *Hemixis* involves unusual behaviour of the macronucleus only. The process has been observed in *Paramecium aurelia*, *P. caudatum*, and *P. multimicronucleatum*. In one type of hemixis there is a precocious division of the macronucleus and the normal nuclear situation is restored in the next fission. In another type, the macronucleus extrudes one or more densely staining masses and then bahaves normally in subsequent fissions. A third type of hemixis combines the elimination of chromatic material with precocious division of the macronucleus.

Physiological Life-Cycle

The description of conjugation by O.F. Müller in 1786 stimulated much interest in the sexual activities of Protozoa. For many years, it was believed that the "ovary" (macronucleus) of ciliates gave rise to "ova" (produces of macronuclear disintegration), while the "testis" (micronucleus) produced "spermatozoa" (chromosomes). In conjugation, two hermaphroditic ciliates were supposed to exchange spermatozoa. In certain cases, small organisms (probably parasites) within the conjugants were interpreted as "embryos" developing within viviparous parents. These interpretations were overthrown by Bütschlii and Engelmann, who showed that the supposed ovary and testis are nuclei and suggested that products of the micronuclei might be exchanged in conjugation. The fusion of pronuclei in conjugation was reported a few years later.

Once conjugation was found to involve nuclear reorganization, and occasionally the reorganization of locomotor structure, the process was interpreted as a sort of rejuvenation. Engelmann suggested that it was unnecessary to suspect any other effect. Bütschli supported a physiological interpretation—ciliates become senescent during continued fission and as a result reproduce less and less frequently until conjugation rejuvenates them and restores the normal reproductive rate. This question was first considered experimentally by Maupas, whose isolation-culture technique involved tracing single ciliates from one generation to the next in order to detect possible senescence. Since all his strains died eventually, Maupas suggested that ciliates, like higher animals, pass through a cycle of youth, maturity, and old age, ending in death.

The characteristic feature of maturity was assumed to be an ability to conjugate normally. Conjugation was believed to rejuvenate ciliates only during the phase of maturity, and therefore was a prophylactic rather than a therapeutic measure. Bütschli maintained that conjugation increased fission-rate after a gradual decline. Hertwig's observations on split-pairs—conjugants separated at the beginning of conjugation and used for starting parallel clones—indicated that fission rate were usually higher in non-conjugation than in exconjugant lines.

As a result, he concluded that conjugation merely regulates metabolism so as to prevent physiological exhaustion. Later investigations were designed to test the theories of Bütschli, Hertwig, and Maupas. Joukowsky, after studying exconjugant and non-conjugant lines of *Paramecium caudatum* and *Pleurotricha lonceolata*, concluded that the degenerative changes described by Maupas were the result of unsatisfactory conditions in cultures. There were no characteristic differences between exconjugant and non-conjugant lines, neither type showed a decreasing fission-rate, and there appeared to be no physiological cycle.

The next important papers were those of Calkins started isolation-cultures of *paramecium caudatum* of February 1, 1901. Four lines were started from each of two ciliates and transferrs were made daily or every other day. After a time, recurrent "depressions" developed. The early depressions, believed to represent the senescence reported by Maupas, were cured by measures other than conjugation. The depression of May, 1901, apparently was cured by jolting during a train ride to Woods Hole; that of August, 1901, by extract of raw beef; that of December, 1901, by beef extract; that of March 1902, by a slight rise in temperature; that of June, 1902, by brain extract. Since the lines were rejuvenated by artificial means, the result were considered analogous to artificial parthenogenesis. In these early papers, Calkins suggested that ciliates have the "potential of endless existence" without conjugation. Later on, however, the depressions became more severe. The "B" lines became extinct after 16 months, the "A" lines in December, 1902. Attempts to rejuvenate the ciliates—treatments with beef extract, pancreas, brain, mutton broth, lecithin, pineapple extract, apple juice, several acid and salts, dried *paramecium*, the electric current and nitroglycerin—were all unsuccessful.

As a result, Calkins was convinced that the final depressions arose from "germinal exhaustion" which could not be prevented by external stimulation. Therefore, strains of *P. caudatum* must pass through a cycle of youth, maturity, and old age unless vitality is renewed by conjugation. A gradual decrease in fission-rate accompanied senescence and the rejuvenation by conjugation was believed to include an increase in fission rate. The observations of Enriques soon questioned the inevitability of senescence. The first important demonstration was that excessive bacterial growth may lead to effect simulating senescence. Later results included the maintenance of *Glaucoma scintillans* without conjugation for almost 700 generations. The ciliates remained healthy so long as fresh medium was supplied; the use of old medium induced depressions.

At this point, Enriques suggested that exhaustion of the investigator's patience in a more important factor than senescence of the ciliates in such investigations. On May 1, 1907, Woodruff started the line of *Paramecium aurelia* which was to deal a more serious blow to the physiological life-cycle. In May, 1908, the strains had passed 490 generations and at the need of four years, had survived for 2,121 generations without conjugation. By this time, the evidence indicated that *P. aurelia* might reproduce indefinitely without conjugation, or else that the "cycle" must be longer than that of any ciliate investigated previously. The conclusion suggested by Woodruff's strain of *P. aurelia* did not remain unchallenged.

Calkins and Gregory maintained that some strains of *Paramecium* are conjugating races while others are non-conjugating and it was argued that Woodruff's strains was a non-conjugating race which should not be compared with the conjugating strain of Calkins. Woodruff met this objection by reporting conjugation in mass-cultures started from his strain at the end of 4,102 generations. The completion of 25 years without conjugation was reported in 1932. Evidence against the physiological cycle gradually accumulated from other sources. *Glaucoma scinillans* showed no senescence after 2,700 generations. Lines of *Paramecium caudatum* lived for ten years without conjugation of decrease in vitality.

The colonial flagellate, *Eudorina elegans*, was maintained for eight years without syngamy or indications of senescence. *Actinophrys sol* passed more than 1,200 generations without syngamy. *Spathidium spathula*, previously credited with a cycle, survived for a thousand generations without conjugation or endomixis. *Didinium nasutum* was maintained by Beers without conjugation or a decrease in vitality so long as the food supply was adequate. However, depressions were readily induced by an inadequate diet. In contrast to various other ciliates, *Uroleptus mobilis* failed to follow the prevailing pattern. Instead, a physiological cycle was reported, with the strains living an average of 350 generations. Attempts to prolong the cycle by varying the environmental conditions were unsucessful, and this species remains one in which the cycle has not been eliminated.

More recently, Jennings concluded that his experience with *Paramecium bursaria* also supports the concept of a physiological cycle, although some clones were maintained for eight years before their health began to decline. In spite of the fact that strains of various ciliates could be grown in the laboratory for long periods without conjugation—and perhaps they could be maintained indefinitely—one question remained unanswered. Does conjugation really have any stimulatory or rejuvenating effect on ciliates? The early investigations had produced little information. A few observations by Hertwig on *Dileptus gigas* and some inconclusive data cited by Calkins represented the available evidence. Some years later, the first convincing experiments were reported by Calkins.

Several strains of *Uroleptus mobilis*, which were entering depressions, showed a higher fission-rate and greater longevity after conjugation than the non-conjugation parental stocks. Comparable effects of conjugation were reported subsequently in *Spathidium spathula* and *paramecium bursaria*. At present it seems clear that conjugation, whether or not it is essential, can produce a physiological stimulation in at least certain strains. However, it is equally evident that conjugation is no universal remedy for senescent ciliates.

In fact, the odds are slightly against survival after conjugation in *Paramecium bursaria*. Records kept for 20,478 exconjugants show that under conditions in which all non-conjugant lines remained vigorous, 29.7 per cent of the conjugating ciliates died before the first post-conjugant fission, and only 47.3 per cent survived for more than four fission. Conjugation between inbred lines is even more dangerous, and mortality often reaches 90-100 per cent in such cases in *P. bursaria*. Conjugation between old stocks which are not closely related also may be almost 100 per cent lethal, although unrelated young stocks may show little or no mortality after conjugation.

11

Life of Amoeba

Amoeba is the simplest living organism, which can perform all its life-activities, such as locomotion, nutrition, digestion, excretion, respiration, reproduction etc. in a single cell. It was first discovered by *Von Rossel Rosenhof* in 1755. *H. I. Hirschfield* (1962) has provided a detailed account of its biology.

Systematic Position

Phylum	—	Protozoa
Subphylum	—	Sarocomastigophora
Superclass	—	Sarcodina
Class	—	Rhizopodea
Subclass	—	Lobosia
Order	—	Amoebida
Genus	—	*Amoeba*
Species	—	*proteus*

Habits and Habitat

Amoeba proteus lives in fresh water ponds, pools, ditches and streams having green plants. It is also found in damp soil. *Amoeba proteus* name is derived from two Greek words (Gr. *amoeba* = change; *proteus* = a mythological sea-god, who could change its shape) because it always changes its shape.

Culture

Amoeba may be obtained for laboratory purpose from a variety of places such as organic ooze from decaying vegetation or the lower surface of the lily pads. Culture of *Amoeba* can be prepared in laboratory. Place some pond water, mud and leaves in 100 ml of water containing a few wheat grains. Amoebae will appear after a few days. To make a pure culture, boil 4-5 grains of wheat in 100 ml of distilled water for about 10 minutes and cool for a few days. To this solution add some amoebae from the previous culture and cover with glass plate. In about 10 days many amoebae will be formed in this culture.

MORPHOLOGY

Shape and Size

Amoeba is unicellular and can be seen clearly with the help of a microscope. It appears as a colourless, transparent and gelatinous mass. Its shape keeps on changing because of presence of many finger like process, the *pseudopodia*. The size varies from 0.2-0.3 mm. in diameter. Some larger forms measure 0.5 mm. in diameter, being just visible to the naked eyes as tiny white speaks.

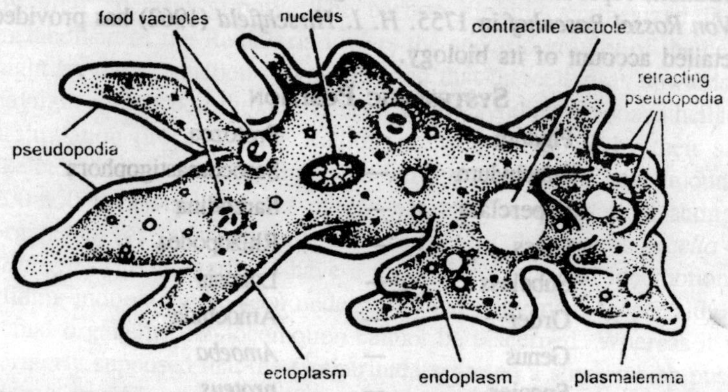

Fig. 11.1. Amoeba proteus.

Plasmalemma

The plasmalemma is a very thin, invisible, elastic external membrane which bounds the cytoplasm. It measures 1-3 μ in thickness. The membrane is selective permeable as water and some small molecules can pass freely across it in both the directions, but larger molecules can not pass through. Under electron microscope it is trilaminar structure, consisting of two, darkly stained, layers separated by a clear layer. Each dark layer is composed of mucoproteins and the light layer of lipid (bimolecular layer). The plasmalemma externally possesses numerous, ridge-like extensions, the *microvilli*. Zoologists believe that these have adhesive properties and serve to bind the animal to the substratum.

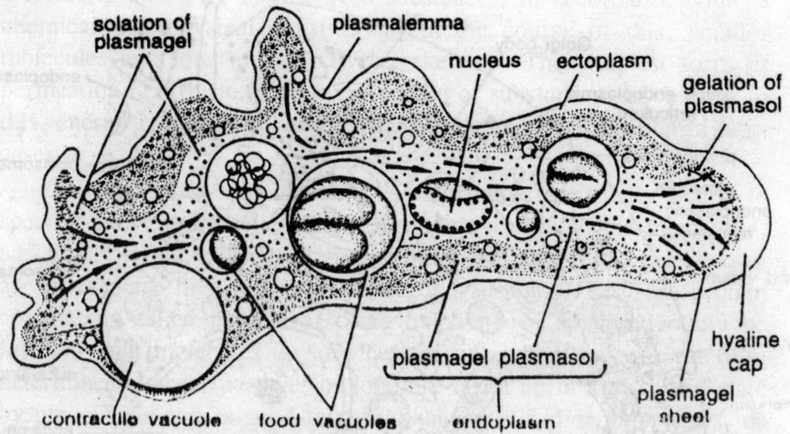

Fig. 11.2. Amoeba proteus showing the solation and gelation of cytoplasm during amoeboid movement.

Cytoplasm

The cytoplasm is differentiated into an outer thin cortical layer called *ectoplasm* and an inner medullary mass called *endoplasm*.

Ectoplasm

The ectoplasm forms the outer and relatively firm layer lying just beneath the plasmalemma. It is thin, non-granular and almost transparent. It is slightly thicker on the advancing side of the body and at the tips of pseudopodia where it forms the *hyaline cap*.

Endoplasm

It is less fibrous and more fluid zone of granular cytoplasm. It occurs in two colloidal states (*Mast*, 1926). The peripheral viscid or *gel* state called *plasmagel* and the central flowing or *sol* state called *plasmasol*. Modern studies have proved that the ectoplasm is gel and the endoplasm is sol state of cytoplasm. The endoplasm shows distinct streaming movements, called *cyclosis*.

Nucleus

Amoeba has a single, disc-like or biconcave nucleus. It is usually located in the central part of plasmasol. It is highly granular and, therefore, refractive to light. The nucleus has a firm nuclear membrane which is double and intercepted by pores. Electron microscopic studies show a honey comb layer under bilayered nuclear membrane. The nucleoplasm contains few nucleoli and about 500 small spherical chromosomes.

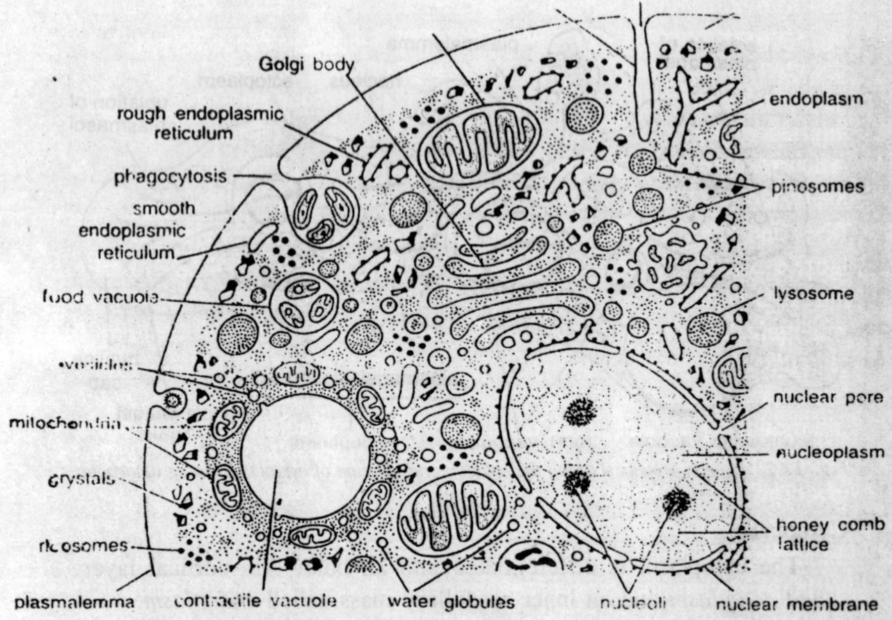

Fig. 11.3. A diagrammatic representation of the structure of a portion as revealed by electron microscopy.

Contractile Vacuole

Lying near the nucleus is a bubble-like spherical body, which is known as contractile vacuole. At regular intervals, it moves to the surface, where it contracts and discharge the water and waste products into the surrounding water. Thus it is *osmoregulatory* in function. The contractile vacuole is bounded by a delicate elastic *condensation membrane*. It is surrounded by many vesicles and mitochondria. Its position is not fixed due to streaming movement of the cytoplasm.

Food Vacuoles

A number of spherical food vacuoles small and large, containing food and water in various phases of digestion occur in the endoplasm. These are not permanent structures, but are formed when *Amoeba* ingest food. They disappear with the egestion of undigestible food from the body.

Water Globules or Vacuoles

Innumerable water vacuoles are scattered in the endoplasm. They are colourless, transparent and non-contractile. Their significance is not well-known.

Other Organelles

Under electron-microscope the endoplasm shows various organelles such as *endoplasmic reticulum*, *mitochiondira*, *Golgi-complex*, *lysosomes* and *ribosomes*. Thin filamentous microfibrils are also present in endoplasm. Besides above organelles numerous minute, regularly shaped crystals called, *biurets* and *triurets* are present.

Polarity

Although the animal is shapeless but *Amoeba* is considered to have a definite polarity i.e. it has definite anterior end and posterior ends. The anterior end possesses pseudopodia while the posterior end is marked by a wrinkled region, called *Uroid*.

PHYSIOLOGY

Locomotion

Amoeba moves from one place to another to capture its food-material and other organisms by forming the temporary finger-like projection called pseudopodia (*Gr. pseudo*= false; *podium* = foot). These pseudopodia are formed at any part of the body. The characteristic irregular movements found in *Amoeba* are termed as *amoeboid movements*. The pseudopodium is formed by pushing out an enlargement of ectoplasm so as to form a blunt projection. The granular endoplasm than flows in it. As a result of which *Amoeba* moves forward in the direction of pseudopodium. This method of pseudopodial formation is called *profluent type* by *Human*. When several pseudopodia are formed by this method it is called *lobose type* and pseudopodia as *lobopodia*.

In some forms, the lobopodia are formed by *eruptive method*. Here the pseudopodia are formed by bursting out of ectoplasm and endoplasm. Sometimes several pseudopodia are formed at the start but usually one survive, which becomes larger and effective, while others disappear. Several theories have been put forward about the formation of pseudopodia. These theories are as follow:

Adhesion theory

According to this theory, amoeboid locomotion is brought about by the force of adhesion between the cytoplasm and the substratum. Just as a drop of water, which spreads irregularly on an uneven glass plate. It was thought that *Amoeba* moves irregularly. Due to adhesive properties pseudopodia generally grow in the path of adhesion. When the surface is clear, adhesion is more perfect, and when it is greasy, adhesion is less suitable. This theory is not satisfactory and does not hold good as the pseudopodia are sometimes given out independently even without any contact with any surface.

Rolling movement theory

This theory was proposed by *Jennings* (1904). If a particle of carmine is placed on the surface of *Amoeba verrucosa*, it is seen that the particle flows forward, rolls over the anterior edge, then it stops on substratum until the entire animal has passed over it, then the particle move upwards at the posterior end and comes on the upper surface and moves forwards. *Jennings* thought that the movement of particle was due to streaming movement of cytoplasm which made the body to roll and move ahead. This theory is also called *Jennings contraction theory*. However, there are two objections to this theory:

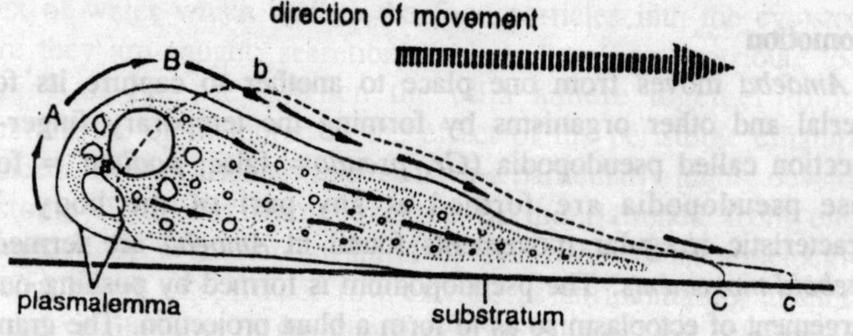

Fig. 11.4. Amoeba verrucosa showing rolling movement.

(a) *Jennings* performed this experiment on *A. verrucosa* only in which pseudopodia are absent. So that this theory could not be apply to *A. proteus.*

(b) He was unable to explain, with certainty, the locomotion of contractile substances, which are responsible for forward movement.

Surface-tension theory

Berthold (1886) proposed that constant internal tension forces the cytoplasm of *Amoeba* to shoot out from its surface in the form of pseudopodium whenever surface tension suddenly lowers due to local effects. This theory has been supported by *Rhumbler* and *Butschli* (1898). Surface tension does exists on a fluid surface, but its existence upon the surface of plasmalemma of *Amoeba* is doubtful. Thus this theory is not supported now-a-days.

Walking movement theory

This theory was proposed by *Dellinger* (1906). He observed *Amoeba* from the side and found that locomotion was a walking process. These movements start by putting out one pseudopodium and then followed

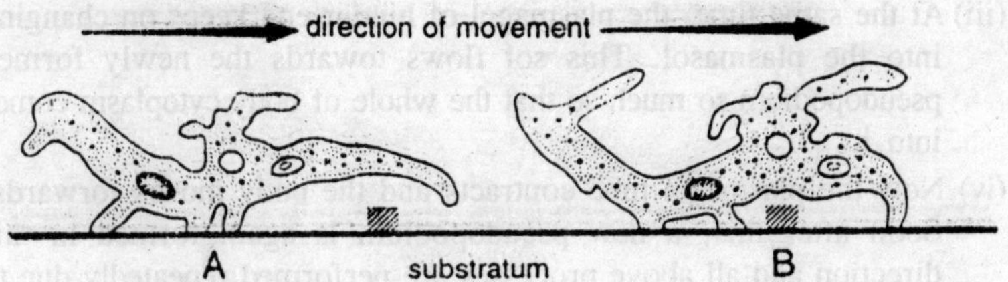

Fig. 11.5. Amoeba proteus in side view showing walking movement due to contraction in the protoplasm.

by another. On making forward movements pseudopodium is projected out, which takes hold and afterwards contracts. During locomotion, the main part of body remains lifted from the substratum and supported on tips of pseudopodia. Thus *Amoeba* virtually walk on pseudopodia just as the animals move on their legs. *Dellinger* pointed out that *Amoeba* can walk on ceiling as well as floor. This theory is also called *Dellinger's Contraction Theory*. This theory is unable to explain how the pseudopodia are formed.

Sol-gel theory

This theory was first proposed by *Hyman* (1917). Later it was supported by *Pantin* (1923-26) and *Mast* (1925). This is the most acceptable theory now-a-days. According to this theory, the pseudopodia are formed by change of cytoplasm from gel to sol and sol to gel. According to *Mast* amoeboid movement is brought about by four process:

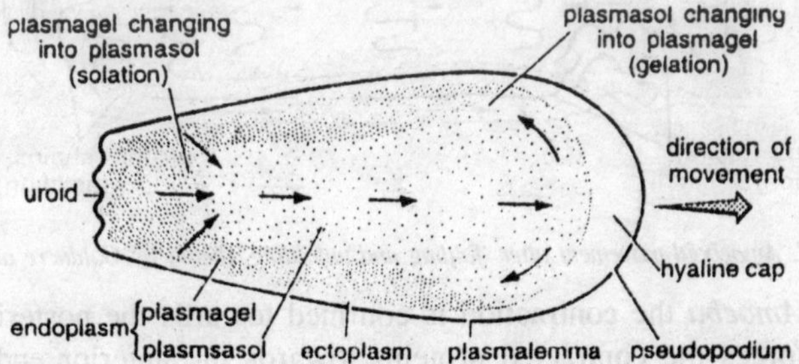

Fig. 11.6. Amoeboid movement after 'sol and gel' theory by Mast.

(i) Plasmalemma attaches to the substratum.

(ii) Plasmasol moves in the direction of locomotion and comes out by rupturing the plasmalemma at the weakest point. In this newly formed pseudopodium, the plasmasol now changes into the plasmagel due to which a strong gelatinous tube is formed in pseudopodium.

(iii) At the same time, the plasmagel of hinder end keeps on changing into the plasmasol. This sol flows towards the newly formed pseudopodium so much so that the whole of body cytoplasm comes into it.

(iv) Now the plasmagel tube contracts and the body moves forwards. Soon after this, a new pseudopodium is again formed in this direction and all above processes are performed repeatedly due to which *Amoeba* moves forward.

Folding and unfolding theory

This theory is proposed by *Goldacre* and *Lorch* (1950). According to them, "There is no doubt that gelation and solution of cytoplasm occur, but it is unlikely that contraction upon gelation would supply the force capable to moving an *Amoeba*." They state that all proteins gelate when their molecules unfold and the solate when their molecules fold. In the fluid endoplasm the protein molecules lie folded compactly, these molecules unfold at the tip of the advancing pseudopodia to form a layer of straightened and attached molecules. Posteriorily the protein molecules begins to fold again and they impart a contraction force.

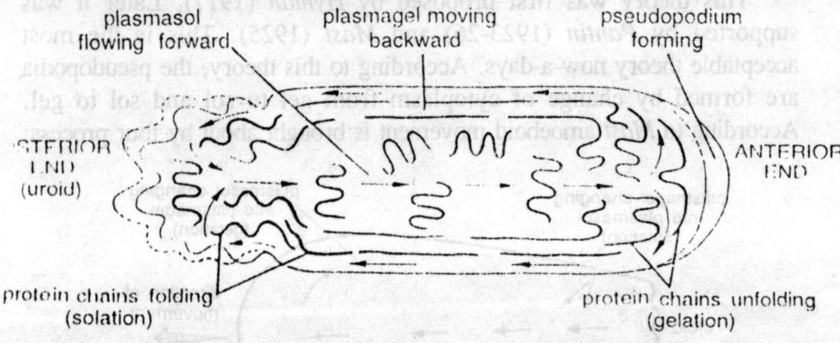

Fig. 11.7. Amoeboid movement after 'folding and unfolding' theory by Goldacre and Lorch.

In *Amoeba* the contraction is confined towards the posterior side which forces the contracted proteins towards the anterior end. They compared contactibility in amoeboid movement with that found in muscle-fibres. It is supposed that the energy for the movement of *Amoeba*, folding and unfolding of protein molecules is provided by *adenosine triphosphate* (ATP).

Fountain-zone contraction theory

Allen (1962) found that on molecular level, amoeboid movement is actually a type of slow contraction similar in many ways to muscle

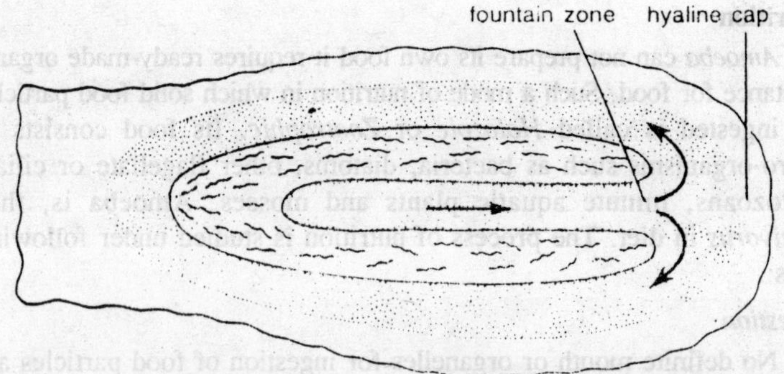

Fig. 11.8. Fountain-zone.

contraction. Endoplasm is regarded to contain long chains of protein which contract at the anterior end. The endoplasm (sol) chains are in contracted state. The protein chains become extended at the posterior end, where the ectoplasm is liquified during its conversion to endoplasm. Thus according to this theory, the body of *Amoeba* is virtually pulled forward by the contraction at the anterior end.

Contraction-hydraulic theory

Rinaldi and *Jahn* (1963) has supported the theory proposed by *Mast* in an improved form. According to them the contraction of posterior plasmagel is responsible for hydraulic pressure on plasmagel. The resistance to the pressure is least anteriorly. The plasmongel always remains under-continuous tension and localized reduction in the elastic strength is responsible for the formation of pseudopodium.

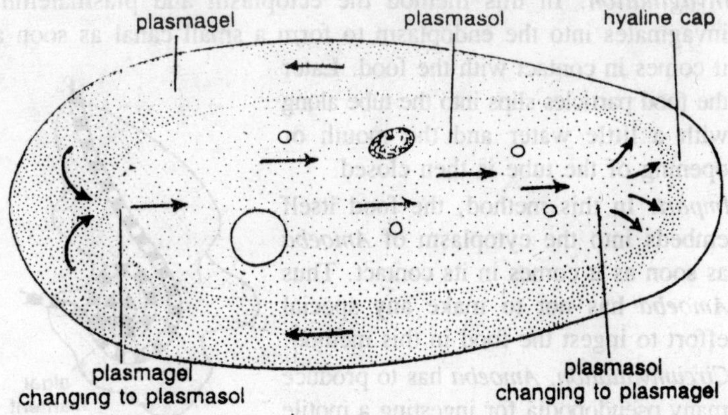

Fig. 11.9. Diagram illustrating the amoeboid movement on the basis of hydraulic theory of Rinaldi and Jahn.

Nutrition

Amoeba can not prepare its own food it requires ready-made organic substance for food. Such a mode of nutrition in which solid food particles are ingested is called *Holozoic* or *Zootrophic*. Its food consists of micro-organisms such as bacteria, diatoms, other flagellate or ciliate protozoans, minute aquatic plants and mosses. Amoeba is, thus *omnivorus* in diet. The process of nutrition is studied under following steps:

Ingestion

No definite mouth or organelles for ingestion of food particles are present. The food captured by pseudopodia. Pseudopodia are formed at the points, where the food in contact with the surface of the body. According to *Rhumbler* (1930) following methods of ingestion are employed according to nature of food particles:

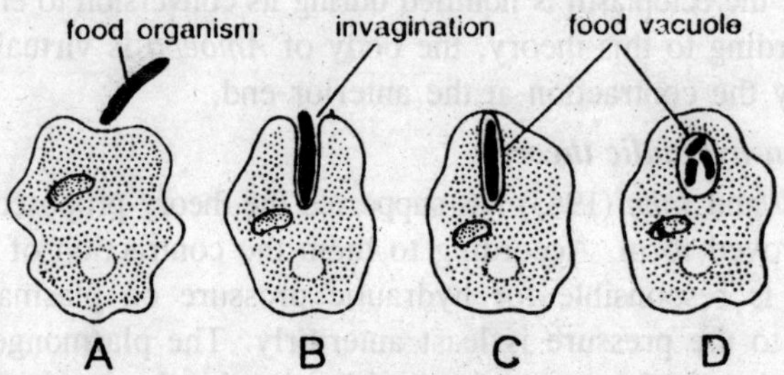

Fig. 11.10. Amoeba ingesting food by invagination.

1. *Invagination*. In this method the ectoplasm and plasmalemma invaginates into the endoplasm to form a small canal as soon as it comes in contact with the food. Later the food particles slips into the tube along with a little water and the mouth or opening of the tube is then closed.
2. *Import*. In this method, the food itself embeds into the cytoplasm of *Amoeba* as soon as it comes in its contact. Thus *Amoeba* has not to make any special effort to ingest the food in this method.
3. *Circumvallation*. *Amoeba* has to produce many pseudopodia for ingesting a motile food particle like ciliates and flagellates. When it approaches near food-particle

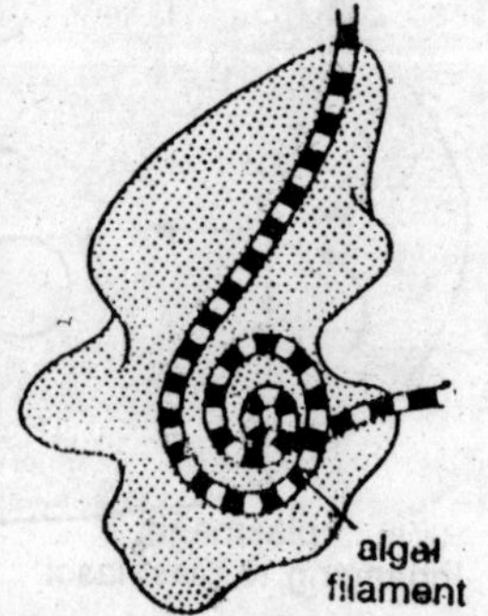

Fig. 11.11. Amoeba ingesting an alga by import.

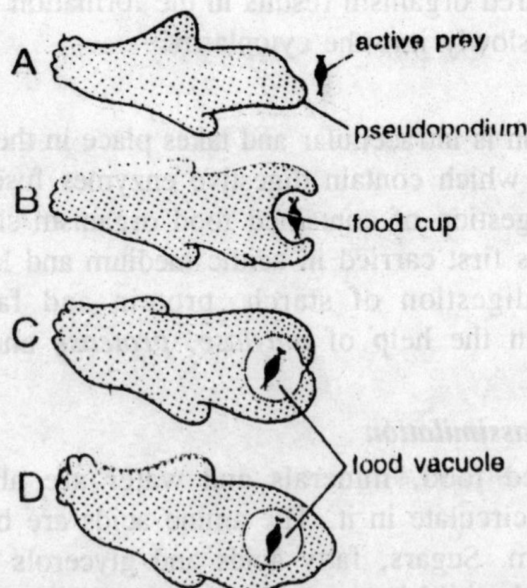

Fig. 11.12. Amoeba ingesting a rotifer by circumvallation.

the part immediately in line with it stops moving, and pseudopodia form a cup-like structure, the *food-cup* around the food-particle. The rim of food-cup narrows and soon closes on the other side of food-particle. The food-particle, thus, gets into the body. Alongwith food-particle a droplet of water is also engulfed. This water droplet surrounds the food-particle and forms the *food vacuole*. The food vacuole is bounded by plasmalemma.

4. *Circumfluence*. This method is applied by *Amoeba* when the prey is less-active or motionless. It extends its pseudopodia around the organism and envelops it completely with cytoplasm. The enclosing

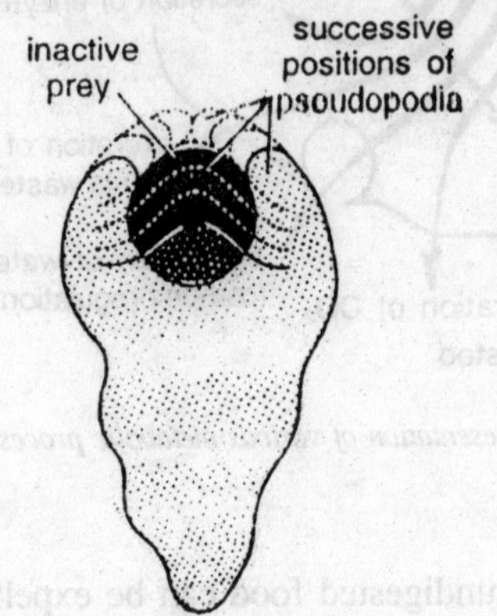

Fig. 11.13. Amoeba feeding by circumfluence.

of the captured organism results in the formation of food vacuole which gets slowly into the cytoplasm.

Digestion

The digestion is intracellular and takes place in the food vacuoles. The lysosomes, which contain digestive enzymes fuse with the food vacuoles and digestion of contained food organism slowly proceeds. The digestions is first carried in acidic medium and latter in alkaline medium. The digestion of starch, protein and fats takes place respectively with the help of *amylase*, *protease* and *lipase* in the alkaline medium.

Absorption and assimilation

The digested food, minerals and water are absorbed by the protoplasm and circulate in it. The amino acids are built up to form living protoplasm. Sugars, fatty acids and glycerols provide energy and are also synthesized into storage products, like glycogen and fats.

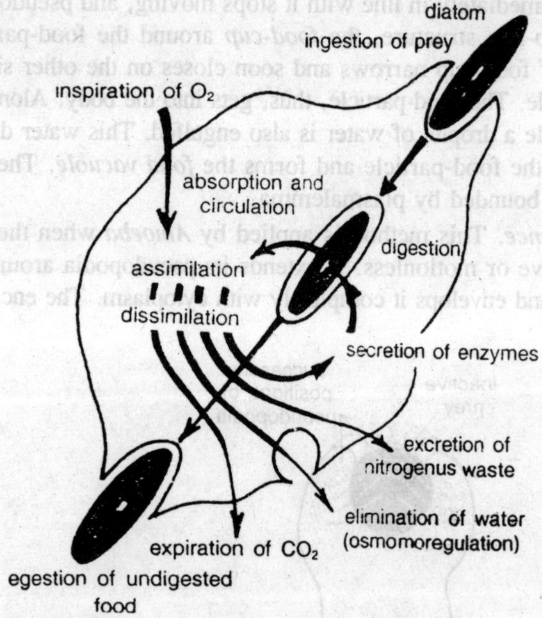

Fig. 11.14. Diagrammatic representation of various metabolic process going on in a living Amoeba.

Egestion

After digestion, the undigested food can be expelled out from any part of body because there is no definite aperture or *cytopyge* for this

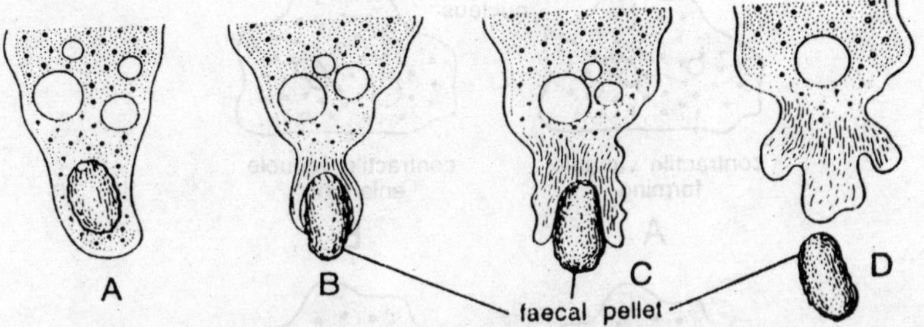

Fig. 11.15. Various stages of egestion.

purpose. The food vacuoles are shifted backwards and finally come in contact with plasmalemma at the posterior end. The plasmalemma ruptures at this point and the undigested food goes out as the animal moves ahead. The plasmalemma soon gets repaired to prevent the outflow of cytoplasm.

Respiration

Respiration in *Amoeba* is aerobic and takes place through general body surface by diffusion, as there is no special respiratory organ or pigments. Oxygen constantly diffuses into the cytoplasm for its concentration in water is always higher than in the body. This oxygen is used in the oxidation of food-stuffs like carbohydrates, fats and even protein to obtain energy. As a result water, carbon dioxide etc. are formed. The carbon dioxide is diffused out in the surrounding water.

Excretion

The nitrogenous waste products consist of compounds of ammonia such as urea and occasionally urates. In the dissolved state, they diffused out from the general body surface. It has been recently claimed that the crystals called *biurets* or *bipyramidal* and *triurets* or *tripyramidal* found in cytoplasm. These crystals are formed of an excretory substance, carbonyl diurea. Some of these crystals are removed with undigested material and some are redissolve and diffused out. Contractile vacuole also helps removing liquid excretion to some extent.

Osmoregulation

Since the cytoplasm of *Amoeba* is hypertonic to the surrounding medium, water continuously enters the body by *endosmosis*. Some of the water also comes in cytoplasm from body vacuoles. If all this water is allowed to accumulate, the body will soon swell and burst. So to regulate the amount of water in cytoplasm the excessive water

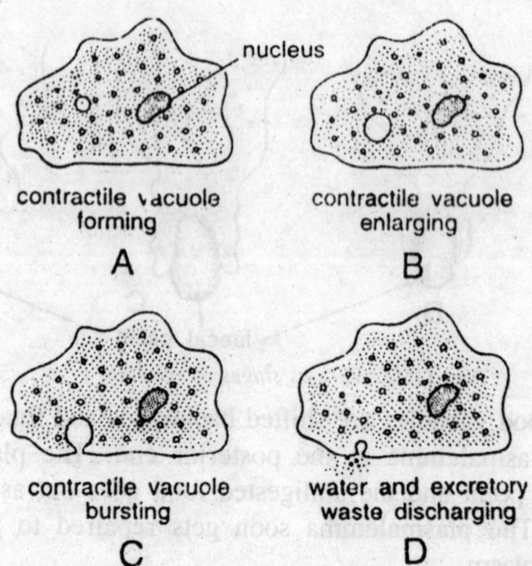

Fig. 11.16. Amoeba showing various stages of osmoregulation and excretion.

is eliminated out from the body and this is brought about by the contractile vacuole. The contractile vacuole regularly pumps out excess water. Its working is divisible into two phases-diastole and systole. The growing phase up to the maximum size reached is called *diastole* and the collapsed condition is referred to as *systole*. How the water is forced out of the protoplasm into the vacuole is not known. There are three theories for the growth of contractile vacuole:

(i) *Secretion theory*. According to this theory the water is first absorbed in the vacular membrane and then secreted into the vacuole.

(ii) *Osmosis theory*. According to this theory, the water passes into the growing vacuole by the simple process of osmosis.

(iii) *Filteration theory*. According to this view the water filters through the vacular membrane into the vacuole because of the hydrostatic pressure of the endoplasm.

It has been found that the growth of the vacuole takes place in spurts and not steadily. From this it is concluded that the fluid enters the vacuole by the fusion of similar smaller vacuoles. On attaining the maximum size (diastole) the contractile vacuole migrates to the surface until it reaches the plasmalemma which subsequently bursts discharging the contents (systole). It is possible that the expelled water may contain some excretory products, but there is no experimental evidence to show such function.

Behaviour

Amoeba responds to both external and internal stimuli. Responses of *Amoeba* to the stimuli constitute *behaviour* or *irritability*. *Amoeba* responds to stimuli mainly by locomotary behaviour. When stimulated,

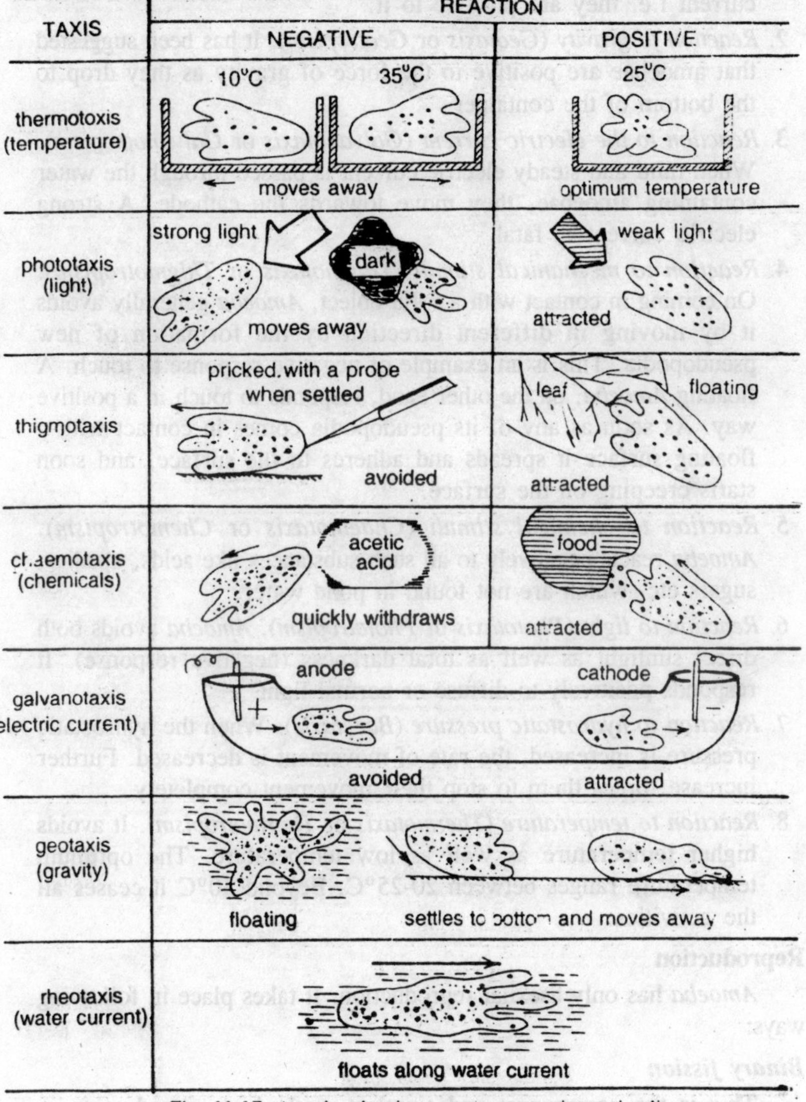

Fig. 11.17. Amoeba showing reaction to various stimuli.

it either continues advancing towards the source of stimulation (positive response), or moves away from it (*negative response*). Its responses are as follows:

1. *Reaction to water current* (*Rheotaxis* or *Rheotropism*). When floating freely in water the amoebae place in line with the water current i.e. they are positive to it.
2. *Reaction to gravity* (*Geotaxis* or *Geotropism*). It has been suggested that amoebae are positive to the force of gravity as they drop to the bottom of the container.
3. *Reaction to the electric current* (*Galvanotaxis* or *Galvanotropism*). When mild and steady electric current is passed through the water containing amoebae, they move towards the cathode. A strong electric current is fatal.
4. *Reaction to mechanical stimuli* (*Thigmotaxis* or *Thigmotropism*). On coming in contact with a solid object, *Amoeba* generally avoids it by moving in different direction by the formation of new pseudopodia. This is an example of negative response to touch. A floating *Amoeba*, on the other band, responds to touch in a positive way. As soon as any of its pseudopodia comes in contact with a floating surface it spreads and adheres to the surface, and soon starts creeping on the surface.
5. *Reaction to chemical stimuli* (*Chaemotaxis* or *Chemotropism*). *Amoeba* reacts negatively to all such substances like acids, alkalies, sugars etc. which are not found in pond water.
6. *Reaction to light* (*Phototaxis* or *Phototropism*). *Amoeba* avoids both direct sunlight as well as total darkness (negative response). It responds positively to diffuse or normal light.
7. *Reaction to hydrostatic pressure* (*Barotaxis*). When the hydrostatic pressure is increased, the rate of movement is decreased. Further increase forces them to stop their movement completely.
8. *Reaction to temperature* (*Thermotaxis* or *Thermotropism*). It avoids higher temperature as well as low temperature. The optimum temperature ranges between 20-25°C, Beyond 30°C it ceases all the activities.

Reproduction

Amoeba has only asexual reproduction. It takes place in following ways:

Binary fission

This is the commonest and simplest method of reproduction in *Amoeba*. In this, the entire body divides, like an ordinary cell, into two daughters by mitosis. It occurs during favourable conditions and takes about half an hour at 24°C. For long such a division was thought to be amitotic, but it has now been shown that the *Amoeba* divides mitotically. Such a mitosis is called *Cryptomitosis* which, however, involves the usual four phases (Prophase, metaphase, anaphase and telophase).

1. *Anaphase*. It is marked by a change in the external form, the pseudopodia become course and thick. The two sets of chromosomes move apart almost entirely by the enormous elongation of the inter-chromosomal region. Anaphase lasts for about 10 minutes.
2. *Telophase*. In telophase, pseudopodia assume normal shape; the body first elongates, then constricts in the middle. Two sets of chromosomes reach their poles where each set in enveloped by a nuclear membrane. The construction finally divides the *Amoeba* into two daughters each having a daughter nucleus. The contractile vacuole is retained by one of the daughter, the other one acquires a new vacuole. The telophase takes 8 minutes. Daughter amoebae may begin to divide after 24 hours.
3. *Prophase*. During prophase the body becomes spherical, studded with short, blunt-pseudopodia and very small sized spherical chromosomes make their appearance. Nuclear membrane and

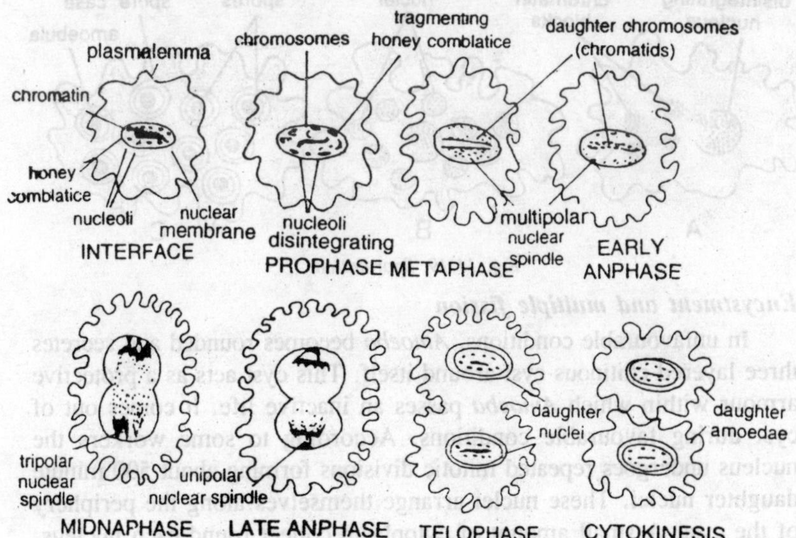

Fig. 11.18. Amoeba showing binary fission.

endosome remain intact, working of contractile vacuole slows down. The prophase lasts for 10 minutes.

4. *Metaphase*. The chromosomes arrange at the equator of the spindle-shaped nucleus to form a *metaphase plate*. The nucleus becomes larger endosome breaks up into pieces and the nuclear membrane breaks down. The chromosomes segregate into two sets. The metaphase lasts for about 4-6 minutes.

Sporulation

It occurs when the conditions become unfavourable. Before the conditions become unbearable, the nucleus divides repeatedly and mitotically into nuclei without the breaking up of the nuclear membrane. Such a nucleus which contains several sets of chromosomes is called *polyenergid nucleus*. Now the nuclear membranes breaks and chromosomes are set free, each set forming a daughter nucleus. Now, the cytoplasm of parent body segregates into small masses, each surround a daughter nucleus. A spore case is formed around this cytoplasm. Thus, about 200 spores are formed. *Amoeba* sinks and disintegrates liberating the spores. On return of favourable conditions spore case bursts open by absorbing water, and a small amoeba is liberated from each spore.

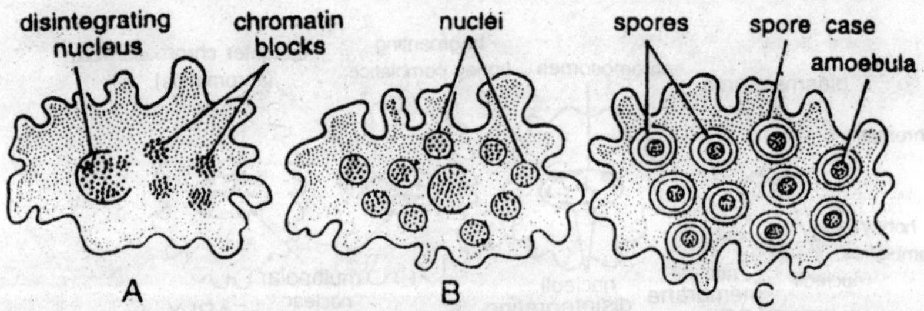

Fig. 11.19. Sporulation.

Encystment and multiple fission

In unfavourable conditions, *Amoeba* becomes rounded and secretes three layered chitinous cyst around itself. This cyst acts as a protective armour within which *Amoeba* passes an inactive life. It comes out of cyst during favourable conditions. According to some workers the nucleus undergoes repeated mitotic divisions forming about 500 minute daughter nuclei. These nuclei arrange themselves along the periphery of the cyst. A small amount of cytoplasm collect round each nucleus, forming a daughter *Amoeba* called *amoebula* or *pseudopodiospore*. When

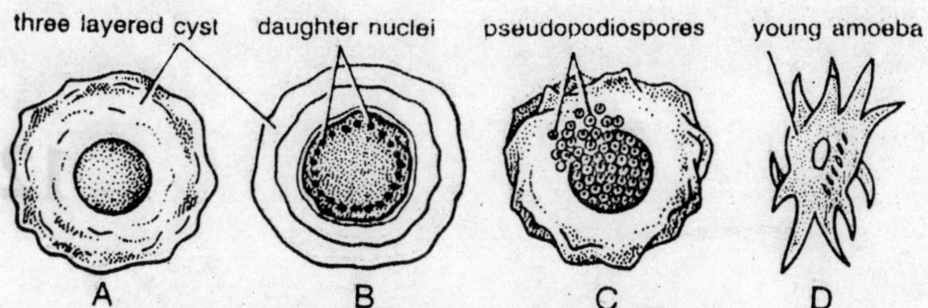

Fig. 11.20. Multiple fission during encysted condition.

favourable conditions arrive, the cyst breaks off liberating *amoebulae*, each with fine pseudopodia. They lead an active life and grow into adults in due course of time. According to modern workers, multiple fission in encysted *Amoeba* is doubtful.

Conjugation

According to some workers, sometimes two amoebae conjugate for sometime and then separate to lead their independent lives. Although there is no definite evidence for conjugation.

Regeneration

Amoeba has a great power of regeneration. If it is cut into small pieces, each piece, containing a part of nucleus, rapidly regenerate into a complete *Amoeba*. A piece without nuclear pat, does not regenerate and soon dies.

12

LIFE OF PARAMECIUM

Paramecium is the widely studied type of the class-Ciliata. The members of class Ciliata are much more developed than the other protozoans. In spite of their bring acellular, special organelles are found in them for ingestion, excretion etc. Thus division of labour is clearly seen in ciliates.

SYSTEMATIC POSITION

Phylum	—	Protozoa
Subphylum	—	Ciliophora
Class	—	Ciliata
Subclass	—	Holotrichia
Order	—	Hymenostomatida
Family	—	Paramecidae
Genus	—	*Paramecium*

Habits and Habitat

Paramecium is cosmopolitan in distribution. It is found in fresh water lakes, pounds, puddles, sewage pipes and rice fields. It is abundant in stagnant water containing decaying organic matter. It is omnivore in diet and reproduces in a variety of ways, which are asexual and sexual both. There are 10 species of *Paramecium*. They differ in shape, size, number of micronuclei etc. The common species are *Paramecium caudatum*, *P. aurelia*, *P. bursaria* and *P. multimicronucleatum*.

Culture

For general laboratory purpose, *P. caudatum* can be grown satisfactorily on a boiled hay medium to which wheat or rice grains have been added. It has also been maintained on several complex sterile media and simplest of which the yeast juice. Paramecia are easily grown in chalky medium (NaCI-80mg, $NaHCO_3$-4mg, $CaCI_2$-4mg, KCI-4mg and CaH_4 $(PO_4)_2$ H_2O-1.6 mg. all are dissolved in one litre of distilled water). Add 6-8 drops of skimmed milk weekly. The jar is kept away from direct light. Other less uniform but very satisfactory method of growing paramecia include a monofloral suspension of bacterium, *Pseudomonas ovalis*, and a suspension of pure, living yeast.

Morphology

Shape and Size

The first description of *Paramecium* was given by *Christian Huygens* (1678). The term *Paramecium* was coined for the first time by *John*

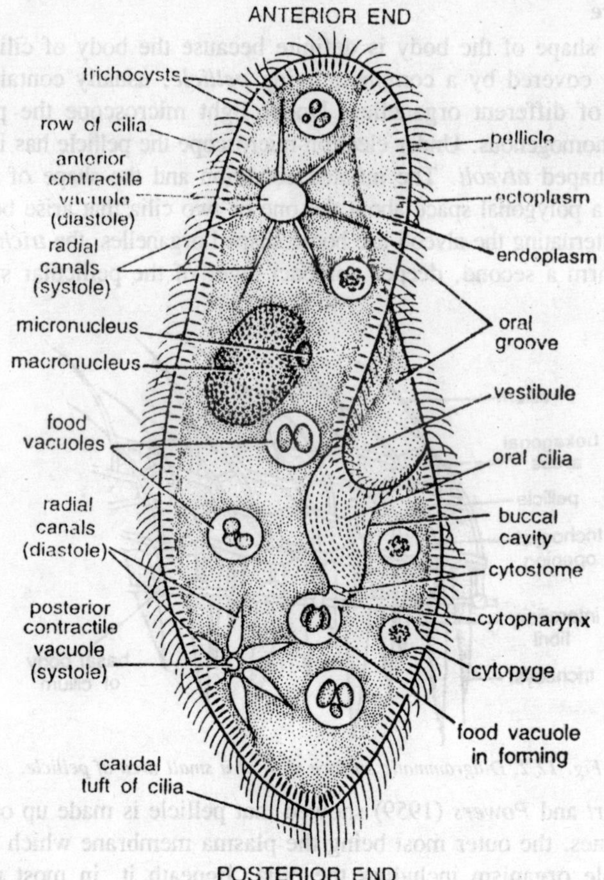

Fig. 12.1. Paramecium caudatum.

Hill (1752) from the Greek word *paramekos* meaning oblong. *Paramecium* in general appearance appears like the sole of the slipper, hence it is popularly known as "*slipper-animalcule.*" One species was named *P. caudatum* (*Joblot*) because of its resemblance to an imprint of human foot. They vary in size. The largest species *P. caudatum* measures 180-300 μ in length; *P. aurelia* is 170-250 μ and *P. trichium* only 62 μ in length.

Colour

Some species are somewhat translucent and colourless, others are vary from light gray, white to pale yellow. *P. bursaria* is green because of the presence of *Zoochlorellae* in the endoplasm.

Structure

The shape of the body is definite because the body of ciliates is typically covered by a complex, living *pellicle*, usually containing a number of different organelles. Under light microscope the pellicle appears homogenous. Under electron-microscope the pellicle has inflated kidney-shaped *alveoli*. The inflated condition and the shape of alveoli produce a polygonal space about the one or two cilia that arise between them. Alternating the alveoli are bottle shaped organelles, the *trichocysts*, which form a second, deeper compact layer of the pellicular system.

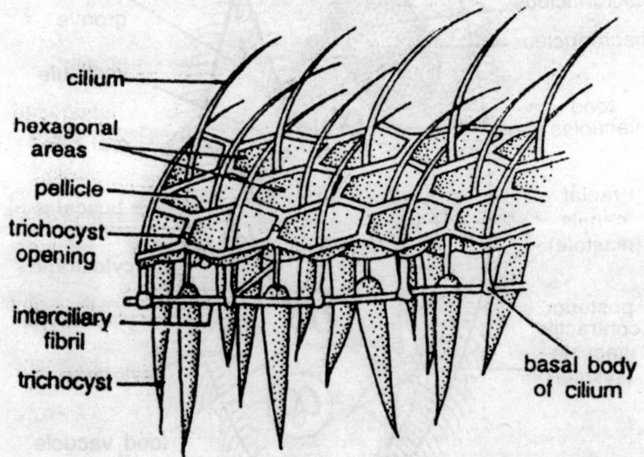

Fig. 12.2. Diagrammatic surface view of a small area of pellicle.

Ehert and *Powers* (1959) showed that pellicle is made up of three membranes, the outer most being the plasma membrane which covers the whole organism including the cilia. Beneath it, in most places, there are two further membranes, the *outer* and *inner membranes* the

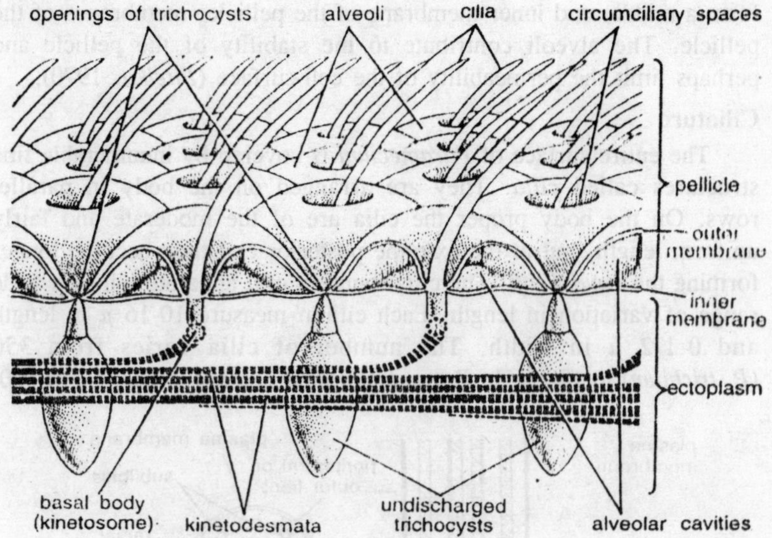

Fig. 12.3. A diagrammatic three-dimensional electron microscopic representation of a portion of pellicle and infraciliary system.

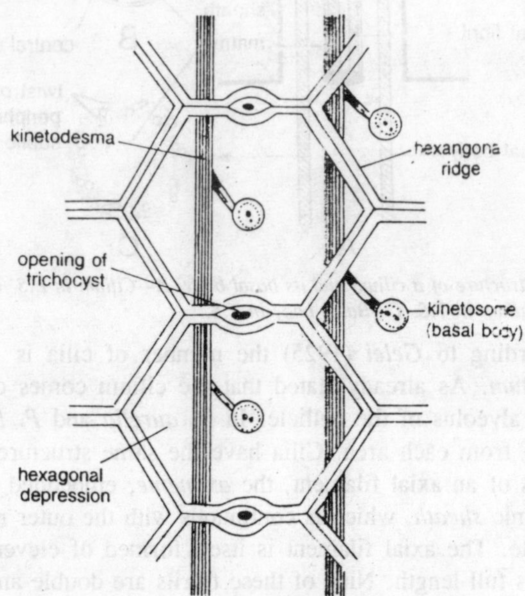

Fig. 12.4. Surface view of a part of Paramecium.

outer and inner membranes bounding a flattened alveolus would thus form a middle and inner membrane of the pellicle. membranes of the pellicle. The alveoli contribute to the stability of the pellicle and perhaps limit the permeability of the cell-surface (*Pitelka*, 1970).

Ciliature

The entire surface of *Paramecium* is covered by innumerable fine structures called *cilia*. They are arranged on the body in parallel rows. On the body proper the cilia are of the moderate and fairly uniform length, but at the extreme posterior end they become longer forming the *caudal tuft*. The cilia found in oral apparatus show a wide range of variation in length. Each cilium measure 10-16 μ in length and 0.1-2 μ in width. The number of cilia varies from 350 (*P. trichium*) to 18000. In *P. caudatum* their number is 10,000- 14,000.

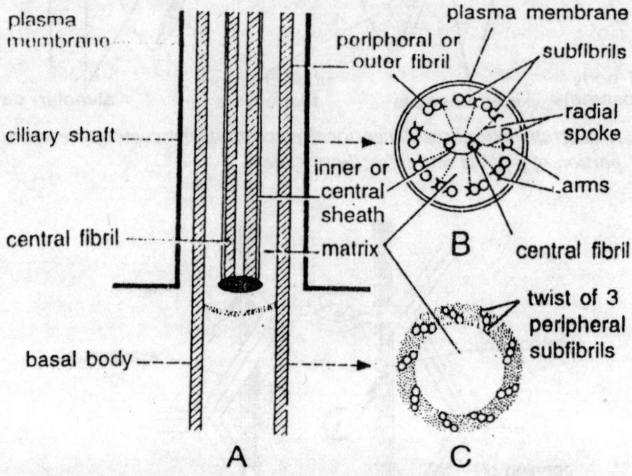

Fig. 12.5. Structure of a cilium and its basal body. A—Cilium in L.S. B—Free part of cilium in T.S. C—Basal body in T.S.

According to *Gelei* (1925) the number of cilia is 18000 in *P. nephridiatum*. As already stated that the cilium comes out from the centre of alveolus of the pellicle. In *P. aurelia* and *P. bursaria* two cilia arise from each area. Cilia have the same structure as flagella. It consists of an axial filament, the *axoneme*, embedded in an elastic protoplasmic *sheath*, which is continuous with the outer membrane of the pellicle. The axial filament is itself formed of eleven fibrils that extends its full length. Nine of these fibrils are double and occur in a ring called *peripheral fibres*. Each is composed of two adhering subfibres, one of which is provided with a pair of arms all disposed in

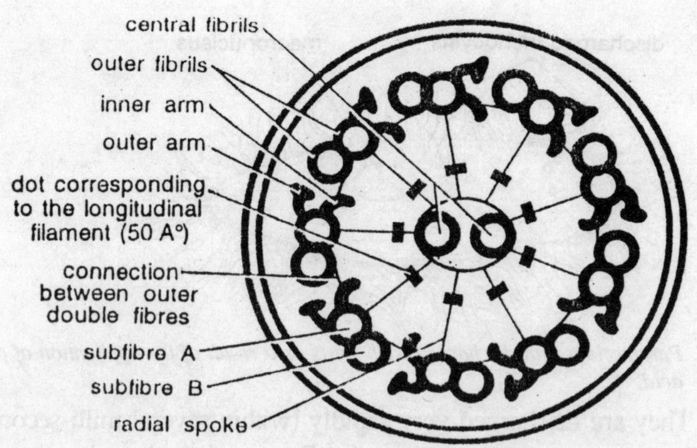

Fig. 12.6. Diagrammatic transverse section of a cilium.

same direction. The remaining two are *central fibres* which are independent surrounded by a thin membranous *central sheath*. Nine radially directed fibres are disposed between the peripheral fibres and the inner sheath like spokes of a wheel. The peripheral fibres continued below in to the basal body or *Kinetosome*.

Cytoplasm

The cytoplasm is differentiated into an outer ectoplasm and inner endoplasm.

Ectoplasm

It is clear and dense part of the cytoplasm which forms its outer layer below the pellicle. In the ectoplasm occurs following three important structures:

Trichocysts

Trichocysts are peculiar, rod-like or oval organelles characteristic of many ciliates. They are oriented at right angles to the body surface being located at the centre of the anterior and posterior walls of the pellicular polygons. The trichocysts are small in size only 4 μ in length. They lie a little (1-2 μ) below the pellicle to which each is attached by a vary delicate fibre. An undischarged trichocyst is a *sac* containing a *swelling substance* (the quelkorper) and closed by a cap. Just within the cap is a conical *spike* resting on a *basis*.

The discharged trichocyst consists of a long, striated, thread-like shaft surmounted by a barb. The shaft is not evident in the undischarged trichocysts and is probably polymerized in the process of discharge. The trichocysts are discharged on mechanical, chemical or electric

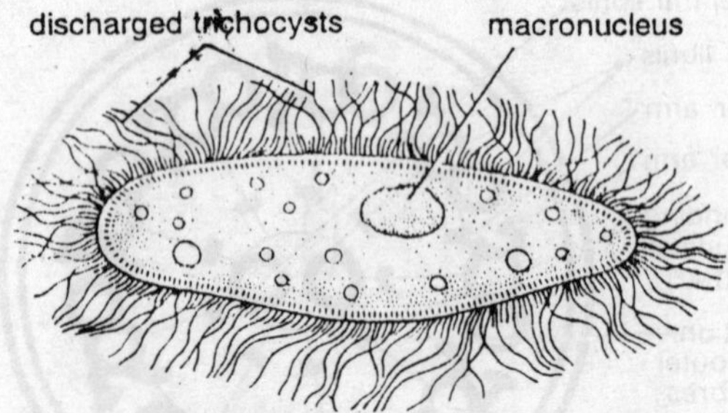

Fig. 12.7. Paramecium with discharged trichocysts as a result of the application of picric acid.

stimuli. They are discharged very rapidly (within several milli-seconds). The trichocysts are *non-toxic* type in *Paramecium* thus they are not defensive organelle. Probably they may be used in anchoring the animal when feeding. Area in the neighbouring of the nucleus is believed to the centre of origin of trichocysts. Recently, it has been established that trichocysts are produced by the kinetosomes. A kinetosome first ejects a *trichocyst granule* or *tricystosome* which forms a budding that develop into a trichocyst.

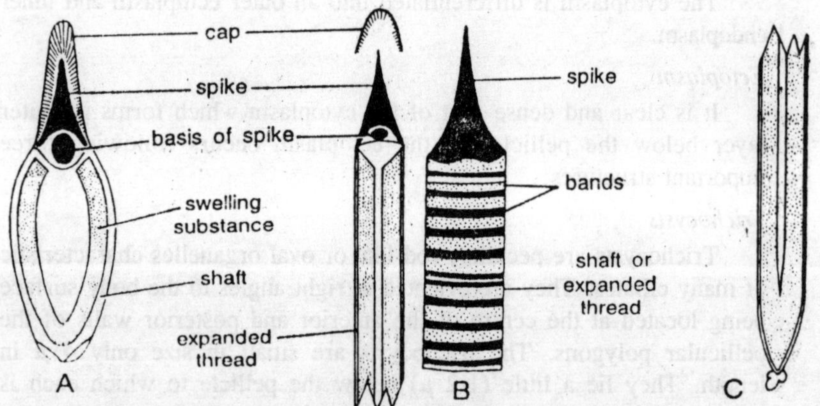

Fig. 12.8. Trichocyst. A—L.S. of hypothetical undischarged trichocyst; B—Tip of trichocyst as seen in electron microscope; C—Discharged trichocyst.

Infra-ciliary system

As mentioned earlier, such cilium arises from a basal body or kinetosome, located in the alveolar layer. The kinetosomes that form a particular longitudinal row are connected by means of fine, striated fibrils, called *kinetodesma*. The kinetosomes plus the fibrils of that

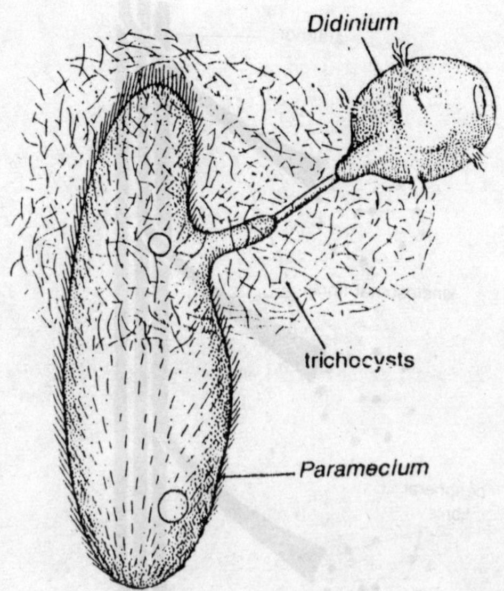

Fig. 12.9. Throwing out trichocysts when attacked by a small Didinium.

row make up a *kinety*. The longitudinal bundle of fibrils run to the right side of the row of kinetosomes, and each kinetosome gives rise to one kinetodesoms (fibril), which joins the longitudinal bundle and extends anteriorly. A single kinetodesmos is tapered and extends for varying distance, as a part of bundle. At the kinetosome, the kinetodesmos is connected to certain of the kinetosome triplets. A kinety system is characteristic of all ciliates although there are variations in details of the pattern. The kineties with their cilia constitute the neurolocomotor organelle. The kineties control and coordinate the beating of cilia.

The fibrillar system or Neuromotor apparatus

The fibrillar system is visible with the help of nuclear stains. According to *Lund* (1933) on the left dorsal wall of the cytopharynx at about the level of posterior margin of the cytostome is a very small, bilobed mass called *neuromotorium*. From neuromotorium, fibrils radiate into the endoplasm. Of these four or more usually pass almost to the dorsal body wall but rest are shorter and not definite in position. Their function is uncertain and apparently function as part of the cilium anchorage system. But the electron microscopic studies reveal only the presence of the alveoli, kinetosomes and kinetodesmata and there is no evidence regarding the existence of this system.

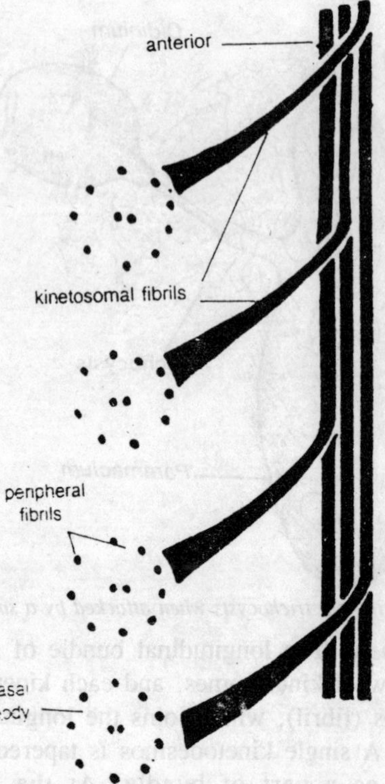

Fig. 12.10. A ciliary kinety.

Endoplasm

The endoplasm of *paramecium* is semi-fluid and less granular. It contains following structures:

Nucleus

Paramecium is *hereokaryotic* having two types of nuclei. The smaller one or the *micronucleus* is lodged in a depression at one side of the large nucleus called *meganucleus* or *macronucleus*. The meganucleus is bean or kidney shaped body lying appoximately in the centre. It is a compact structure, containing fine threads and tightly packed discrete chromatin granules of variable size. The meganucleus controls the vegetative activities of the animal. The micronuclei vary in number in different species. These are smaller structure and may be compact (*P. caudatum*) or vesicular (*P. aurelia*). *P. cardatum* and *P. bursaria* each has one micronucleus. *P. aurelia* has two and *P. multimicronuclealum* has 3-9 micronuclei. The micronuclei are concerned with reproduction.

Contractile vacuoles

In *Paramecium* there are two contractile vacuoles occupying somewhat fixed positions in the endoplasm. One vacuole is located at the posterior, and anterior end of the body. (In *P. multimicro-nucleatum* there are 2-7 contractlie vacuoles). The anterior vacuole is sometimes called the *nuclear vacuole* being near the nucleus and the posterior one is called the *peristomial vacuole* being in vicinity of peristome. Each contractile vacuole is surrounded by 5-12 *radiating canals*. Each radiating canal consists of (a) a tabular terminal part (b) an ampulla and (c) an injector canal. The injector canal opens into the vacuole. Electron Microscopy reveals that the outer or terminal part is surrounded by a network of fine coiled *nephridial tubules*.

The nephridial tubules collect water from the endoplasm and pass into the terminal part. This fluid is collected in the ampulla which becomes bulb-like when distended, but when the fluid passed out it is of same diameter as the terminal part. When the ampulla is fully distended (diastole) it collapses (systole) and the fluid is passed through the injector canal to form the contractile vacuole. Thus it is apparent that the systole of ampulla becomes the diastole of the vacuole. In *P. trichium* the contractile vacuole is without radial canals. The contractile vacuole opens outside through *discharge canal*. Bundles of fibres surround

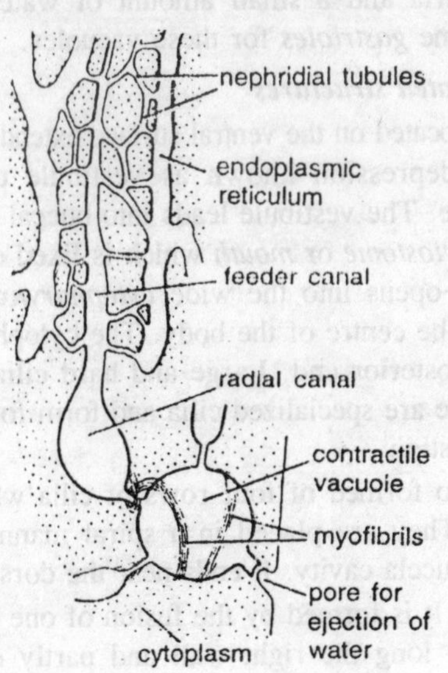

Fig. 12.11. Contractile apparatus.

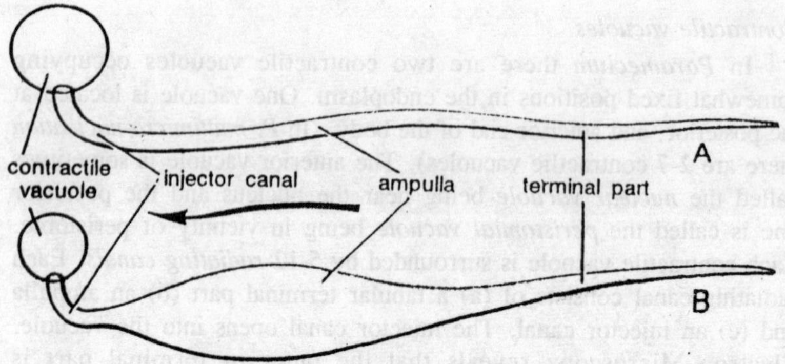

Fig. 12.12. Radial canal. A—Empty; B—Full.

the ampullae and the main vacuoles, and concerned in the production or control of the pressure required to control the vacuole. The contractile vacuoles pulsate at different rates depending on their position. The posterior vacuole faster than the anterior vacuole because of large amount of water being delivered into the posterior region by the cytopharyns.

Food vacuoles

These are roughly spherical, non contractile bodies varying in size and number lying in the endoplasm. They contain ingested food materials, mainly bacteria and a small amount of water. *Volkovsky* (1934) proposed the name *gastrioles* for these vacuoles.

Oral groove and associated structures

The oral groove is located on the ventral surface extending backward into a funnel shaped depression known as vestibule consisting of invaginated body pellicle. The vestibule leads into buccal cavity which directly leads into the *cytostome* or *mouth* which is fixed oral opening. The cytostome directly opens into the wide *cytopharynx* or *Pharynx* which extends towards the centre of the body. The cytopharynx forms a food vacuole at the posterior end. Large and hard cilia are present in these structures. These are specialized cilia and form four organelles which help in food ingestion:

(a) *Quadrulus*. It is also formed of four rows of cilia which are less densely arranged. They are placed in a spiral manner along the dorsal wall of the buccla cavity. It ends near the dorsal peniculus.

(b) *Endoral membrane*. It is formed by the fusion of one row of cilia. It runs transversely long the right wall and partly encircles the opening of vestibule into cytostome.

(c) *Ventral peniculus.* It is also formed by four rows of cilia. It extends backwards to a distance on the left side of buccal cavity and ends near the cytostome.

(d) *Dorsal peniculus.* This is formed by four rows of densely arranged cilia which are placed in a spiral manner on the left side of buccal cavity. They end behind the cytostome.

The quadrulus and peniculi control food passage while endoral membrane guards the entrance of buccal cavity. A group of long cilia, *post buccal cilia* behind the lateral wall of cytopharynx.

Cytopyge

Just behind the cytostome on the ventro-lateral surface there is a minute pore called *cytopyge*. Through this cytopyge the undigested food particles are eliminated out.

PHYSIOLOGY

Locomotion

The locomotion in *Paramecium* takes by two methods:

1. Ciliary locomotion
2. Metaboly.

Ciliary locomotion

Paramecium swims in water by beating of its cilia and has put on speed because of development of a large number of these locomotary organs (10,000-18,000). It can swim very slowly, perhaps 5 metres per hour. A cilium performs whip-lash movement by which water is propelled. In its beat each cilium performs an *effective* and a *recovery stroke*. During the effective stroke the cilium is outstretched and moves from a forward to a backward position. In the recovery stroke the cilium is bent over to the right against the body and is brought back

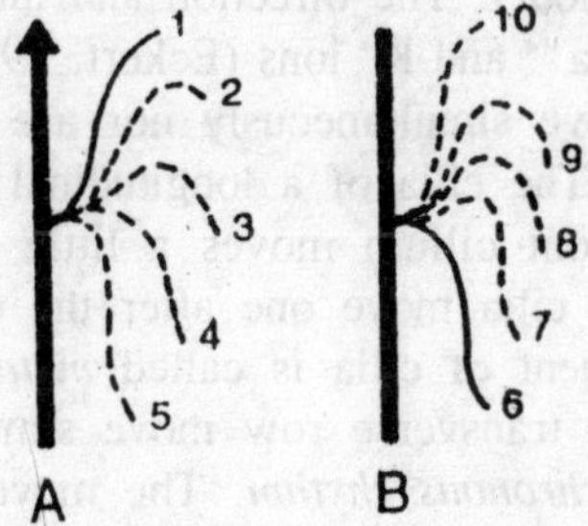

Fig. 12.13. Diagram illustrating ciliary movement of a single cilium. A—Effective stroke; B—Recovery stroke.

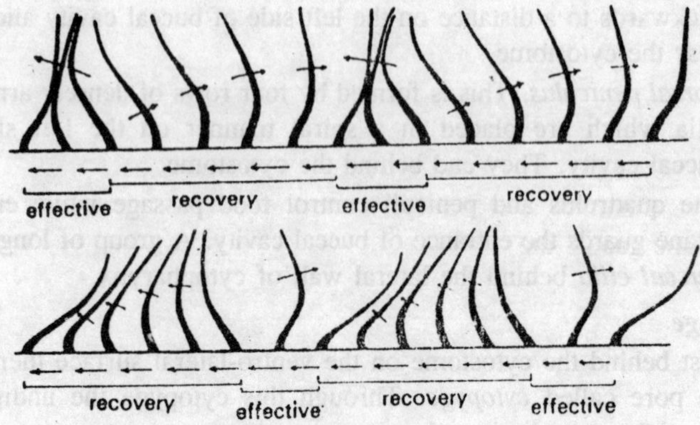

Fig. 12.14. Cilia indicating effective and recovery strokes.

to the forward position in a counterclock-wise movement. The recovery position offers less water resistance and is somewhat analogous to feathering an oar. The direction of effective stroke is oblique to the long axis of the body. This causes the animal to swim in a spiral path and at the same time to rotate on its longitudinal axis.

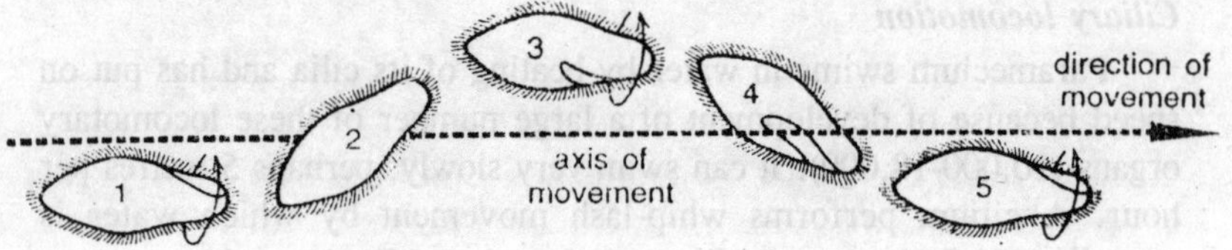

Fig. 12.15. Anticlockwise spiral path followed by a swimming Paramecium.

The ciliary beat can be reversed, and the animal can move backward. This backward movement is associated with the so-called avoiding reaction (behaviour). The direction and intensity of beat is controlled by levels of Ca^{++} and K^{+} ions (Eckert, 1972). All the cilia of body can neither move simultaneously nor are they capable of independent movement. The cilia of a longitudinal row oscillate in such a way that the front cilium moves a little earlier than the one behind it. Thus the cilia move one after the other from front backwards. This movement of cilia is called *metachronal rhythm*. However, the cilia of a transverse row move simultaneously. The movement is called *synchronous rhythm*. The movement of cilia is controlled by the infraciliary system and ATP plays an important role (*Serawin*, 1967).

Metaboly

The body of *Paramecium* possesses elasticity. It can pass through a passage narrow than its body, after which it regains its normal size. Such type of movement is called *metaboly*. This type of body movement is caused by the cytoplasm.

Nutrition

The free-living ciliates are almost entirely *holozoic*. It feeds on bacteria, small protozoan, yeast, algae etc, but its favourite food is *Bacterium coli*. It can devour 2-5 million *Bacterium* in 24 hours. Thus *Paramecium* is a selective feeder, showing considerable choice of food, and often rejecting undigestible non-nutritive particles.

Feeding Mechanism

It feeds while stationary or swimming slowly. In feeding, the cilia of the oral groove produce a current of water that sweeps in an arc-like manner down the side of the body and over the oral region. The food particles are sucked with water current into the oral groove and then to the vestibule. Ciliary tract of vestibule direct the food particles into the buccal cavity. Some of these particles are discarded and left out. Quadrulus and peniculi regulate the passage of food into the cytopharynx. At the bottom of cytopharynx the food particles are gathered and form a rounded or ball-like mass, the *food vacuole*. It

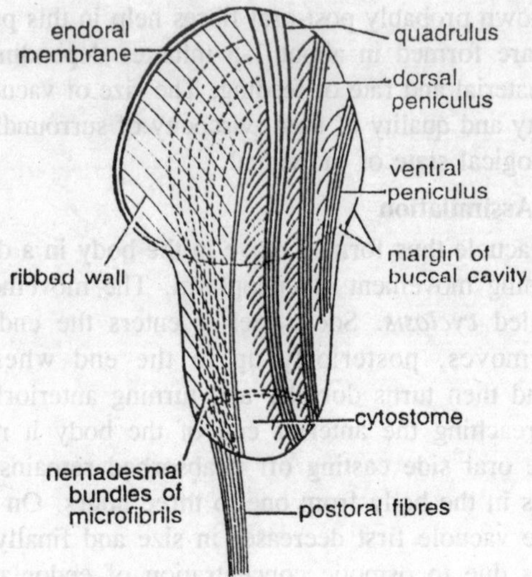

Fig. 12.16. Oral apparatus showing buccal ciliature.

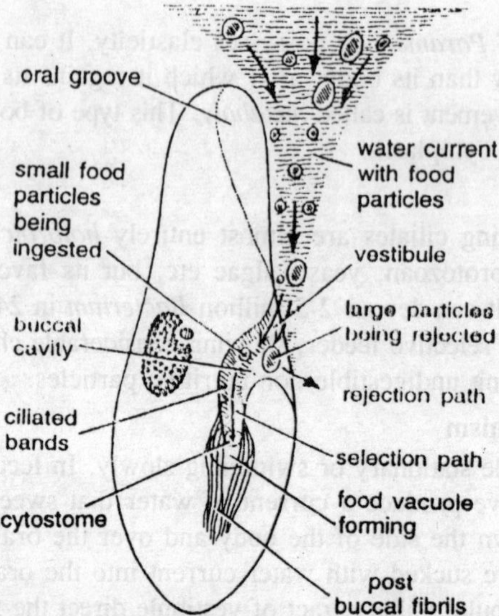

Fig. 12.17. Paramecium receiving food particles with water current drawn into buccal cavity by ciliary action.

grows rapidly and when the vacuole attains a maximum size it is pinched off and takes its course in endoplasm. The food vacuole besides food particles also contains a little amount of water.

The exact mechanism by which the food vacuole is liberated is not properly known probably post-oral fibres help in this process. The food vacuoles are formed in about 1-5 minutes depending upon the abundance of material and rate of feeding. The size of vacuoles depend upon the quantity and quality of food, viscocity of surrounding medium and the physiological state of individual.

Digestion and Assimilation

The food vacuole thus formed move in the body in a definite path with the streaming movement of endoplasm. The movement of food vacuoles is called *cyclosis*. Soon after it enters the endoplasm the food vacuole moves, posteriorly up to the end where it stops momentarily and then turns dorsally and turning anteriorly it moves forwards. On reaching the anterior end of the body it moves back again along the oral side casting off unabsorbed remains. The food vacuole remains in the body from one to three hours. On leaving the cytopharynx the vacuole first decreases in size and finally increases. The decrease is due to osmotic concentration of endoplasm as it is

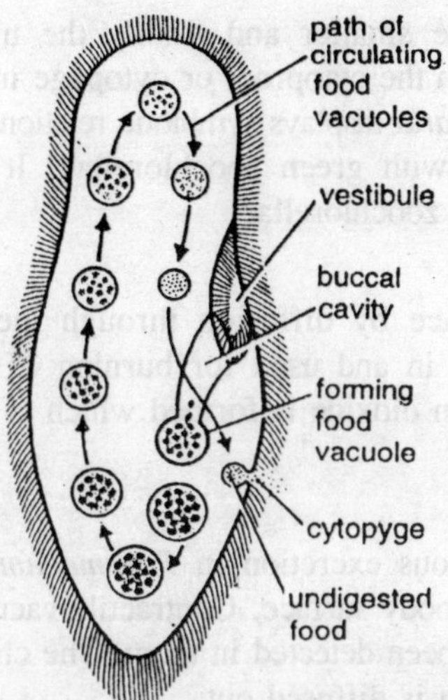

Fig. 12.18. Paramecium showing cyclosis and path of food vacuoles in endoplasm.

higher than the vacuole, the water is lost from the vacuole and it becomes small.

The digestion of food takes place in the food vacuoles in which the medium is first acidic and later alkaline. As a result of digestion the proteins are hydrolysed into amino acids by *proteases*, carbohydrates into glucose by *carbohydrases*, and probably fat is also digested. The process of cyclosis can be demonstrated experimentally in *Paramecium*. If milk, coloured with *Congo Red* stain, is given to *Paramecium*, the fat particles of milk first become red due to the acidic reaction of the enzymes in food vacuoles, then they gradually turn purple and finally blue due to alkaline medium in the food vacuoles.

Complete digestion takes place mainly during the later alkaline phase of food vacuole. The digested food materials are thoroughly distributed to all parts of body during cyclosis. Reserve food material is glycogen and fat-droplets which remain scattered throughout the endoplasm. If *Paramecium* does not get food, it feeds upon the reserve food materials in the endoplasm. When even these are exhausted the macronucleus becomes smaller, the animal becomes transparent, vacuoles appear in the cytoplasm, cilia disappear and degeneration of specialized structures set in. This phenomenon is called *inanition*. Its restores it normal shape when supplied with rich food such as bacteria.

Egestion

The vacuoles become smaller and finally the undigested food particles are released from the cytoproct or cytopyge into surrounding medium. *Paramecium bursaria* displays symbiotic relationship with algae. The endoplasm is filled with green zoochlorellae. It takes its food material manufactured by zoochlorellae.

Respiration

Respiration takes place by diffusion through the general body surface. Oxygen is taken in and used for burning of food to obtain energy. As a result carbon dioxide is formed which is diffused out in surrounding water.

Excretion

Most of the nitrogenous excretion in *Paramecium* is carried by diffusion through general body surface. Contractile vacuoles also serve for excretion as urea has been detected in them. The chief nitrogenous product is ammonia which is diffused out.

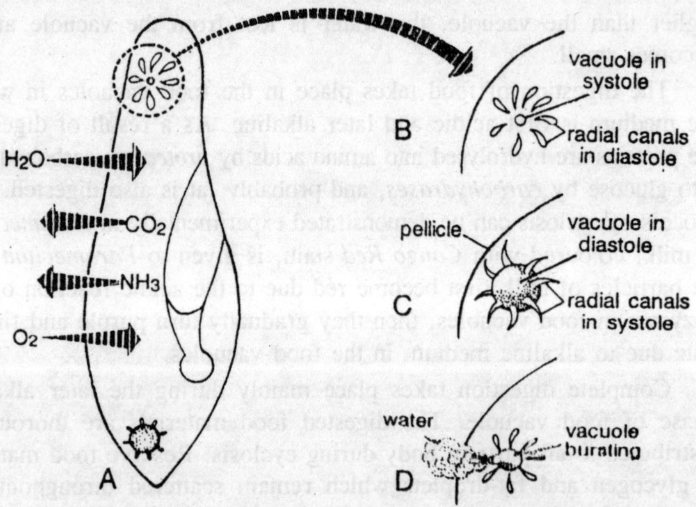

Fig. 12.19. Diagrammatic sketch to show the process of respiration, excretion and osmoregulation.

Osmoregulation

Two contractile vacuoles regulate the water content of the body. They occupy rather fixed position, one being anterior and the other posterior. Each vacuole is surrounded by 5-12 radiating canals. Both vacuoles derive energy to work from the mitochondria crowded around them.

Behaviour

Paramecium is extremely sensitive to different type of stimuli and shows its reaction to them. The response is positive when the animal moves towards a stimulus and negative when it moves away. To an adverse stimulus the animal continuous to give the avoiding reaction until it escapes. Various stimuli and their response given by *Paramecium* are:

1. *Reaction to gravity* (*Geotaxis* or *Geotropism*). *Paramecium* generally exhibit a negative response to gravity as seen in culture contained in a test tube where they gather close to the surface film with their anterior ends pointed upward. If paramecia are introduced in an inverted water filled 'U' shaped tube stoppered at both the ends, they immediately move upward into the horizontal part of the tube.
2. *Reaction to electric current* (*Galvanotaxis* or *Galvanotropism*). When two electrodes are placed opposite each other in a shallow dish containing paramecia and a constant current applied all the organisms swim towards the cathode.
3. *Reaction to hydrostatic pressure* (*Barotropism*). When the hydrostatic pressure is increased, the rate of movement is decreased.
4. *Reaction to mechanical stimuli* (*Thigmotaxis* or *Thigmotropism*). Response to contact is varied. If the anterior end lightly touches a fine point, the animal shows strong avoiding reaction. A slow moving individual responds positively to contact with an object by coming to rest upon it.
5. *Reaction to light* (*Phototaxis* or *Phototropism*). It shows negative response to bright light as well as dark. It responds to the diffused light. However, *P. bursaria* is positively phototactic.
6. *Reaction to water current* (*Rheotaxis* or *Rheotropism*). In a gentle water current the paramecia will mostly move with the flow with their anterior ends upstream thus showing positive response.
7. *Reaction to temperature* (*Thermotaxis* or *Thermotropism*). The optimum range of temperature for paramecia is 24-28°C, but they can withstand considerably lower and higher temperature than the optimum range. Extreme cool and hot water is avoided.
8. *Reaction to chemical* (*Chemotaxis* or *Chemotropism*). It shows avoiding reaction to chemical like bases. If a drop of weak salt solution (0.5%) is introduced in a *Paramecium* culture on a

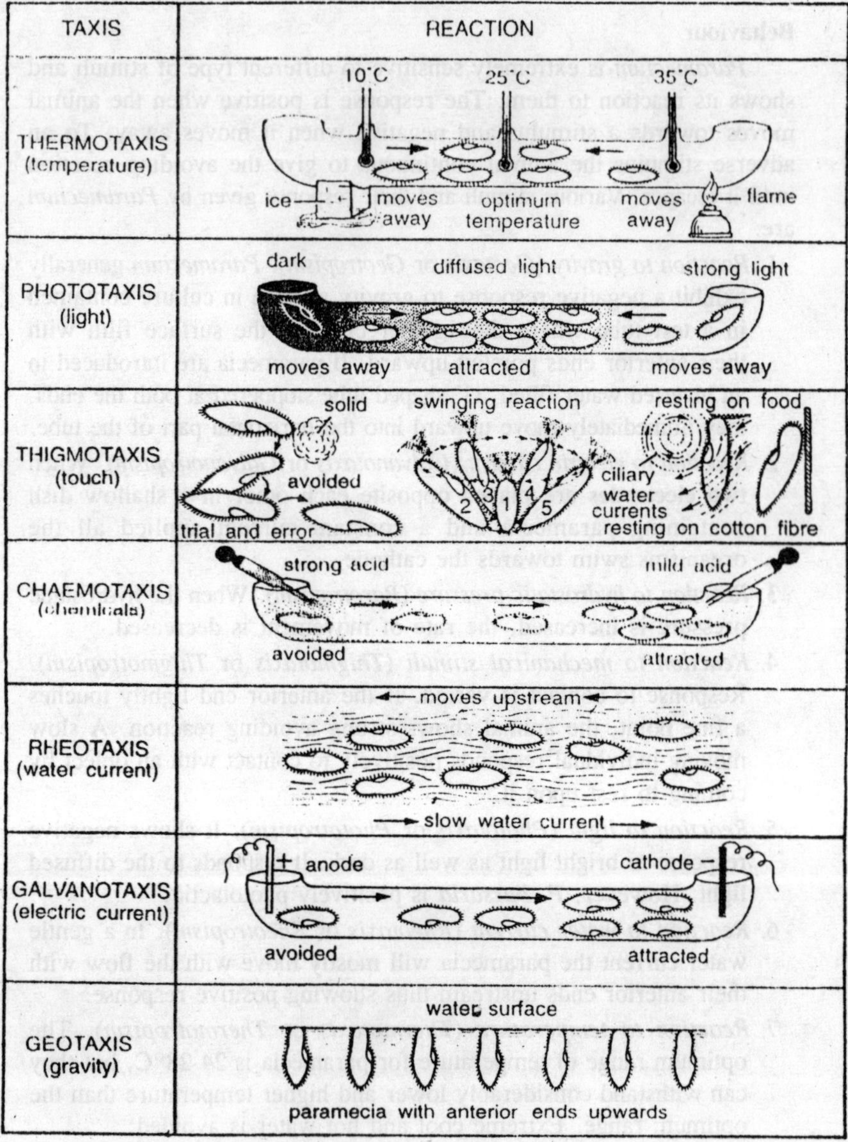

Fig. 12.20. Responses to different stimuli.

microslide, the animal show negative response. However, they show a positive response to the weak acids.

REPRODUCTION

Ciliates differ from almost all other organisms in possessing two distinct types of nuclei-usually one large macronucleus and one or more small micronuclei. They are diploid with little RNA. The micronucleus, a store of genetic material, is responsible for genetic exchange and nuclear organization, and also gives rise macronucleus.

In *Paramecium* two types of reproduction is reported:

1. Asexual Reproduction
2. Sexual Reproduction

Asexual Reproduction

Asexual reproduction takes place during favourable conditions. Asexual reproduction is always by means of binary fission, which is typically transverse. More accurately, fission is described as being homotheogenic, with the division plane cutting across the kineties—the longitudinal rows of cilia. Before the beginning of fission, *Paramecium* stops feeding and the body elongates to become spindle-like. The oral groove disappears. The micronucleus divides by the complicated process of mitosis and the resulting two daughter micronuclei move towards opposite ends of the cell. All the stages of mitosis i.e. prophase, metaphase, anaphase and telophase take place inside the nuclear membrane. The number of chromosomes in 36-150 depending upon the species. When this is happening, the macronucleus also elongates, divides transversely by amitotic process in which the number of chromosomes

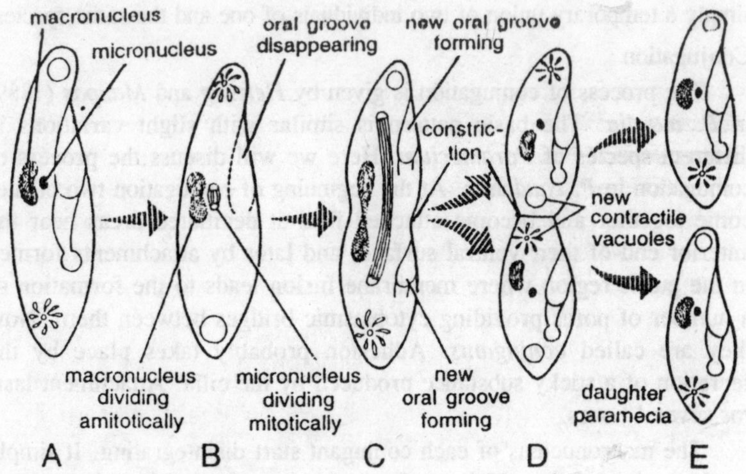

Fig. 12.21. Stages showing binary fission.

may differs. A transverse constriction appears in the middle of *Paramecium*. The constriction deepens gradually.

In the mean time two oral grooves begin to form one in the anterior half and another in the posterior half. Two original contractile vacuoles of parent remain, one in each half of the dividing parent individual. Two contractile vacuole are later formed. Two new buccal structures also appear. Now the constriction completely divides the parent into two daughter paramecia of equal size and with a complete set of cell-organelles. These grow to full size before dividing again. The new formed daughter paramecia neither resemble each other nor the typical parent, so that for a while they can be distinguished as the anterior daughter (*Proter*) and the posterior daughter (*Opisthe*). Soon they assume the typical form.

The entire process of binary fission is completed with 30-120 minutes depending upon the availability of food and temperature. In *P. caudatum* it may occur 2-3 times a day and in *P. aurelia* about 5 times a day, so that a large number of them can be obtained in a short time. All the individuals which are produced from a single parent, collectively, form a *clone* and all the members of a clone are genetically alike. The binary fission cannot go on indefinitely. After several generations, the rate of fission declines.

Sexual Reproduction

Paramecium reproduces sexually in which nuclear material of two individuals is interchanged. This method is called *conjugation*, but is simply a temporary union of two individuals of one and the same species.

Conjugation

The process of conjugation is given by *Hertwig* and *Maupas* (1889) in *P. aurelia*. The basic pattern is similar with slight variations in different species of *Paramecium*. Here we will discuss the process of conjugation in *P. caudatum*. At the beginning of conjugation two ciliates come together and become attached first at deciliated areas near the anterior end of their ventral surface, and later by attachments formed in the gullet region where membrane fusion leads to the formation of a number of pores providing cytoplasmic bridges between them. Now they are called *conjugants*. Adhesion probably takes place by the secretion of a sticky substance produced by the cilia. Attachment lasts for several hours.

The macronucleus of each conjugant start disintegrating. It simply breaks up into pieces which are digested in cytoplasm. The micronucleus

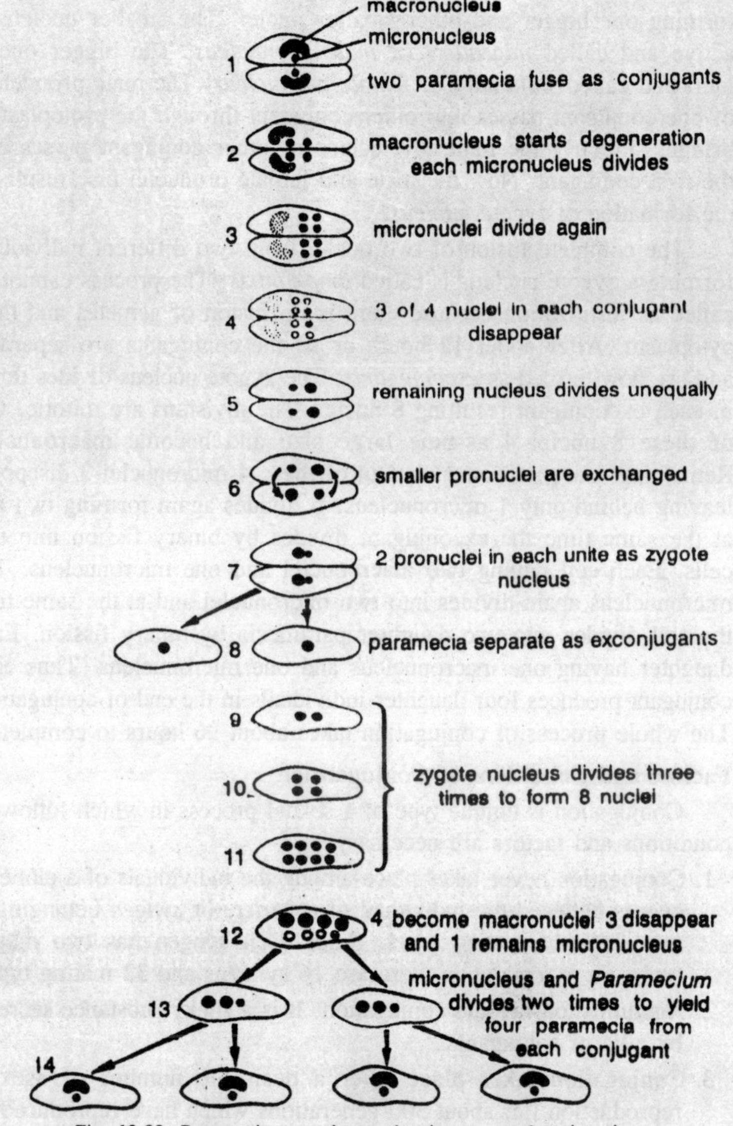

Fig. 12.22. Paramecium caudatum showing stage of conjugation.

of each conjugant now grows in size and divides twice, the first division being meiotic result in the formation of 4 haploid nuclei. Three micronuclei degenerate and disappear from cytoplasm of each conjugant living behind one nucleus. The remaining nucleus divides again unequally forming one bigger and other smaller nuclei. The smaller nucleus is active and called *migratory* or *male pronucleus*. The bigger one is inert and called *stationary* or *female pronucleus*. The male pronucleus of one conjugant passes into other conjugant through the protoplasmic bridge, similarly the male pronucleus of other conjugant passes into the first conjugant. Now the male and female pronuclei fuse results in the formation of *zygote nucleus*.

The complete fusion of two nuclei from two different individuals forming a zygote nucleus is called *amphimixis*. The process cannot be called as fertilization because there is no fusion of gametes and their cytoplasm. After about 12 hours or so the conjugants are separated and are now termed as *exconjugants*. The zygote nucleus divides thrice in each exconjugant resulting 8 nuclei. The divisions are mitotic. Out of these 8 nuclei 4 assume large size and become macronuclei. Remaining 4 are micronuclei. Out of these 4 micronuclei 3 disappear leaving behind only 1 micronucleus. It divides again forming two and at the same time the exconjugant divides by binary fission into two cells. Each cell having two macronuclei and one micronucleus. The micronucleus again divides into two micronuclei and at the same time the cell divides into two daughter paramecia by binary fission. Each daughter having one macronucleus and one micronucleus. Thus each conjugant produces four daughter individuals in the end of conjugation. The whole process of conjugation takes about 96 hours to complete.

Factors and Conditions of Conjugantion

Conjugation is unique type of a sexual process in which following conditions and factors are necessary:

1. Conjugation never takes place among the individuals of a clone. It occurs between the individual of a *variety* or *syngen* belonging to two *different mating types*. Since each syngen has two mating types, in *P. caudatum* there are 16 syngens and 32 mating types.
2. Agglutination favours conjugation. It is a sticky substance secreted by cilia of conjugants.
3. Conjugation takes place after a desirable number of asexual reproduction i.e. about 300 generations which have reproduced by binary fission or it occurs after a long continuance of binary fission so that the rejuvenation may take place in the exhausted or perishing clone.
4. Conjugation starts early in the morning and is continued till the afternoon. It suggests that dark favour conjugation and light supress it.

5. Conjugation occurs when changes take place in the physiological condition. The individuals are usually smaller in size (210 μ in length) than the normal individuals.
6. It occurs during unfavourable conditions such as scarcity of food, light etc.

According to *Sonneborn* (1957) in *P. aurelia*, there are 14 syngens and 28 mating types. These are as follows:

Table 12.1. Showing the mating types of P. aurelia complex.

Syngen number	*Mating type*	*Other mating types with which conjugation occurs (and %)*	*Group*	*+ or – type*
1	2	3	*4*	*5*
1.	I	II (95%), X (40%)	A	–
	II	1 (95%), IX (40%), XIII (10%), V (1%)		+
2.	III	IV (95%)	B	
	IV	III (90%)		
3.	V	VI (95%), XVI (40%), II (1%)	A	–
	VI	V (95%)		+
4.	VII	VIII (95%), XVI (95%)	B	–
	VIII	VII (95%), XV (60%)		+
5.	IX	X (95%), II (40%)	A	–
	X	IX (95%), I (40%)		+
6.	XI	XII (95%)	B	–
	XII	XI (95%)		–
7.	XIII	XIV (95%) II (10%)	B	–
	XIV	XIII (95%)		+
8.	XV	XVI (95%), VIII (60%)	B	–
	XVI	XV (95%), VIII (95%), V (40%)	+	–
9.	XVII	XVII (95%)	A	–
	XVIII	XVII (95%)		–
10.	XIX	XX (95%)	B	–
	XX	XIX (95%)	A	–
11.	XXI	XXII (95%)	A	
	XXII	XXI (95%)		
12.	XXIII	XXIV (95%)	B	–
	XXIV	XXIII (95%)		+
13.	XXV	XXVI (60%)	C	–
	XXVI	XXV (60%)	C	–
14.	XVII	XXVIII (95%)	A	–
	XXVIII	XXVII (95%)		–

Significance of Conjugation

The significance of conjugation has been discussed in details, but it still remains uncertain. The following significances are attributed to conjugation.

1. It is the method of rejuvenation by which the vigour of the species is maintained. If conjugation does not occur for a long time than paramecia become weak an ultimately die.
2. Conjugation is a temporary union in which there is no mixing of cytoplasm of the conjugants but a zygote nucleus is formed by the fusion of their nuclei, therefore, the zygote nucleus of each conjugant possesses the hereditary material of both the conjugants i.e. new characters come in them.
3. In conjugation, the macronucleus is replaced by the material of zygote nucleus. The number of chromosomes is restored in macronucleus by conjugation due to which the vigour of daughter paramecia is renewed.
4. The function of micronucleus is to maintain balanced chromosomes and gene complex.

ABERRANT BEHAVIOUR IN REPRODUCTION

Paramecium shows certain variations in its nuclear behaviour during fission and conjugation these deviations are *Autogamy*, *Endomixis*, *Cytogamy* and *Hemixis*.

Autogamy

Diller (1934,36) described a process of *self-fertilization* or *autogamy* occurring in a single individual of *P. aurelia*. About once a month *P. aurelia* undergoes this process. During autogamy the macronucleus increase in size to form an irregular structure and later breaks up into small fragments are absorbed in the cytoplasm. The two micronuclei divide twice forming eight nuclei. As a rule the first division is meiotic thus all the eight nuclei are haploid. Out of these eight nuclei, seven disintegrate and only one nucleus is left. It again divides to form to nuclei. These nuclei are called *gametic nuclei*. By this time a peculiar cone-like bulge appears on the surface of autogamous animal, slightly to the right and anterior to the cytostome. Because of its position it is called the *oral cone* or *protoplasmic cone*. The gametic nuclei are at first elongated and spindle like bodies.

Often there is slight size variation between them so that they may be termed to as male (smaller) and female (larger) nuclei. They approach each other and then move into the cone and fuse to form zygote nucleus. Later it comes out of cone and the cone disappears. The zygote nucleus divides twice forming four nuclei. Two of these nuclei increase in size to become macronuclei and remaining two as

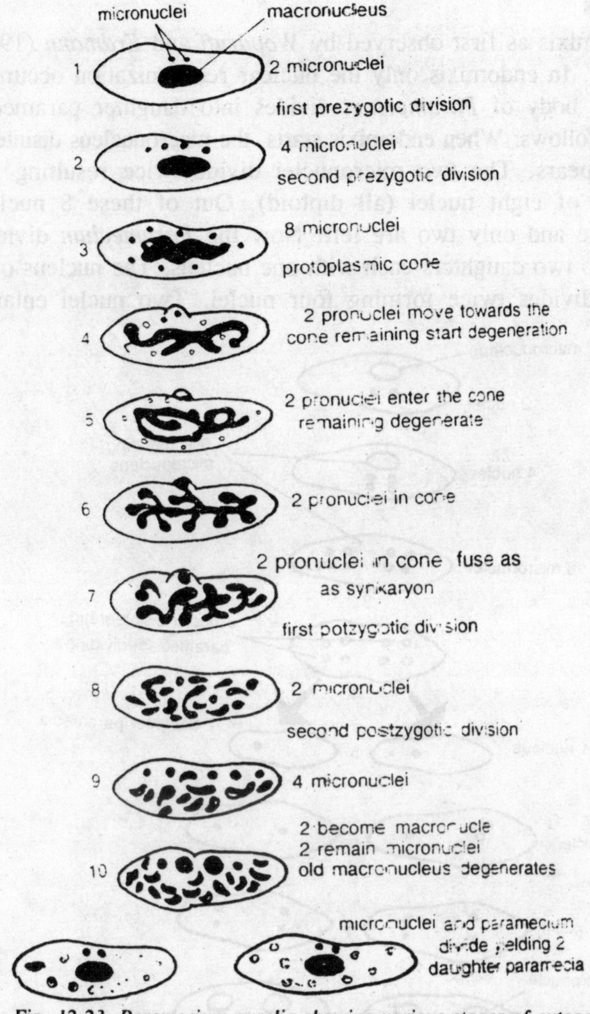

Fig. 12.23. Paramecium aurelia showing various stages of autogamy.

micronuclei. The cell body and micronuclei divide by binary fission to form two daughter paramecia each with a macronucleus and two micronuclei.

Significance of autogamy

1. Rejuvenation in the species is accomplished by autogamy.
2. A new macronucleus is formed from the material of micronuclei.
3. Paramecium regains its lost vigour by this process.

Endomixis

Endomixis as first observed by *Woodruff* and *Erdmann* (1914) in *P. aurelia*. In endomixis only the nuclear reorganization occurs after which the body of *Paramecium* divides into daughter paramecia. It occurs as follows: When endomixis starts, the macronucleus disintegrates and disappears. The two micronuclei divide twice resulting in the formation of eight nuclei (all diploid). Out of these 8 nuclei six disintegrate and only two are left. Now the *Paramecium* divides by fission into two daughters each with one nucleus. The nucleus of each daughter divides twice forming four nuclei. Two nuclei enlarge to

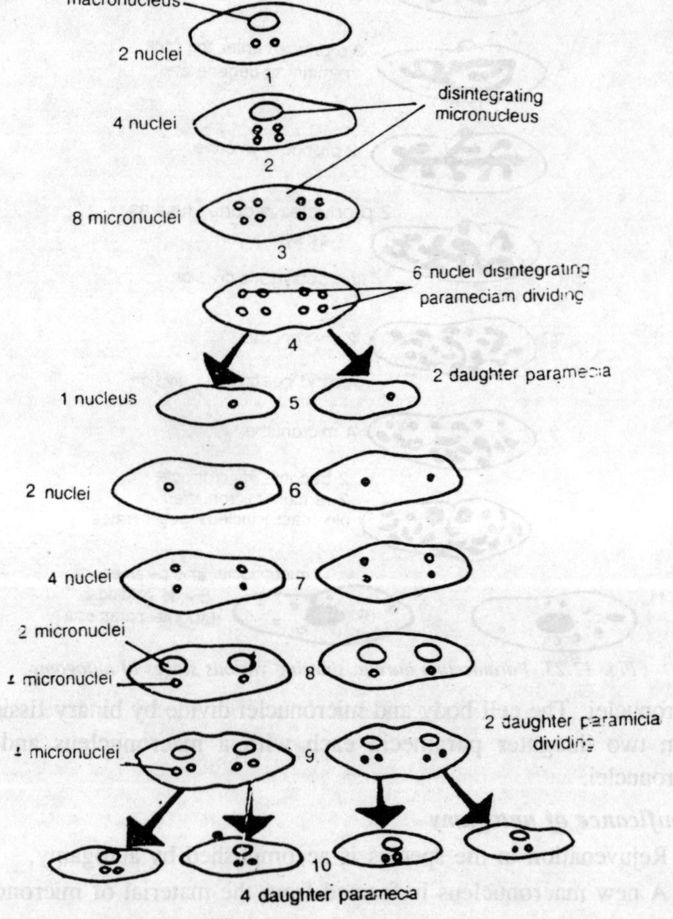

Fig. 12.24. Paramecium aurelia showing endomixis.

form macronuclei and remaining two micronuclei. The two micronuclei, again divide to form four micronuclei in the mean time the cell divides again by fission into two daughters each containing one macronucleus and two micronuclei. In the end four daughter paramecia are formed from one individual. According to *Diller* (1936), *Sonneborn* (1947) and *Whichterman* (1953) the endomixis does not occur in *Paramecium*.

Cytogamy

Whichterman (1940) observed the process of cytogamy in *P. caudatum*. The process is something intermediate between conjugation and autogamy. Two individuals stick by their oral surfaces, the micronucleus divides twice (first division in meiotic) of these three disintegrate into the cytoplasm, the remaining one divides once producing two gametic nuclei. Self-fertilization takes place. The conjugants now separate and called exconjugants. Exconjugants behave and multiply the same manner as in conjugation. The process is completed in about 13 hours. The genetical consequence of cytogamy is always homozygosity.

Hemixis

This method of nuclear reorganization was observed by *Hartman* in *P. aurelia* and was supported by *Diller* (1936). During hemixis the micronuclei do not participate. The macronucleus throws out two or more equal or unequal fragments to be absorbed in the cytoplasm. The remainder of macronucleus and micronuclei behave normally at the cell division. *Diller* classified hemixis into four types.

1. *Type A.* It is the simplest form of hemixis characterized by a division of the macronucleus into two or more parts in the cytoplasm.
2. *Type B.* It is characterized by the extrusion of 1-20 or more chromatin balls from the macronucleus.
3. *Type C.* It is characterized by the simultaneous splitting of macronucleus into 2 or more major portions and extrusion of balls into cytoplasm.
4. *Type D.* It is considered to be the pathogenic conditions in which micronucleus undergoes complete fragmentation into chromatin balls that disappear from the cell later on. Micronuclei generally disappear before the dissolution of macronucleus.

The hemixis is considered as an act of purification on the part of the macronucleus and the process of chromatin material cast off by it are held in contain waste and harmful substances.

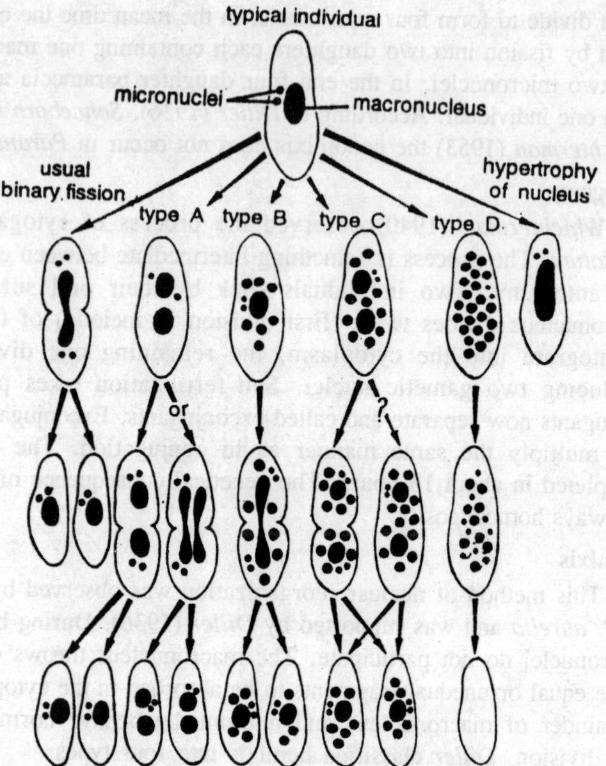

Fig. 12.25. Paramecium aurelia showing hemixis.

Killer Paramecia

Sonneborn discovered that certain strains of *P. aurelia* liberate a toxic substance into the medium in which they live. This toxic substance kills the other strains of *Paramecium*. He called the former strain as *killers* and later as *sensitives*. The toxic substance is named as *paramecin*. The killing takes place slowly, the sensitive taking about two days to die. Symbiotic bacteria (*Caedibacter*) have been found living in the cytoplasm of *P. aurelia* (*Soldo*, 1974; *Preer*, 1981) and have been given such names as *kappa*, *mu*, *pi*, *lambda* etc;. Lambda possess bacteria flagella, and several types of symbiont have bacterial cell walls, Killer paramecia were found to contain kappa, but the persistence of *kappa* in the cytoplasm was dependent upon the presence of dominant macronuclear gene. Only *kappa bodies* that contained refractile inclusions (tightly-coiled ribbons associated with virus-like granules) were able to kill, but other *kappa* could develop these inclusions.

Other symbionts act in a comparable manner, paramecia that possess *mu* kill sensitive at conjugation (they are mate killers), while *pi* may be inactive kappa. Paramecia that contain *lambda* produce lethal granules which are released into the medium. It is assumed that these symbionts derive nutrients from the paramecia. A killer stock has an advantage over a sensitive stock in ecological competition (*Landis*, 1981). *Lambda* apparently synthesises folic acid, and paramecia with this symbiont have no external requirement for this vitamin.

13

LIFE OF EUGLENA

Euglena belongs to the Order-Euglenoidida (Class-Mastigophora), which comprises relatively large forms, usually of a conspicuously green colour. In the members of this order one or two flagella are present, arising from the invaginated anterior end of the body called *reservoir*. All chlorophyll bearing members possess a light-sensitive organelle called *stigma*. Several non-pigmented forms e.g. *Peranema* etc. are also placed in this order because of their basic structure.

SYSTEMATIC POSITION

Phylum	—	Protozoa
Subphylum	—	Sarcomastigophora
Superclass	—	Mastigophora
Class	—	Phytomastigophorea
Order	—	Euglenida
Family	—	Euglenoidae
Genes	—	*Euglena*

Habits and Habitat

Euglena is a common, solitary and free-living flagellate found in stagnant water containing nitrogenous organic matters. Such water with high organic contents is called *polysaprobic*. Water with moderate organic contents is called *mesosprobic*, and with low organic contents is *oligosaprobic*. The pure water is described as *kathrobic*. There are about 100 species of *Euglena* widely distributed all over world in polysaprobic fresh water. It is known to occur in soil, mud and often has been found on the bark of trees. *Pringsheim* (1953) mentioned that some euglenae are marine and some are parasite.

Euglenomorpha lives in the rectum of tadpoles, where they play an unknown symbiotic role. The most common species of *Euglena* is *E. viridis*. The common Indian species are *E. agilis*, *E. orientalis* and *E. fusiformis*. The presents description is based on the biology of *E. viridis* (Gr. *eu*=true; *glene*= eyeball; and *L. viridis*=green). They are found in such a large number that they produce a green scum on the surface of ponds.

Culture

They can be cultured in laboratory. For preparing culture, some cow or horse dung is boiled in water in a jar, and allowed to cool it for two days. Now pour some weeds from the pond containing euglenae into the jar, and place the jar near the light source. In a few days euglenae develop in the medium. Another method can be utilized for the preparation of culture. Boil the pond or tap water for some times

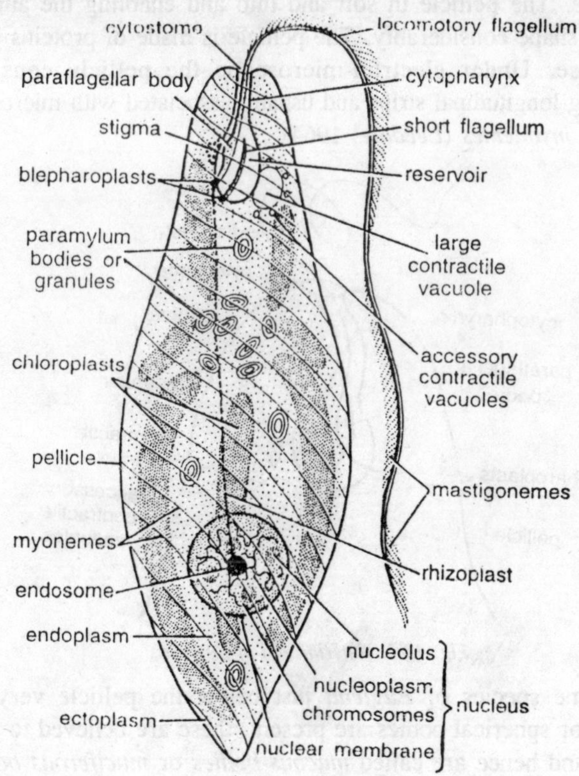

Fig. 13.1. Euglena viridis.

and allow to cool it. Add 20 boiled wheat grains to this water and inoculate with euglenae. Keep the jar in sunny place. Wheat grains may be added monthly to maintain the culture.

MORPHOLOGY

Shape and Size

Euglena viridis is a minute elongated and spindle-shaped. It is pointed at the posterior end and blunt at the anterior end. The posterior end is called *tailpiece*. It measures 60 μ in breadth at the thickest part of the body.

Pellicle

The shape of the animal is maintained by the pellicle which presents parallel or spiral striations. The striations may or may not be present on the tail piece. According to *Chadfaud* (1937) the pellicle is two layered, having an outer thin *epicuticle* and deeper and thick inner *endocuticle*. The pellicle in soft and thin and enabling the animal to change its shape considerably. The pellicle is made of proteins but not of cellulose. Under electron-microscope the pellicle consists of interlocking longitudinal strips and usually associated with microtubular fibrils, the *myonemes* (*Leedale*, 1964).

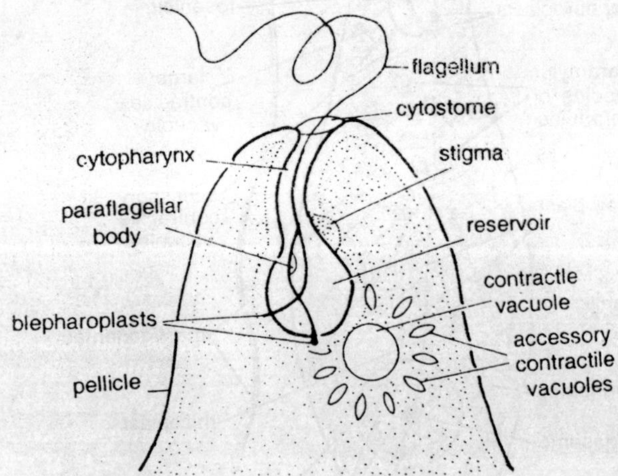

Fig. 13.2. Anterior end of Euglena.

In some species of *Euglena* just below the pellicle very small elongated or spherical bodies are present, these are believed to secrete mucilage and hence are called *mucous bodies* or *muciferous bodies*.

Cytostome and Cytopharynx

Shortly to one side of the middle line at the anterior end, is a small opening called *cytostome* or *cell mouth*. The cytostome leads into a tube called *gullet* or *cytopharynx* which in turn opens at its other end into a more or less spherical chamber called the *reservoir*. The pellicle invaginates into the gullet through the cytostome and lines the reservoir. The pellicle here contains only the epicuticle layer. The cytostome and gullet are not used for the ingestion of food but as a canal for the escape of fluid from the reservoir.

Contractile Vacuole

A large contractile vacuole lies towards one side of the reservoir. It is surrounded by several small accessory vacuoles. The contractile vacuole receives water from the *accessory vacuoles* and discharges the fluid into the reservoir from where it goes out through the cytostome.

Stigma

A conspicuous structure in *Euglena* is the red eye spot or *stigma*. This is placed near the inner end of the gullet close to the reservoir. It consists of protoplasm in which clusters of lipid globules containing orange or red carotinoid pigments. The stigma is cup-shaped with a

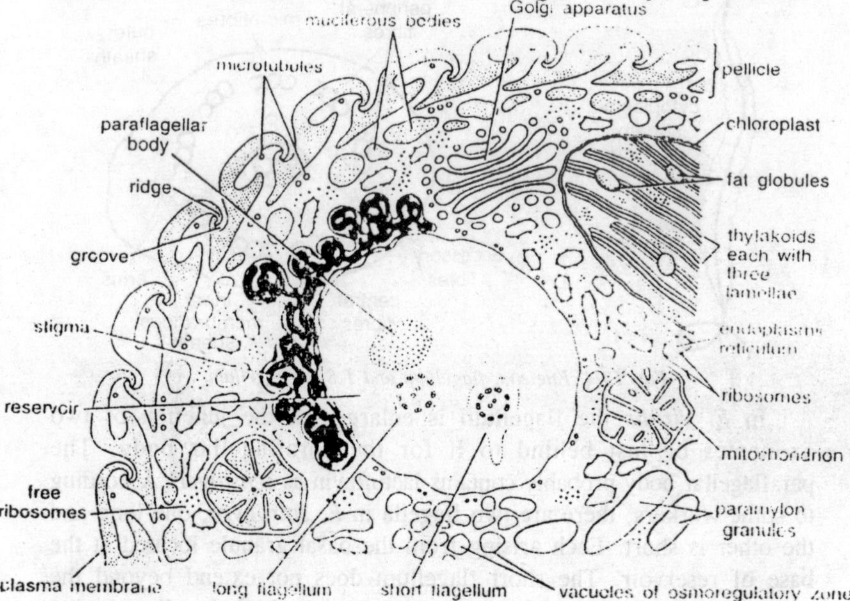

Fig. 13.3. An electron micrograph (diagrammatic) of a portion of the T.S. body, passing through the reservoir.

colourless mass of oily droplets in its concavity which functions as a lens. The stigma and paraflagellar body, together form a photoreceptor apparatus.

Paraflagellar Body

A small swelling lies either on one root or at the junction of two roots of the flagellum. It is known as *paraflagellar body*. *Chadefaud* and *Provasoli* have shown that paraflagellar body and stigma together form the photoreceptor apparatus.

Flagellum

A single, long whip-like flagellum emerges out of the cytostome through cytopharynx. This whip-like process, by its lasting movements, causes the organism to move in water. The flagellum of *Euglena* is also called *tractellum* because it does not propel the body from behind but draws its forward. The flagellum arises from the bottom of the cytopharynx, is forked at the base and is connected by two *rhizoplasts* to the *basal granules* or *blepharoplasts*. Each blepharoplast is a compact granule that gives rise to a separate filament (*axoneme*) and in most of the species the two axonemes soon unite to form a definite flagellum.

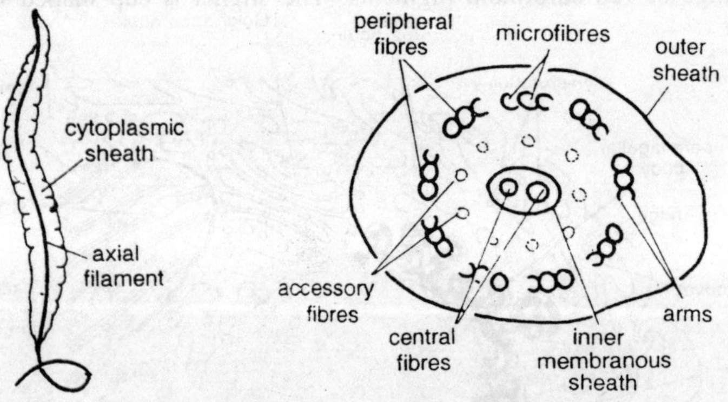

Fig. 13.4. Euglena, flagellum and T.S. of flagellum.

In *E. viridis* the flagellum is enlarged at the junction of two axonemes or just behind to it for the paraflagellar body. The paraflagellar body probably contains lactoflavin as *sensitiger*. According to some workers, there are two flagella in *E. spirogyra*, one long and the other is short. Each arising from the basal granule located at the base of reservoir. The short flagellum does not extend beyond the neck of the reservoir and is often adherent to the long flagellum giving the appearance of bifurcation. Each flagellum is cylindrical structure about 0.25μ in diameter. It is composed of a longitudinal bundle of microtubular fibres (axonemes) enclosed within a unit membrane which is continuous with the cell membrane.

The microtubular fibrils of the axoneme are arranged in a precise pattern, best seen in a T.S. Two microtubules occupy the central position. They are called *central fibres* and remain surrounded by inner membraneous sheath. The central fibres are surrounded by nine double microtubular fibrils called *peripheral fibres*. Each peripheral fibre is made of two microfibres. One of the microfibres bears a double row of short *arms* or projections all pointing in the same direction.

The distal portion of flagellum contains numerous minute fibres, known as *mastigonemes*, which project laterally (*Manton*, 1959). Such a flagellum that bears mastigonemes is called *stichonematic type*. Mastigonemes are believed to project rigidly from the flagellum in two rows arranged in the plane of flagellar undulation. The structure of blepharoplast is similar to that of the centriole found in animal cells. There are evidences which show that basal body contains DNA and may have some power of self-replication.

Cytoplasm

Two types of cytoplasm can be distinguished in *Euglena*, a dense outer layer, the *ectoplasm* and the central mass, the *endoplasm*, which is more fluid. The endoplasm contains nucleus, chromatophore and paramylum bodies.

Nucleus

Euglena has a single oval nucleus lying in a definite position near the centre of the body. There is a distinct nuclear membrane. The nucleus contains a central body known as *endosome* (Syn. *nucleolus*, *karyosome*).

Chromatophores

Nearly all species of *Euglena* have small or large coloured bodies called *chromatophores*. They are present in the form of little plates or discs, sometimes spindle-shaped band-like or radiating star-like. The number, shape, size and arrangement of chromatophores vary from one species to another. The colour of chromatophore is green due to the presence of green pigment called *chlorophyll*. The chlorophyll consists of two pigments, the green *chlorophyll* a and a yellowish b *carotene*. Under electron-microscope the chromatophore is swollen in the centre where it contains two non-pigmented hemispherical bodies called *pyrenophores*. Attached to one of each are protein bodies called

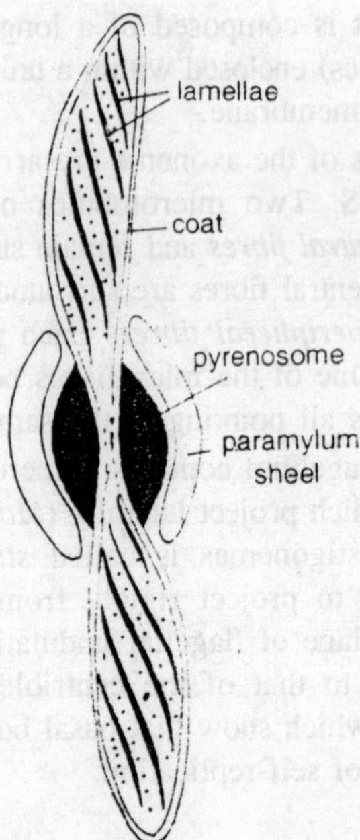

Fig. 13.5. A chloroplast.

pyrenosome. Each pyrenosome is covered with a *paramylum shell*. The two pyrenosomes constitute a *pyrenoid*. Pyrenoids are associated with the formation of a kind of animal carbohydrate called *paramylon*. Paramylon granules thus produced may be found scattered through the cytoplasm. The chromatophore is surrounded by two membranes. The outer membrane is similar to plasma membrane. The inner membrane is similar to internal membrane system of the chloroplast. The *matrix* or *stroma* contains the chlorophyll-bearing bundles or *thylakoids*. The matrix contains the photosynthetic carbon cycle enzymes, ribosomes and DNA.

Paramylon

Here and there are scattered light bluish-green refractile granules called *paramylon*. It is a polysaccharide similar to starch but it does not give violet colour with iodine solution.

Other Endoplasmic Organelles

The endoplasm outside the nucleus clearly has a system of double-membranes forming the *endoplasmic reticulum*. Numerous small granules, termed *mitochondria*, are scattered throughout the cell. The mitochondria are important as a location of many of the biochemical reactions of the cell during respiration. *Golgi complex* are also found in abundance. The *ribosomes* also scattered and remain attached to the endoplasmic reticulum. In some species of *Euglena*, a fine fibril called *rhizoplast*, extends from one of the blepharoplast to a small granule on the nucleus. This connection suggests that the flagellum is under the direct control of nucleus. The flagellum, blepharoplast, rhizoplast and the granule on the nucleus are said to form *infra-ciliary system* or *neuromotor system*.

PHYSIOLOGY

Locomotion

There are two methods of locomotion is *Euglena*:

1. Flagellar movement
2. Euglenoid movement.

Flagellar movement

The process of the flagellar movement is not clearly understood because, except a few, its movements are not easy to observe in most animals. The whip-like movement of flagellum moves the animal forward and at the same time rotates it around the axis. Thus the animal moves forward in a spiral manner at a fair speed of 3.6 mm per minute.

Various theories have been put forward to explain the mechanism of flagellar movements in *Euglena*. *Butschli* was first to describe the flagellar movement. He observed that the flagellum undergoes a series of lateral movement and in doing so, a pressure is exerted on the water at right angles to its surface. This pressure creates two forces: one directed parallel and other at right angles, to the main axis of

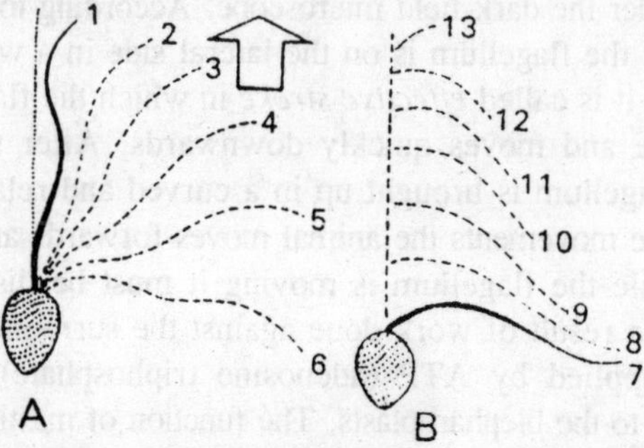

Fig. 13.6. Action of flagellum. A—Recovery stroke, successive stages from 1 to 7; B—Effective stroke, successive stages 8 to 13.

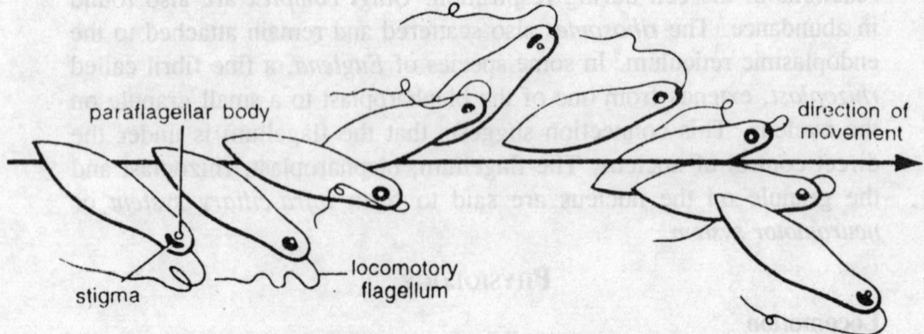

Fig. 13.7. Successive stages in flagellar movement.

body. The parallel force will drive the animal forward and the other force would rotate the animal on its own axis. According to *Gray* (1928), the flagellum beats sideways in such a way that its movement proceeds from one end to other. These waves create two types of forces, one in the direction of the movement and the other in the circular direction with the main axis of body. The first force will drive the animal forward and the later would rotate the animal. The posterior end of body describes a smaller circle than the anterior end while the body is propelled forward through the water like a rotating plane which always remains inclined at an angle of 30° to the axis of movement.

Lowends (1941-43) also studied the locomotion in *Euglena*. His views are a little different from those of *Gray*. According to him, the flagellum is always held backwards during swimming and the movement in it occurs in a spiral manner from base to its tip. The flagellum beats, on an average at the rate of 12 beats per second with increasing velocity and amplitude. *Krijisman* (1952) observed the movement of flagellum under the dark-field microscope. According to him the normal movement of the flagellum is on the lateral side in a whip-like manner due to which it is called *effective stroke* in which the flagellum remains on outer side and moves quickly downwards. After this in *recovery stroke* the flagellum is brought up in a curved and relaxed state. As a result of these movements the animal moves forwards and rotate around its axis. While the flagellum is moving it must be dissipating energy (*Brokaw*) as a result of work done against the surrounding water. The energy is supplied by ATP (adenosine triphosphate) present in the mitochondria to the blepharoplasts. The function of mastigonemes present on the flagellum is unknown. *Euglena* is unable to swim backwards as the direction of motion can not be reversed.

Euglenoid movements or metaboly

Sometimes a creeping movement is found in *Euglena*. It is bought about by the contraction and expansion of body. A peristaltic wave of contraction and expansion passes over the entire body from anterior to posterior end and the animal moves forward. The body becomes shorter and wider first at the anterior end, then in the middle and later at the posterior end. The contraction and expansion of the body is due to thin, elastic pellicle and the myonemes present beneath it. This is a very slow movement. *Lowends* found that an *Euglena* of (50-64μ long) travels 168μ per minute.

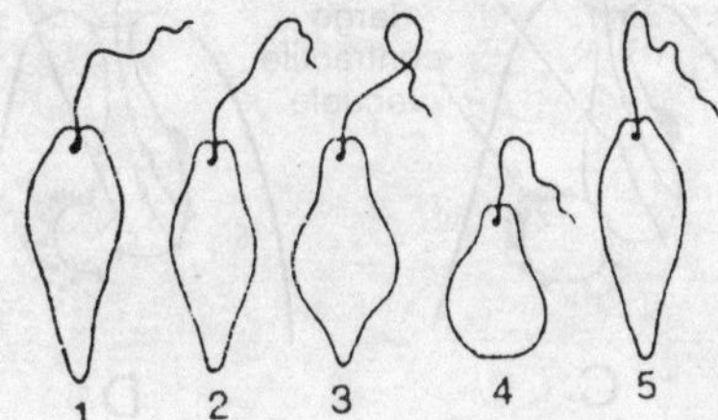

Fig. 13.8. Successive stages in euglenoid movement.

Nutrition

The mode of nutrition in *Euglena* is *mixotrophic* i.e. the nutrition is accomplished either by *holophytic* or *saprozoic* or by both the modes.

Holophytic or autotrophic nutrition

Although *Euglena* has a mouth (cytostome) and gullet, it is doubtful if any food is ingested. The food is manufactured photosynthetically, as in plants, with the aid of light, carbon dioxide and the chlorophyll present in chromatophores. This mode of nutrition is known as *holophytic* or *autotrophic*. Photosynthesis enables the oxygen and a water molecule to be released, while hydrogen of the water molecule reduces the carbon dioxide present through a series of steps to form carbohydrates and more water. The carbohydrates are stored as a starch-like substance called *paramylon*.

Saprozoic nutrition

When the animal is kept in dark for a long time or the chlorophyll is destroyed by treatment with certain drugs, the colourless *Euglena* resulting can continue to live if provided with a medium rich in organic nutrients. This seems to indicate that organic substances in solution are absorbed through the surface of the body, that is, *saprophytic* nutrition.

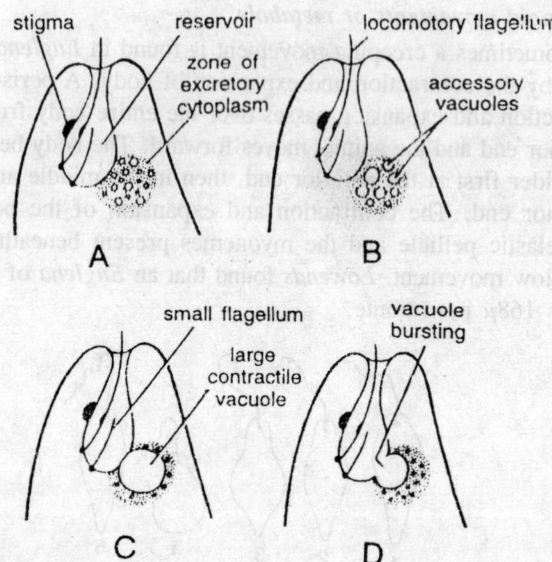

Fig. 13.9. Successive stages of diastole and systole of contractile vacuole during excretion and osmoregulation.

Myxotrophic nutrition

The euglenae are sometimes holophytic and sometimes saprozoic depending upon the external conditions for maintaining life. According to recent information, the *Euglena* is *amphitrophic*. It is holophytic in sun light and saprozoic in dark. In the sunlight, it is holophytic but not completely as it needs Vitamin B_{12} or B_1. In the absence of these vitamins it will not multiply. This nutritional versatility is called *amphitrophy*.

Respiration

Respiration in *Euglena* is *aerobic* i.e. occurs in the presence of oxygen. The exchange of gases takes place through general body surface. Oxygen diffuses into the cytoplasm from the surrounding water and carbon dioxide diffused out. During active photosynthesis in the organism diffusion of carbon dioxide inwards and the output of oxygen mask the respiratory diffusion.

Excretion

Chemical processes keep on taking place in the body all the times, some of which produce materials that are of no use to the animal and may even prove harmful. During the catabolism of amino acid ammonia is formed. It is highly toxic and dissolved in water. This ammonia must be excreted as soon as it is formed. Some of the ammonia is diffused out through general body surface. Some of the ammonia is excreted out by contractile vacuole. The excretory role of contractile vacuole has gained support in the past few years. (*Chaudefaud*). The contractile vacuole is surrounded by specialized granular cytoplasm called *excretory cytoplasm*. Ammonia with large quantities of water first collect in the region of excretory cytoplasm as small droplets called accessory vacuoles which later on fuse to form the contractile vacuole. Periodically, the vacuole reaches its maximum size (*diastole*) and then burst (*systole*) so as to discharge its contents in reservoir, from where waste passes out through the cytostome.

Osmoregulation

The primary function of contractile vacuole is *osmoregulation* or removal of excess of water from the body. The cytoplasm is denser than the surrounding water and thus has a relatively higher osmotic pressure. The surrounding water constantly enters the body by endosmosis through the pellicle. Moreover, water is also formed in the body during respiration. The excess of water from all over the body flows towards the excretory cytoplasm (described above) and collects in small accessory vacuoles. These water droplets unite to form the contractile vacuole (diastole) and during systole the water is discharged in the reservoir.

Behaviour

Like other protozoans, *Euglena* possesses a good degree of irritability i.e. the power of responding to external stimuli. This power enables it to avoid unfavourable environmental conditions. While swimming forward rotating and gyrating on its long axis *Euglena* may reach an unfavourable area. When this happens it turns violently usually heading into favourable area again. For this the gyration of the anterior end is suddenly widened and a new spiral path is taken. This behaviour gives the appearance of an animal trying a particular direction and correcting the error, if misled. Thus, such a behaviour is called *trail* and *error* behaviour, but it is better to call it *avoiding. reaction*.

Reaction to light or phototaxis

Euglena is sensitive to light. It swims towards an ordinary light such at that from a window and avoids strong light. The light is perceived only at the anterior end by the specialized eye-spot. This specialization is important since they depend upon photosynthesis for their nourishment. *Euglena* orientates itself parallel to rays of light whenever the paraflagellar body is shaped by the eye spot or stigma. The animal adjusts its position to the direction of light moving either

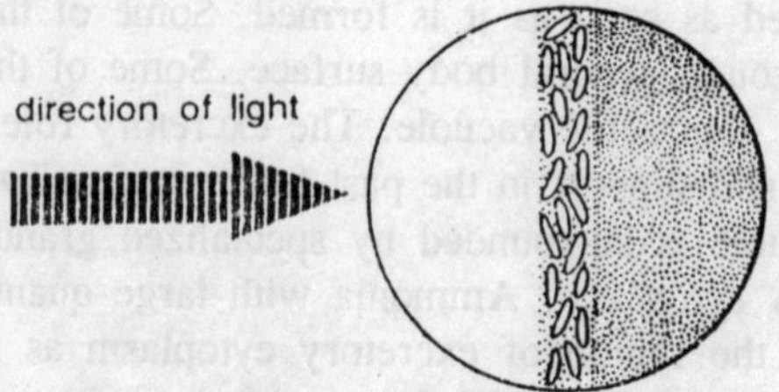

Fig. 13.10. Euglena showing reaction to light.

towards or away from it. It is now know after the experiments of *Tchakotine* (1936) that if the stigma is experimentally destroyed by ultraviolet light, the light-reactions disappear. *Euglena* gives avoiding reaction to mechanical, thermal and chemical stimuli on a trial and error pattern.

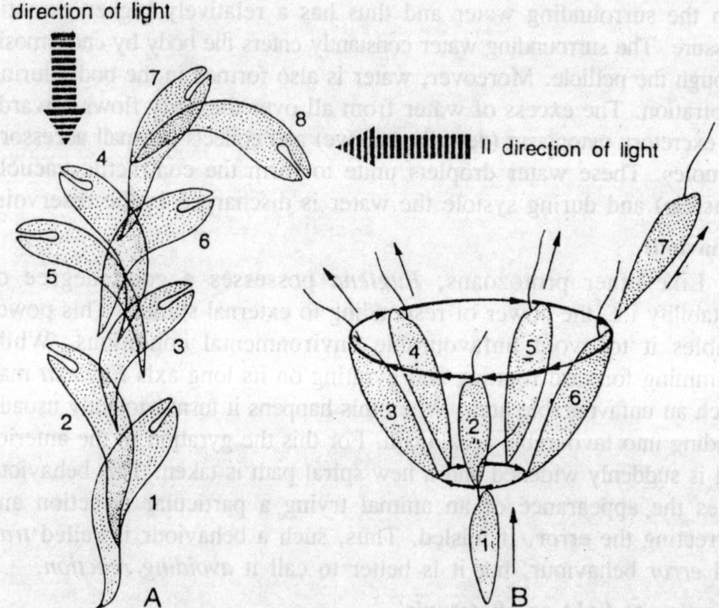

Fig. 13.11. A—Shock reaction and reorientation to light; B—Avoiding reaction on a trial and error pattern.

Reproduction

In *Euglena* only asexual reproduction occurs. Sexual reproduction is not reported. The asexual reproduction takes place by longitudinal binary fission and multiple fission. Encystment also takes place.

Longitudinal Binary Fission

Longitudinal binary fission takes place during favourable conditions. Nuclear division takes place within the nuclear membrane. The chromomeres in the vegetative stage, forms pairs of chromosomes each of which divides longitudinally into two. The endosome becomes constructed into two approximately equal parts. The nucleus before its division migrates from the centre of body to lie just beneath the bases of the flagellum.

During nuclear division, duplication of anterior extranuclear organelles such as flagella, blepharoplasts, cytostome, cytopharynx,

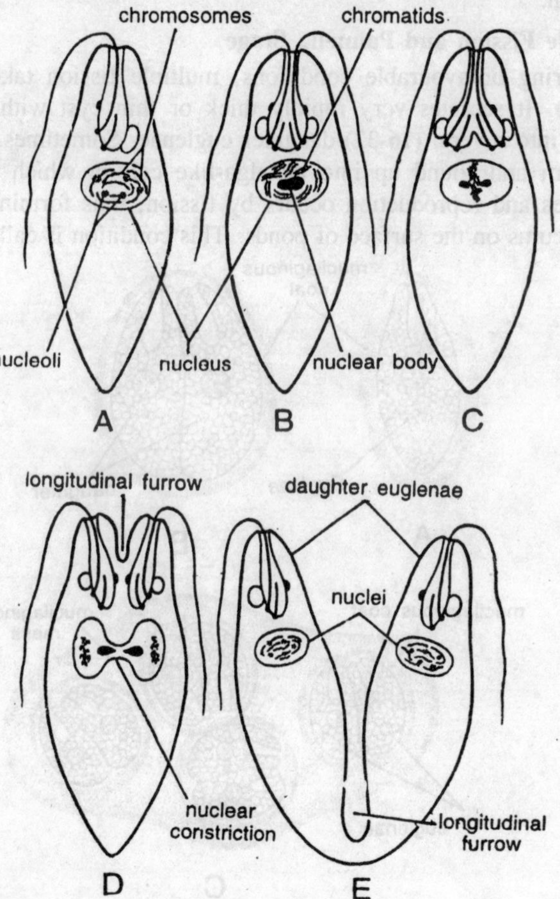

Fig. 13.12. Stages of longitudinal binary fission. A—Interphase. B—Prophase. C—Metaphase. D—Anaphase. E—Telophase and Cytokinesis.

reservoir, contractile vacuole, chromatophores etc. also take place. Soon after this, a longitudinal groove appears at the anterior end and gradually proceeds backwards as a result of which the body of animal is divided longitudinally into two halves. This division is *symmetrogenic* as one daughter is the plane mirror image of the other. In some species, it has been observed that the paraflagellar body disappears before the onset of division. In others, the stigma breaks up into small pieces prior to binary fission. According to some workers the flagellum is completely lost and each daughter cell develops its own new flagellum. Often division takes place while the animals are in encysted condition.

Multiple Fission and Palmella Stage

During unfavourable conditions, multiple fission takes place in *Euglena*. It secretes very rapidly thick or thin cyst within which it divides into several (16-32) daughter euglenae. Sometimes, it loses its flagellum and round up into an alga-like cell in which metabolism continues and reproduction occurs by fission, thus forming extensive green scums on the surface of ponds. This condition is called palmella

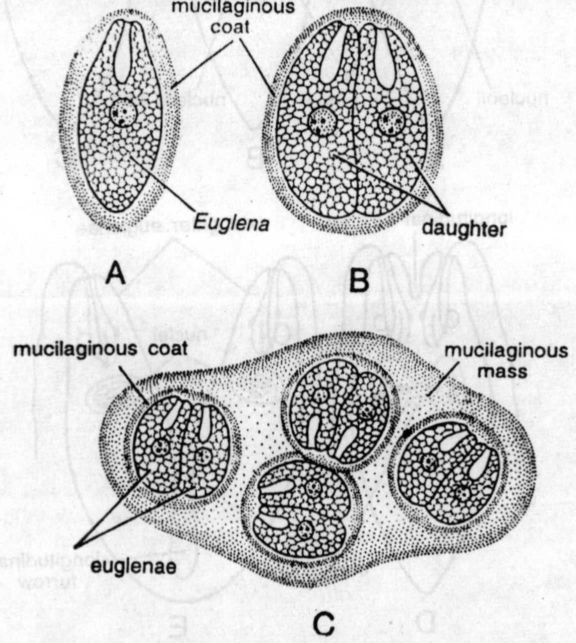

Fig. 13.13. Multiple fission and encystation. A—Encysted individual. B—Fission in encysted condition. C—Palmella stage.

stage. Later flagella grow in these daughter euglenae which become free and develop into adult.

Encystment

This is not a method of reproduction but is a mean to overcome unfavourable conditions. In the process of encystment, the cysts are formed by a special type of yellowish-brown carbohydrates. Their shape is generally rounded and their walls are formed of three concentric layers. During encystment binary fission may occur one or more times. On return of favourable conditions cyst wall breaks, the animal becomes active and emerge from the cyst to lead a normal life. In different species of *Euglena*, the cyst may be thick, stalked or operculated and the organism lying centrally or eccentrically.

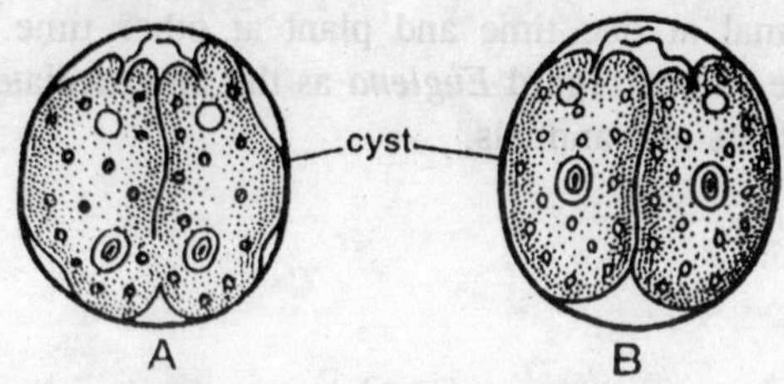

Fig. 13.14. Binary fission in a cyst.

Position of *Euglena*

Euglena is considered to be an animal by zoologists and plant by botanists.

Animal characters

The animal characters of *Euglena* include:

1. Presence of cytostome and gullet which resembles similar structure in the relatives of *Euglena* that are distinctly animals.
2. Pellicle is made up of protein and not of cellulose as in plants.
3. Nutrition is heterotrophic, partial or complete.
4. Presence of contractile vacuole.
5. Reproduction by binary fission.
6. Utilization of amino acids, pentoses or polypeptides as sources of nitrogen.

Plant characters

The plant characters of *Euglena* are:

1. Pyrenoid bodies are present.
2. Presence of chromatophore and chlorophyll.

3. Existence of palmella stage.
4. Photosynthesis takes place in sun-light.
5. Mode of nutrition is holophytic.

The main plant-like character is presence of chromatophores and pyrenoids. But are present only in some species and these can survive even without them. Moreover, the chlorophyll of this organism contains fewer pigments then the chlorophyll of plants. Only few species of *Euglena* are autotrophic and they too only partly. In the absence of light the mode of nutrition becomes saprozoic. During sun light the carbohydrate formed is paramylon which differs from the plant starch as it does not give violet colour with iodine solution. Palmella stage is an adaptation rather than affinity with plants. This is appears that some of the species of *Euglena* are true animal like while few others behave as animal at one time and plant at other time. Its position suggest that we should regard *Euglena* as the intermediate stage in the evolution of plants and animals.

14

LIFE OF POLYSTOMELLA

Elphidium is a shelled protozoan belongs to the order-Foraminiferida, class-Rhizopodea. Its shell is perforated by numerous minute pores hence the name Foraminifera (L. *forare*-pores; *feree*-to bear) is given to this order.

SYSTEMATIC POSITION

Phylum	—	Protozoa
Subphylum	—	Sarcomastigophora
Superclass	—	Sarcodina
Class	—	Rhizopodea
Subclass	—	Granuloreticulosia
Order	—	Foramaniferida
Genus	—	*Elphidium*
Species	—	*crispum*

Habits and Habitat

Elphidium is a free-living and marine protozoan. It is a bottom dwelling form found at the sea floor to depth of 540-600 metres. It slowly creeps along the sand and sea weeds. It is omnivorous. It shows dimorphism and the life-cycle shows alternation of generations.

MORPHOLOGY

Body of *Elphidium* is covered with a hard and translucent; biconvex and oval or spherical shell made up of calcium carbonate and silica. The shell is perforated with several chambers (about 45-50). All the chambers are filled with cytoplasm (*Polythalamous*). The shell is pale-yellow in colour and measures about 1 mm in diameter. The surface of the shell is beautifully chiselled bearing tiny tubercles. The central

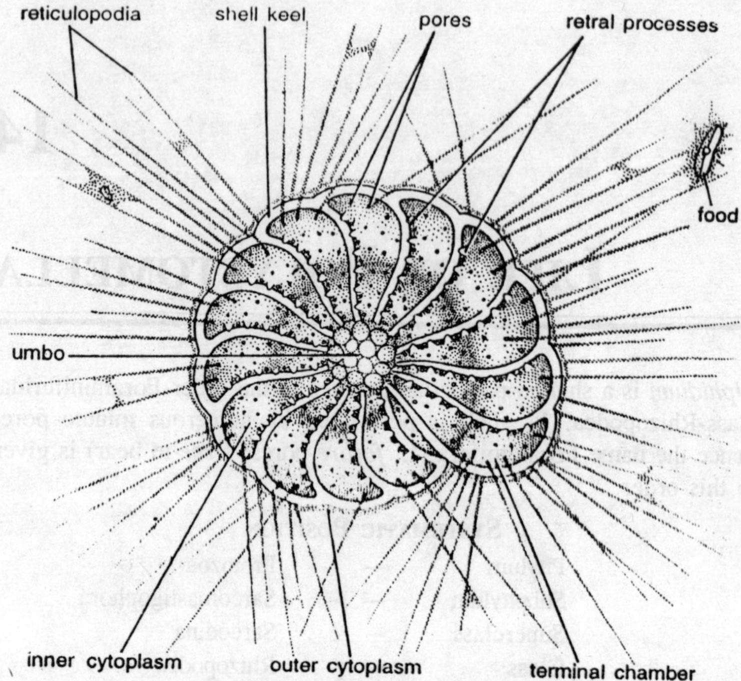

Fig. 14.1. Elphidium crispum (living).

part of the shell forms the rounded *umbo* while peripheral part the *keel*. The chambers of the shell are V-shaped, laid down serially and arranged in a flat spiral in which each whorl of chambers overlaps the previous one i.e. *equitant*. The overlapping portions are called

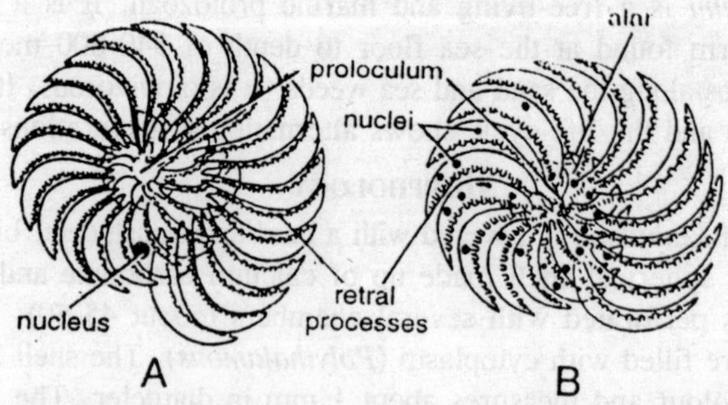

Fig. 14.2. Decalcified and stained specimens. A—Megalospheric individual. B—Microsperic individual.

alar processes. Due to overlapping of the chambers only the last chamber is visible from outside. The hinder margin of each chamber bears a row of projections, the *retral processes*.

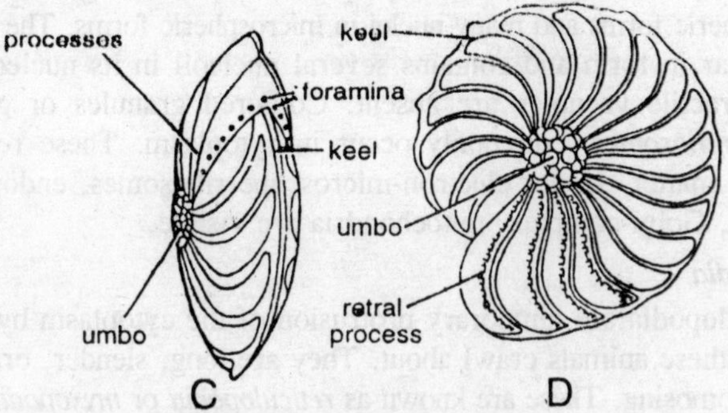

Fig. 14.3. Elphidium cirspum. A—Endon view of the shell. B—Lateral view of the shell.

The adjacent chambers remain separated from each other by perforated septa through which cytoplasmic continuity is maintained. The first or initial chamber formed by the animal is known as *proloculum*. As the young grows, its cytoplasm overflow through the aperture of proloculum. The extruded cytoplasm rounds itself off and secretes a new chamber. By the repetition of this process a multi-chambered shell is formed. Each new chamber is large than the predecessor. In *Elphidium* there are two types of shells differing in relative sizes of the proloculum. If the proloculum is small and is formed by the individual developing from zygote, the shell is called *microspheric*. In others the initial chamber or proloculum is proportionately larger and they have produced by individuals resulting from schizogony. Such shells are called *megalospheric*. Thus the animals show dimorphism. The differences between two forms given in the table:

Megalospheric	*Microspheric*
1. Proloculum large but thin-walled diameter 60-100μ.	1. Proloculum small but thick-walled diameter about 10μ
2. Second chamber horn shaped.	2. Second chamber spherical.
3. Uninucleate.	3. Multinucleate.
4. Gamont-haploid.	4. Schizont-diploid.
5. More common.	5. Less common.
6. Retral processes present in all the chamber.	6. Retral processes found in later chambers only.
7. Rotation of chamber counter-clockwise.	7. Rotation of chambers clockwise.

Cytoplasm

The shell-chambers are filled with cytoplasm. The cytoplasm is viscous and granular in nature. There is no marked distinction of ectoplasm and endoplasm. The cytoplasm contains single nucleus in megalospheric forms and many nuclei in microspheric forms. The nucleus is vericular in form and contains several nucleoli in its nucleoplasm. The contractile vacuoles are absent. Coloured granules or globules called *xanthosomes*, commonly occur in cytoplasm. These represent excretory matter. Under electron-microscope ribosomes, endoplasmic reticulum, Golgi-complex, mitochondria are visible.

Pseudopodia

Pseudopodia are temporary protrusion of the cytoplasm by means of which these animals crawl about. They are long, slender, branching and anastomosing. These are known as *reticulopodia* or *myxopodia*. Each reticulopodium consists of an inner fibril-like *axis* and an outer fluid-like *cortex*. They are very active. They are withdrawn and put out very quickly, and freely wave about in water. The recticulopodia are arranged in bundles and help the organisms in locomotion as well as in nutrition.

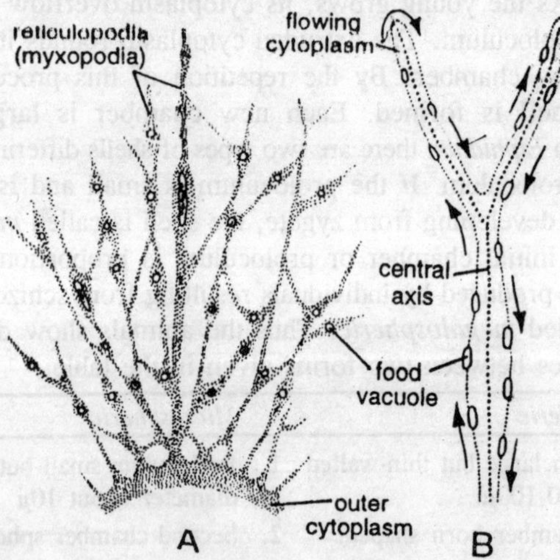

Fig. 14.4. Reticulopodia. A—A group of reticulopodia. B—A reticulopodium showing streaming circulation of cytoplasm.

Physiology

Locomotion

The animal slowly creeps along the sand or sea-weeds at the sea floor. The locomotion is performed by reticulopodia, which help the animal to drag its shell along. The movement is very slow i.e. 4-6 mm. per hour.

Nutrition

The mode of nutrition is *holozoic*. The animal is omnivorous as it feeds on other protozoans, algae and minute crustancean larvae. The food is traped by pseudopodial net, and is promptly surrounded by cytoplasm. Digestion occurs in food vacuole and is completed before the food reaches the shell.

Respiration and Excretion

The respiration takes place through general body surface by diffusion. Oxygen is taken in and carbon dioxide is eliminated out. The nitrogenous waste products also get rid off by diffusion. The xanthosomes are excretory granules or globules which are passed out by the retracting pseudopodia, as the animal creeps along.

Osmoregulation

As the animal is marine so that the osmotic pressure of cytoplasm and its surrounding medium is almost equal. No water enters the body in excess. Thus there is no necessity of osmoregulation hence the contractile vacuole is absent.

Reproduction and Life-Cycle

The first attempt at the study of life-cycle was made by *Schaudinn* (1990). Later *Lister* (1903) added some information on the reproduction in *Polystomella*. *Myers* (1930-40) made further observations and most of the earlier findings were confirmed by *Le Calvez* (1938, 1950) who provided the first complete account of the life-cycle of this genus. *Jepps* (1962) gave a detailed account on the life-cycle. *Grell* (1967) found that gametes are haploid and the amoebulae produced asexually by the microspheric forms are diploid. Typically there are two generations in the life-cycle of *Elphidium* a sexual one (*gamont*) which alternates with an asexual one (*schizont*). This alternation of generations is further combined with dimorphism in the adult condition.

Asexual Phase or Microspheric Form

The microspheric form reproduces asexually which is therefore known as the *agamont* or *schizont*. The microspheric are smallest forms as described by *Schaudinn*. It has nine chambers and 28 nuclei. The nuclei are irregularly scattered, although they are absent in terminal chambers. Nuclei multiply by simple division. Nuclei give off irregular strands of darkly staining substances, which are called *chromidia* by

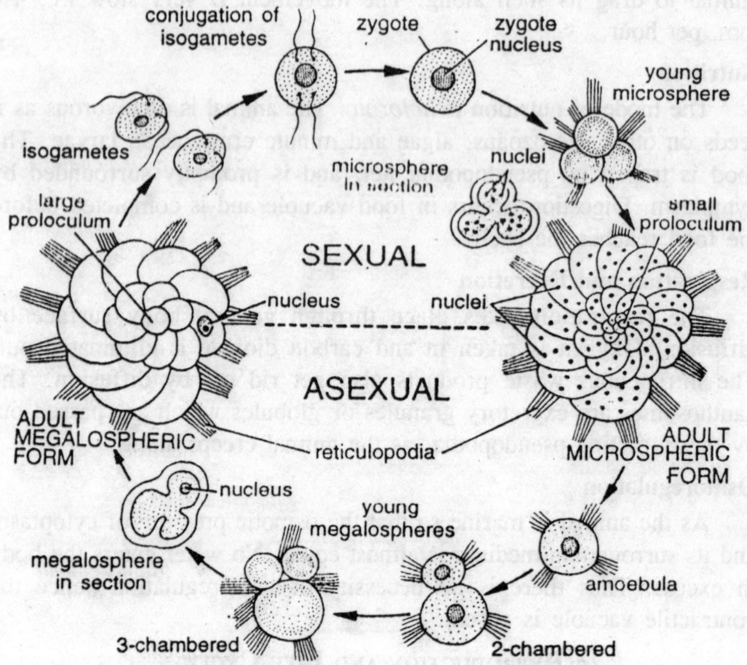

Fig. 14.5. Reproduction and life cycle.

some authors. The first indication of reproduction starts with the great increase in the number of pseudopodia the protoplasm streams out and forms a *halo* around the brown shell. Within a short time, coarse brown granules, the *chromidia* pass out. A large number of new nuclei are recognized from the chromidia. A bit of cytoplasm gathered around each nucleus. These give rise to small amoeboid bodies called *amoebulae* or *pseudopodiospores*. Each amoebula secretes a shell that later becomes proloculum of macrospheric form. New chambers are added as the animal grows in size. Thus a macrospheric form is formed from the microspheric form by multiple fission.

Sexual Phase or Macrospheric Forms

The sexual reproduction occurs in macrospheric forms so that they are called *gamonts*. The nucleus of gamont divides into numerous nuclei and also the cytoplasm. Each nucleus is surrounded by a bit of cytoplasm, which acquires a pair of flagella. This minute biflagellated gamete is called *flagellula* or *zoospore*. The formation of gametes occur within the parental cell. The gametes are released in the surrounding water through the foramina of parent shell, they swim about freely for sometime and then become somewhat elongated. The gametes from different macrospheric forms unit in pairs to form the *zygote*. The zygote loses its flagella. Each zygote secretes a shell around itself and becomes a young microspheric form. The first chamber is small and called proloculum. New chambers are laid down.

Nucleus divides several times to form several nuclei. The first division is meiotic so all the nuclei are haploid. With the division of nucleus, the cytoplasm does not divides so a multinucleated condition is found. The microspheric form on becoming full-grown, undergoes asexual reproduction i.e. multiple fission. According to *Jepps*, it takes about two years for a complete dimorphic cycle in *Elphidium crispum*. The slow phase of the cycle is related to the amount of growth, the young gamont must undergo growth before maturity. The gamont develops a test with about 4 to 5 chambers in about eight days. The 6th chamber is completed after about eleven days, the 15th chamber in about one month and usually 40th almost four months. From the above description it is evident that each phase takes a sufficiently long time to mature, and that is why the life-cycle takes a long period.

Alternation of Generations

The life-cycle shows the phenomenon of alternation of generations. The asexual form (the microspheric) gives rise to several amoebulae or pseudopodiospores, which forms macrospheric forms (sexual forms). These, in turn, give rise biflagellate isogametes by multiple fission. These gametes fuse to form zygote. The zygote forms macrospheric form. In this manner two generations alternate with each other

15

LIFE OF RADIOLARIA

Radiolaria are exclusively marine and chiefly pelagic. They are found equal well in warm and cold water; littoral or oceanic; surface dweller or inhabiting deep down upto 500 metres. West areas of oceanic floor are covered by the ooze of radiolarians. Many of the radiolarians show bioluminescence. Their zoological history gets back from the *Salurian Period* and they are one of the earliest known animals. The body is generally spherical and ranges from 50μ to several mm., and the colonial forms to several cms.

The prevailing symmetry is radial or bilateral although asymmetrical forms are also found. The cytosome is divided into two regions: an *extra-capsular* and an *intra-capsular* by the central *capsule*. The capsule is a thin, delicate structure of a single or double membrane of pseudo-chitinous or mucoid nature. In shape it is spherical, ovoid, lobate or branched. The capsule may increase in diameter with the growth of animal and it dissolves during fission in simple forms. Perforations either distributed uniformly or concentrated in one or more groups. The perforations permit continuity between the extra-capsular and intra-capsular cytosome.

The extra-capsular cystosome is primarily considered with floatation and digestion of food. Three zones are distinguished in it *sarcomatrix*, *calymma* and *sarcodictyum*. The sarcomatrix, or assimilative layer immediately surrounds the central capsule. It contains pigments, excretory substances and undigested food material. In some Tripyleans aggregation of the excretory substances and food material produces greenish or brownish mass surround the aperture of capsule called *phoedium*. The calymma or the vacuolated layer lies in a wide zone next to sarcomatrix. It is filled with vacuoles containing water, saturated

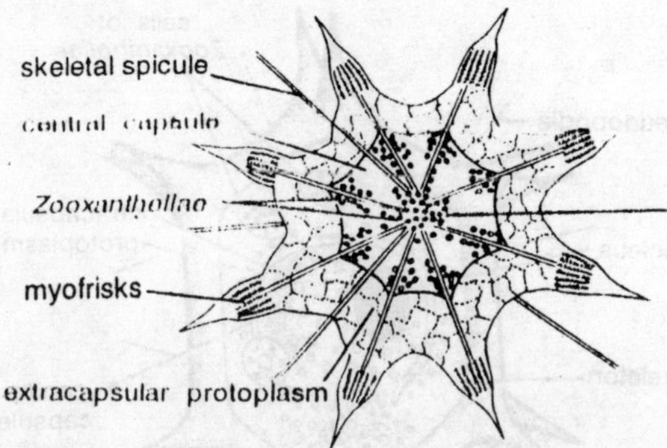

Fig. 15.1. Acanthometra (a radiolarian with skeleton).

with carbon dioxide which gives buoyancy. When the vacuoles burst the specific gravity of individual increases and it sinks; when the vacuole reappears the specific gravity decreases and the animal rises.

In the calymma are usually symbiotic zooxanthellae are present, which are modified dinoflagellate. Zooxanthellae are absent in deep sea individuals. Fat drops, oil spheres and pigments are also present in the calymma. The sarcodictyum is in a thin film on the out side of calymma, from it extends slender filamentous pseudopodia. The typical pseudopodia are *axopodia* which arise from inside the central capsule and radiate in all directions, some pseudopodia are provided with myoneme, which produce circular group of short rod like body clustered round each radial spine.

The intra-capsular part is the seat of reproductive activity, it is granular often vacuolated. It contains one or more nuclei pigments, oil-droplets, fat-globules and crystals. Zooxanthellae are also found in it sometimes. The skeleton is absent in certain forms as *Thalassicola*. In others the skeleton is present in various stages of development. It is formed of strontium sulphate while other consists of siliceous substances. In *Actipylina* the skeleton is composed of 20 rods radiating from centre and emerging from body in five circle comparable one *equitoreal*, two *tropical* and two *circumpolar regions*. This arrangements after *Muller* is known as *Muller's Law*. The siliceous skeleton may be in the form of rods and spines which usually lies outside the capsule. The rods and spines may be intracapsular or inter-nuclear in some forms like *Acantheria*.

The skeleton may be in the form of one or more layers of lattice work peripheral and concentric with central capsule. In the later case the lattice work, may lie in the nucleus, in the intra-capsular cytosome

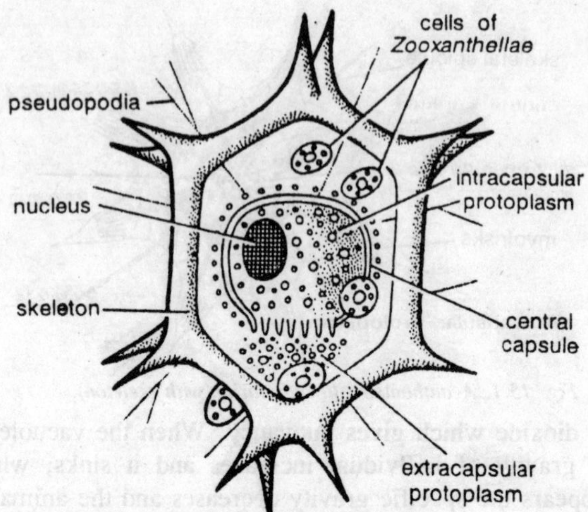

Fig. 15.2. Lithocircus.

and extra-capsular cytosome. Radiolarian are not swimmers. Some of them rise and sink in response to changes it the environmental conditions by regeneration or collapse of vacuoles, such a mechanism allows animals living near the surface to sink when the sea is rough or temperature in the upper region of water is unfavourable. The filamentous, cylinder pseudopodia which come out from extra-capsular cytosome help in creeping along hard surface.

Radiolarians feed on microplankton like diatoms and various protozoa. The food is captured by pseudopodia and by their contraction is drawn in sarcomatrix where it is digested in food vacuole. Radiolarian, can live without food if kept in sun-light due to zooxanthellae which occur in calymma except in *Actipylina* where it is found in intracapsular cytosome. In the absence of organic food the radiolarian lives on the reserve made by zooxanthellae through holophytic nutrition.

Asexual reproduction by *binary fission* occurs in many forms or forms with simple skeleton. In any individual (with a simple skeleton), the central capsule is divided and the skeletal element are passed to daughter organisms. Fission is also known to occur in helmet-shaped skeleton bearers, in such forms one daughter organism retains the old-shell and the other leaves the parent shell and forms a new skeleton. *Multiple fission* is also known from forms like *Thalassophysa*. The central capsule becomes irregular and the nucleus divides into a number of globules around a number of which cytosome gathered to form small ovoidal masses.

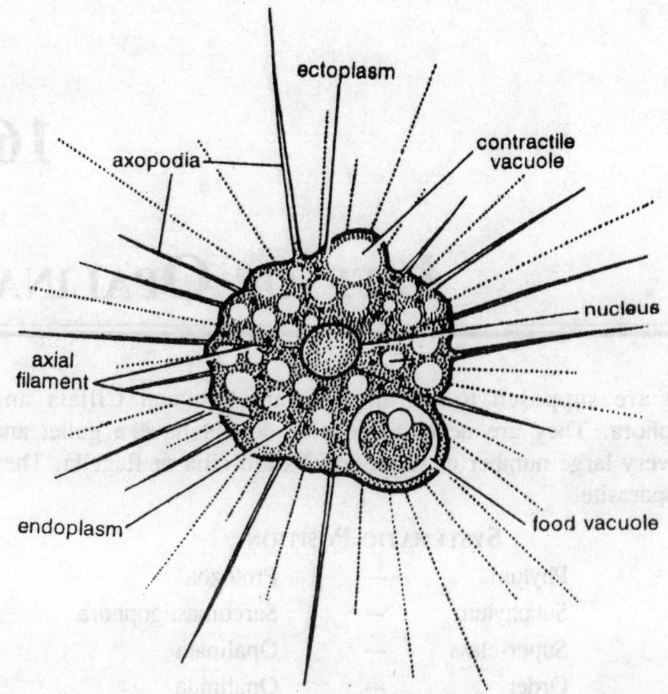

Fig. 15.3. Actinophrys sol.

Finally the extra-capsular and intra-capsular cytosome breaks up into numerous small multinucleated body. Multiplication by budding also takes place in some forms. The multinucleated individual gives off buds each with a number of nuclei. Buds separate and develop into new individuals. Sexual reproduction by swarm formation is known in some forms. In *Thalassicola* the central capsule becomes separates from the body and its nucleus divides into vary many nuclei. Around each nucleus cytoplasm gather and each mass develops a flagellum. The capsule now descended to the depth of several hundred metres where it ruptures liberating the swarmers both *isoswarmers* (isospores) and *anisoswarmers* (anisospores). Each isoswarmer is said to contain a crystal and a fat globule. The anisoswarmers contain refractile granules in cytoplasm. Further development is not known. According to *Kudo* the isoswarmers belonging to asexual generation and anisoswarmer to the sexual. *Hyman* considered isoswarmer as isogametes and anisogametes as escaped zooxanthellae. The complete life-history in any case is unknown.

16

LIFE OF OPALINA

Opalina are supposed to be intermediate between Ciliata and Mastigophora. They are never amoehoid, do not have a gullet and possess very large number of small equal-sized cilia or flagella. They are endoparasite.

SYSTEMATIC POSITION

Phylum	—	Protozoa
Subphylum	—	Sarcomastigophora
Super-class	—	Opalinata
Order	—	Opalinida
Genus	—	Opalina
Species	—	*ranarum*

Habits and Habitat

Opalina ranarum is an endoparsite found in the rectum of frogs and toads.

MORPHOLOGY

Its body is much flattened like a disc and oval in outline. It is almost colourless and transparent measuring about 1 mm in size. The body is covered over by tough but thin pellicle. The pellicle in term is covered over by innumerable uniform small sized cilia arranged in rows over the entire surface. The dorsal rows follow sigmoid paths while ventral rows are straight. About half of the rows pass completely around the body. Between the rows of cilia the pellicle is thrown into narrow longitudial folds, each supported by a simple longitudinal ribbon of 12-24 microtubules (*Noirot-Timothee*, 1959 and *Wessenberg* 1966).

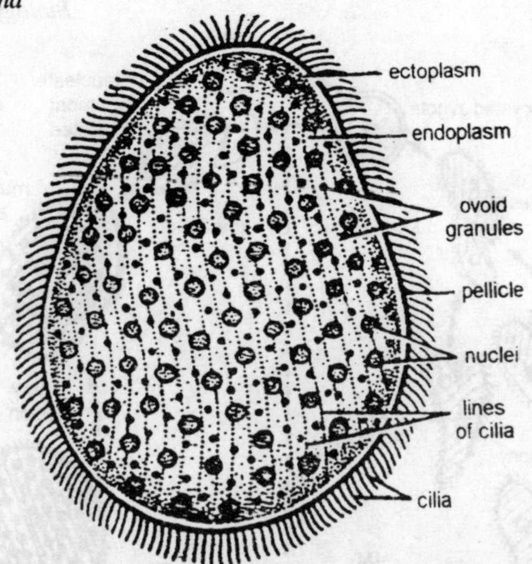

Fig. 16.1. Opalina.

Mouth, gullet, cytopyge, food vacuoles and contractile vacuoles are absent. The cytoplasm is differentiated into outer clear *ectoplasm* and inner granular *endoplasm*. The number of nuclei is very large. These are small, hyaline, spherical and similar (*Monomorphic*). They are evenly distributed in the endoplasm. Each nucleus has both *trophochromatin* and *idiocnromatin*. Endoplasm also contains numerous spherules of chromatin-like material but of unknown function. These are known as *endospheres*.

PHYSIOLOGY

Nutrition

The mode of nutrition is *saprozoic*. It absorbs the liquid food contents of host's intestine through the general body surface by *pinocytosis*.

Life-Cycle

Opalina reproduces by binary fission throughout the year. It is most active during the breeding season of its host i.e. frog or toad. The line of cleavage passes between the ciliated lines. This type of fission is known as *plasmotomy*. But in spring, unusually rapid division produces smaller individuals, each with few nuclei (4-6) and few rows of cilia. Each daughter secretes a cyst wall around itself and passes out with the faecal matter of host. The cysts so formed are transparent and spherical to ovoid and usually left over water-plants. Cysts are

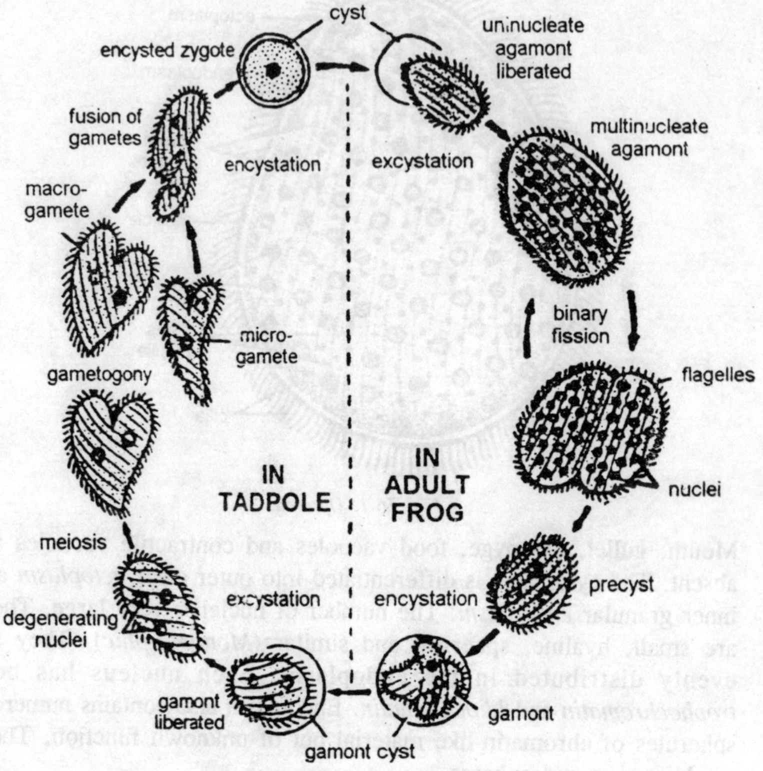

Fig. 16.2. Life cycle of Opalina.

swallowed by young tadpoles as they scrap water-plants. These cysts hatch in the rectum as gametocytes.

The gametocysts divide by a series of longitudinal fiss on to form uninucleate gametes. The gametes are anisogamous as they differentiated into microgametes and macrogametes. Gametes fuse and the zygote is formed. The zygote may encyst for a while as *zygocysts*. With growth and rapid nuclear division each zygote grows into young *Opalina* called *proto- opalina stage*.

Systematic Position of Opalina

Opalina was once classified as a member of Ciliata. They were regarded as primitive ciliates and called *Protociliata* by *Metcalf* (1918). It was based on the basis of numerous nuclei of *Opalina* which considered to be macro- and micro-nuclei of Euciliata. In *Opalina* the nuclei are monomorphic i.e. they can not be differentiated into micro- and micro-nuclei (dimorphic). In true ciliate, the nuclei are dimorphic and sexual reproduction is by conjugation, there being neither gametes nor syngamy.

In *Opalina* as earlier said the nuclei are monomorphic. The sexual reproduction by fusion of anisogametes (micro- and macro-gametes) and never by conjugation. Except cilia, most of the characters of *Opalina* are those of flagellates. The behaviour of infraciliature during division is of the flagellate pattern i.e. the cleavage plane is parallel to the kinetia (longitudinal fission). This type of division is called *symmetrigenic*. In *Opalina* the centrioles are absent. They are also absent from some mastigophores like *Euglena*. Thus *Opalina* shows greater affinities with the flagellates rather than ciliates. Some earlier worker placed them in class Mastigophora. The position of *Opalina* is still uncertain. Some workers include them in Ciliata and some in Mastigophora. Some authors have put them in a separate class Opalinata, equal in rank with Mastigophora and Sarcodina. Probably they are highly simplified metazoans and in past related to the *Turbellaria*.

17

LIFE OF VORTICELLA

Vorticella has over 200 species described by *Kahl* (1953), most of them are fresh water. Some species are marine, some are epizoic and some are parasitic. The most common species are *V. nebulifera*, *V. campanula* and *V. microstoma*.

SYSTEMATIC POSITION

Phylum	—	Protozoa
Subphylum	—	Ciliophora
Class	—	Ciliata
Subclass	—	Peritrichia
Order	—	Peritrichida
Family	--	Vorticellidae
Genus	—	*Vorticella*

Habits and Habitat

It is generally found in lakes, rivers, ponds and streams with aquatic vegetation. It is not free swimming like *Paramecium* but is a sedentary ciliate. Most of them are abundant in stagnant water containing decaying organic substances. *V. campanula* and *V. nebulifera* are found in clean water. It is found attached to the stems and leaves of the aquatic plants, mollusc-shells, living oligochaetes and tadpoles.

MORPHOLOGY

Shape, Size and Colour

Vorticella possesses a bell-shaped or asymmetrical body. It is, therefore, usually called a *bell-animalcule*. To its base is attached a long contractile stalk. Many individual variations are found in the

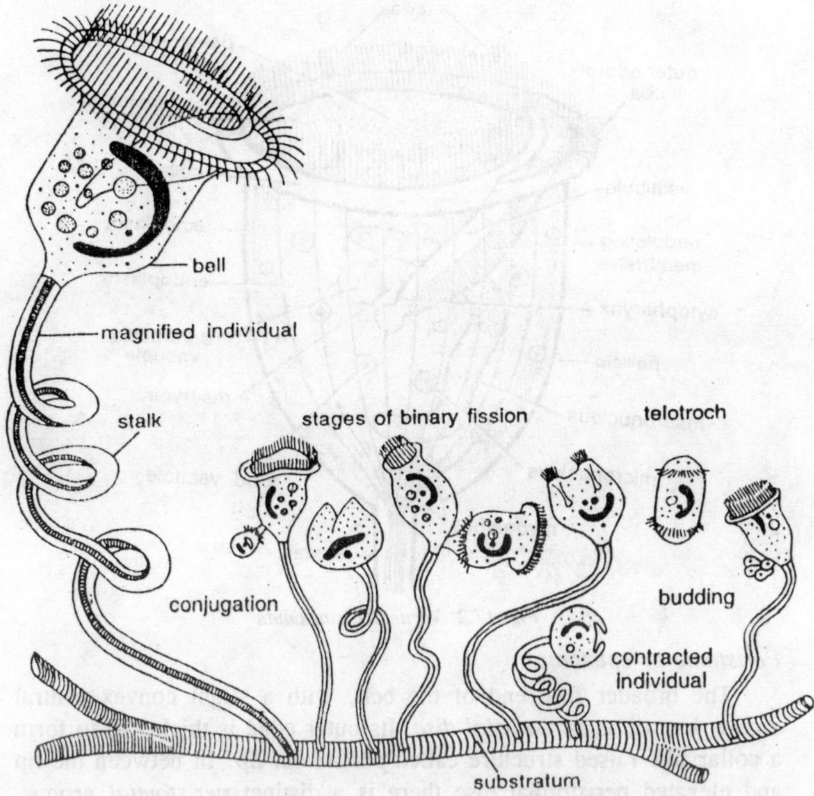

Fig. 17.1. Vorticella. A group of individuals in different states.

shape of the body and stalk. The size of the bell of the largest species *V. canpanula* is about 157´99μ and the length of the stalk is from 53-4150μ. The body of smaller species, *V. microstoma* measures about 55×35μ. The Colour of the *V. nebulifera* is greenish while *V. campanula* is bluish.

Structure

The body is divisible into two parts

1. Bell
2. Stalk.

Bell

Its shape is like that of a solid inverted bell. Its structures can be studied under following headings:

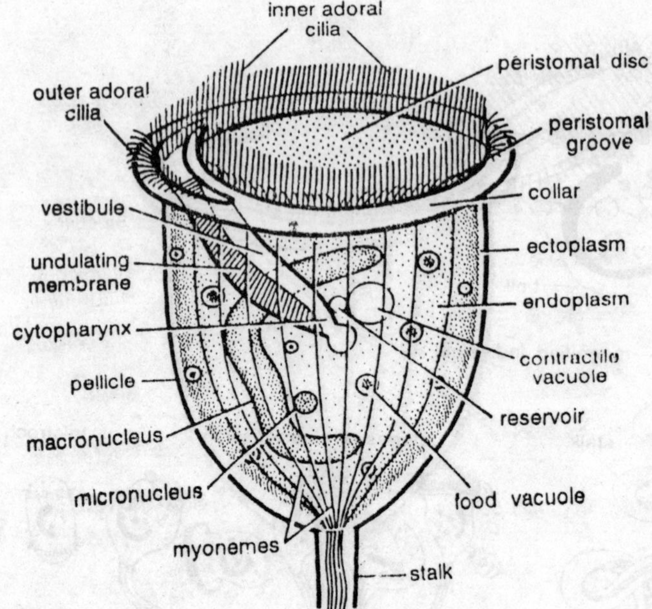

Fig. 17.2. Vorticella campanula

Peristome or epistome

The broader free end of the bell, with a slight convex central area is known as *peristomial disc*. Its outer edge is thickened to form a collar like raised structure called *peristomal lip*. In between the lip and elevated peristomial disc there is a distinct *peristomial groove*.

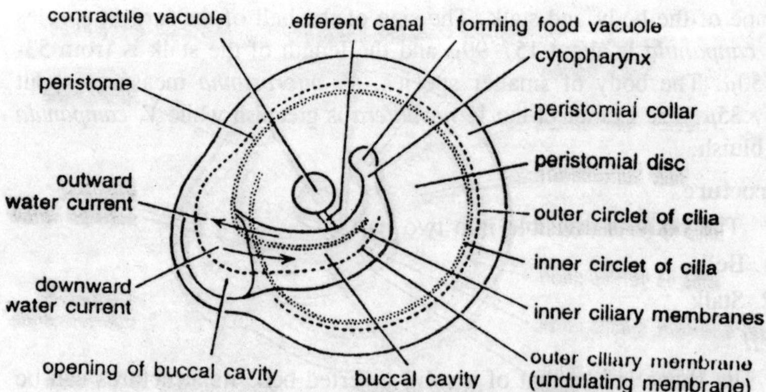

Fig. 17.3. Peristome as seen from above.

At the posterior end of the vase a clear area, the *scopula*, is visible. This secretes the inert matrix of the stalk.

Vestibule

On one side of the disc the peristomial groove is deeply invaginated forming the *buccal cavity* or *vestibule*. At the bottom of the vestibule is an opening, the cytostome that leads into a narrow tube, the *cytopharynx*. The cytopharynx remains collapsed, except when distended with food.

Cilia

Ciliation in *Vorticella* has been studied by *Noland* and *Finely* (1931). The cilia are restricted to a certain area that is known as the *adoral zone*. On the body there are no cilia. Three concentric rows of adoral cilia are found in the peristomial groove. Two rows of cilia are found in the inner circle. These cilia remain straight and always moving. Only one row of cilia is found in outer circle. The cilia of this row hang horizontally towards the outer side on the lip. They direct the food particles towards the vestibule. The adoral cilia are fused with each other at their base, but are free at their distal ends.

Both the spirals of cilia are curved in such a way that they form about one and a half anticlockwise spiral on the peristomial disc. Both the spirals of cilia extend upto the vestibule. The two inner rows of cilia extend together along the inner wall of the vestibule, where as the outer row of cilia is elongated and its cilia unite together to form an *membrane* along the outer wall of vestibule. The cytopharynx is devoid of cilia.

Pellicle

The pellicle of the bell is uniformly thin and delicate (thicker than that of *Paramecium*). It is often marked with circular parallel striations and in some species of *Vorticella* it is tuberculate. It covers the body so closely that it becomes difficult to distinguish it. It is devoid of cilia.

Cytoplasm

The body cytoplasm is differentiated into an outer layer of clear and firm *ectoplasm* and an inner fluid *endoplasm*.

(a) *Ectoplasm.* The ectoplasm is modified to form a *myoneme system*. Innumerable myoneme fibres are found. They are arranged in both straight and oblique manner and extend across the length and breadth both. All the myonemes of the bell are converged at its base forming a *spasmoneme*. The longitudinally myonemes reduce the length of the body (*Schoder*, 1906). The oblique mynonemes pull the disc of bell inwards. The circularly placed myonemes pull the peristomial edge over the cilia and the disc to close the mouth of bell. The parallel myonemes of the sides of the bell are more distinct in its basal part where they enter the stalk.

(b) *Endoplasm.* The endoplasm is granular and fluid like. It contains:

(i) *Nuclear apparatus*. There are two nuclei, a *macronucleus* and a *micronucleus*. Macronucleus is large band-like, often horse-shoe shaped. The micronucleus is small and lying close to the macronucleus.

(ii) *Contractile vacuole*. A single contractile vacuole with a definite vacuolar membrane of the ectoplasm is placed very close to the gullet and communicates with the gullet through a reservoir.

(iii) *Food vacuoles*. Many food vacuoles are found scattered in the cytoplasm.

(iv) *Cytoproct*. It is a temporary or permanent aperture opening into the vestibule. Undigested food is sent out of body through this aperture.

Stalk

The stalk is long, slender and contractile. It is produced by the body itself. It is made up of a central axil rod known as *spasmoneme* or *stalk muscle*, which consists of a band of spiral myonemes ensheathed in delicate transparent ectoplasmic sheath.

Recently, it has been reported that this sheath also has *myonemes*. This sheath is continuous with the pellicle of the bell. *Kolzoff* was of the view that the stalk was a pseudo-chitinous tube enclosing an inner tube packed with granulated *ectoplasm* and a central rod composed of *kinoplasm*. Recently, *Faure-Fremeit* described that the stalk consists of a wall having a liquid and a contractile cord. The wall is double and thickened due to the presence of rodless having cytoplasm and a specialized myoneme called *spasmoneme*, which is continued into the myonemes of the bell.

PHYSIOLOGY

Locomotion

Vorticella is a sedentary creature and normally does not move from place to place. However, the cilia remain very active and the movements of the stalk are frequent and rapid. When the animal is feeding, its stalk remains fully extended and the bell sways to and fro.

A light disturbance excites *Vorticella* and immediately its actions are stopped. The stalk contracts to become spiral-shaped. The cilia of disc contract and stop working. The disc is pulled inwards and the peristomial lip contracts to close its upper part. Due to contraction of the stalk, the whole structure is pulled towards the bottom of water where it remains in this state till the danger is over.

Nutrition

The mode of nutrition is holozoic in *Vorticella*. It feeds on small protozoan, bacteria and organic pieces.

Ingestion

A current of water is created with the help of adoral cilia present in the peristome and the disc. The water current brings water along with food materials to the disc from where they are carried to the vestibule and then into cytopharynx. On reaching the bottom of cytopharynx, the food particles along with water take the form of food vacuole in the endoplasm. The food vacuoles do not move in the endoplasm in a regular manner as in *Paramecium*, but show irregular cyclosis.

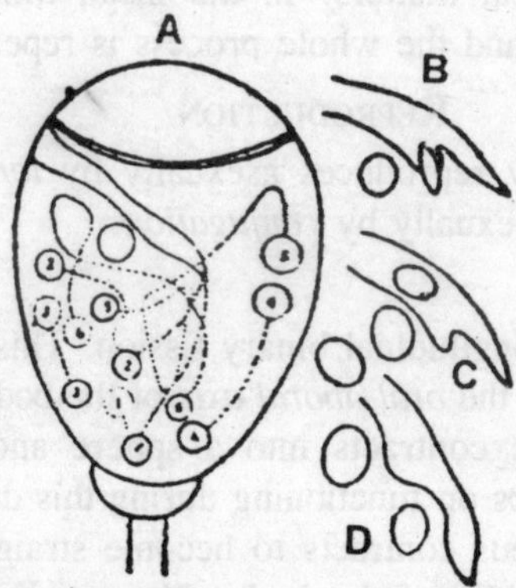

Fig. 17.4. A—Diagram showing movements of food vacuoles in Vorticella. B-D—Stages in removal of food vacuoles.

Digestion and Assimilation

Digestion of the food takes place during movements of food vacuoles. At first the effect is acidic, then neutral and finally alkaline. *Volkovsky* (1931) was of the opinion that certain neutral red granules gather on the food. He named them digestive granules. The digested food materials are diffused in endoplasm. The reserve food material is glycogen.

Egestion

Undigested food-materials and old food vacuoles are removed out by the cytoproct into the cytopharynx on its lower wall. From here they pass out.

Respiration

Respiration is aerobic. The oxygen found dissolved diffuses into the body through pellicle and carbon dioxide diffused out in surrounding medium.

Excretion

The nitrogenous waste product formed as a metabolic process are diffused out through general body surface.

Osmoregulation

The excess of water from the endoplasm collects in the contractile vacuole which enlarges in size (diastole). When full, the vacuole contracts quickly and disappears (systole), discharging its contents into the reservoir situated close by. The reservoir then contracts slowly, deriving its contents into the vestibule. From the vestibule the water goes out with the faecal matters. In the mean time the contractile vacuole appears again and the whole process is repeated.

Reproduction

Vorticella normally reproduces asexually by *longitudinal binary fission* and sometimes sexually by *conjugation*.

Asexual Reproduction

It takes place by longitudinal binary fission. This type of division passes lengthwise along the *oral-aboral axis* of the body. During binary fission, the animalcule contracts into a sphere and broadens. The contractile vacuole keeps on functioning during this division. The long and curved macronucleus contracts to become straight and is placed transversely in the middle of the body. Now it divides amitotically and forms two daughter nuclei. The micronucleus divides mitotically into two. With the division of nuclear apparatus, a longitudinal groove appears in the middle of the free end due to which the peristome is divided into two parts. This groove gradually progresses downward to divide the bell into two parts which are unequal in size the smaller part is devoid of stalk. Whereas the parental remains attached to the larger part. A new spiral of above cilia and a contractile vacuole is formed in the smaller part. Along with this, a spiral of cilia is also formed in the hinder region of the smaller part. This part now becomes

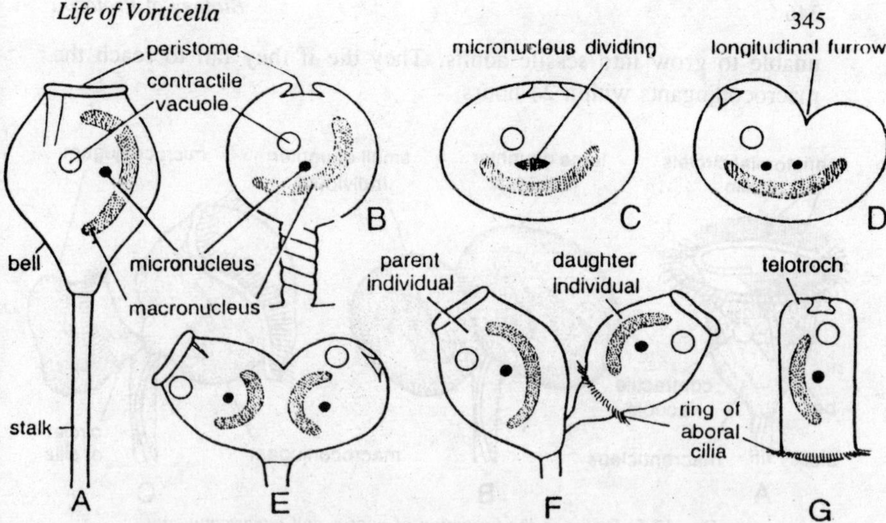

Fig. 17.5. Stages of binary fission.

cylindrical and breaks off from the parental stalk. It is called *telotroch*. It swims in the water with the help of cilia present on posterior part.

After swimming for some times the telotroch attaches itself and undergoes metamorphosis into a normal sessile adult by producing a stalk shortly before developing a stalk, it resembles another ciliate, the *Scyphidia*, hence often called *Scyphidia stage*. According to *Finally* (1939) the binary fission is completed in about 20-30 minutes. During unfavourable conditions the normal *Vorticella* develops a posterior ring of cilia to become a telotroch which breaks from its stalk and swims away to some favourable place and then grows a stalk. At times *Vorticella* encysts in a double cyst wall while still fixed to the stalk, then the cyst fall from the stalk, on excystment it swims away as a telotroch.

Sexual Reproduction

Maupas (1888) has described sexual reproduction in *Vorticella nebulifera*. It takes place by conjugation from time to time. The conjugation taking place in *Vorticella* is different to that of conjugation in *Paramecium*. The process of conjugation in *Vorticella* takes place as follows:

Formation of microconjugants

The *Vorticella* divides twice forming four smaller individuals. Each of them develop posterior circlet of cilia and detached from one another and they swim as independent microconjugants. These are similar to telotrochs, but differ from them in being smaller in size and are unable to grow into sessile adults. They die if they fail to reach the macroconjugants within 24 hours.

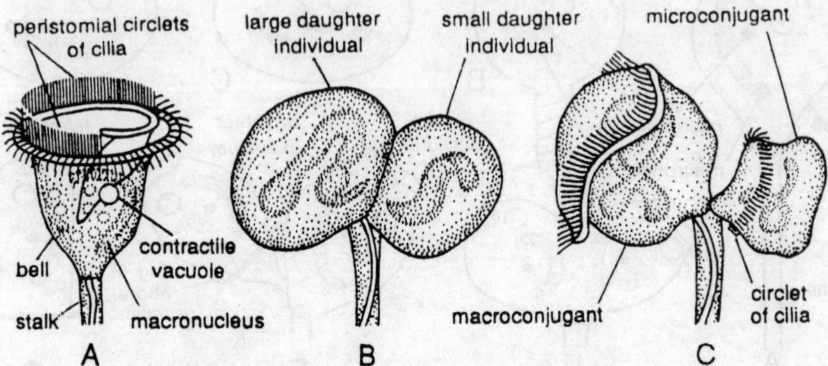

Fig. 17.6. Stages in the formation of micro- and macroconjugant.

Formation of macroconjugants

The vegetative individual becomes a macroconjugant after certain nuclear modifications in it. In structure, it is similar to adult individual but has physiological differences from the later. It attracts the microconjugant towards itself.

Pairing of conjugants

The microconjugant attaches itself to the posterior part of macroconjugant. The pellicle and posterior cilia are lost. After attachment the macronucleus of each conjugant disintegrates and is absorbed in the cytoplasm. The micronucleus of microconjugant divides thrice so as to form 8 nuclei while the micronucleus of macroconjugant divides twice to form 4 nuclei 7 nuclei in microconjugant and 3 nuclei of macroconjugant disintegrate. Thus only one nucleus is left in each conjugant. Both the nuclei more toward the wall separating them. In the mean time the wall separating them dissolved and two conjugants become one. The nuclei of both the conjugants now divide meiotically to form four nuclei. One nucleus of each conjugant remains while other disintegrates.

The remaining nucleus of the microconjugant is called *male pronucleus* or *migratory pronucleus*, while that of macroconjugant is called *female pronucleus* or *stationary pronucleus*. Now both male and female pronuclei unite to form *zygote nucleus* or *synkaryon*. The macroconjugant is now called the *zygote*. The zygote nucleus divides thrice to form 8 nuclei. All the divisions are mitotic. Out of these 8

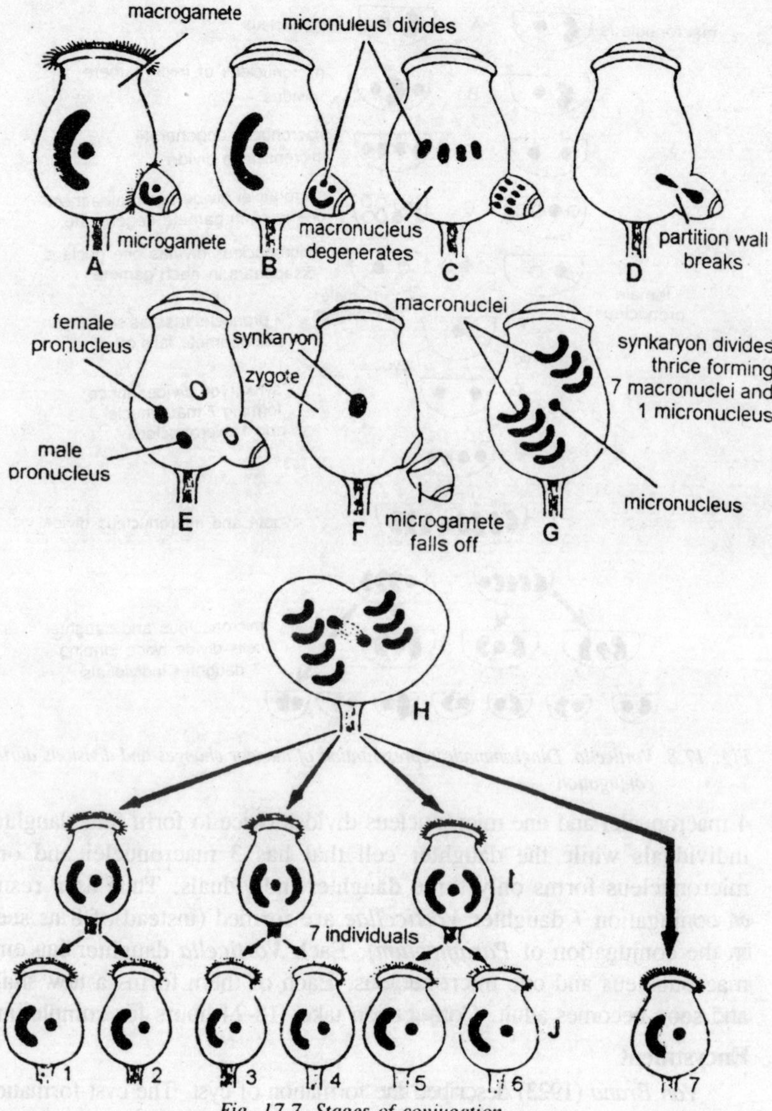

Fig. 17.7. Stages of conjugation.

nuclei, 7 increase in size and become macronuclei. The remaining one is the mircronucleus. The micronucleus and the body of zygote divide in such a manner that one of the daughter cell possesses 4 macronuclei and one micronucleus and other daughter possesses 3 macronuclei and one micronucleus. Now the daughter cell which has

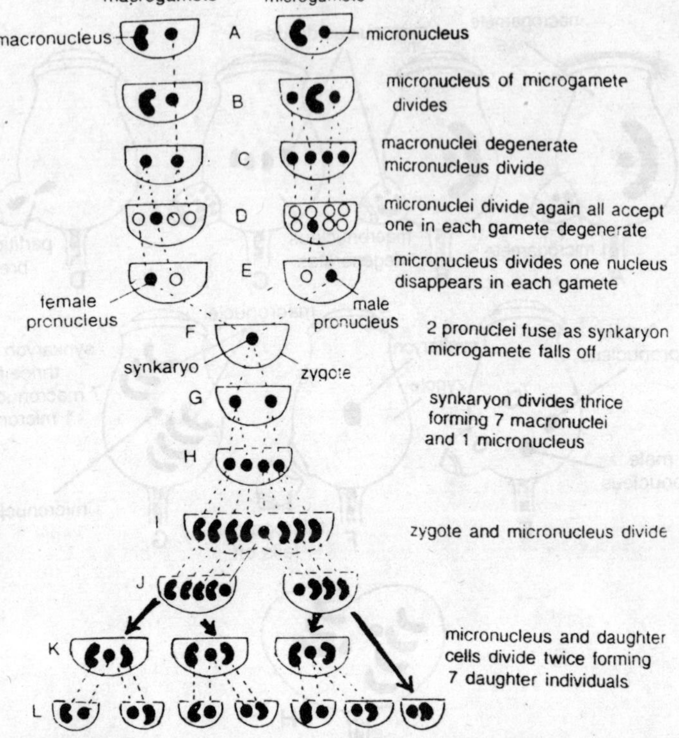

Fig. 17.8. Vorticella. Diagrammatic representation of nuclear changes and divisions during conjugation

4 macronuclei and one micronucleus divides twice to form four daughter individuals while the daughter cell that has 3 macronuclei and one micronucleus forms only three daughter individuals. Thus as a result of conjugation 7 daughter *Vorticellae* are formed (instead of 8 as seen in the conjugation of *Paramecium*). Each *Vorticella* daughter has one macronucleus and one micronucleus. Each of them forms a new stalk and soon becomes adult. Conjugation takes 18-24 hours for completion.

Encystment

Von Brand (1923) described the formation of cyst. The cyst formation takes place during unfavourable conditions. The body becomes rounded and is surrounded by an outer gelatinous *ectocyst*, which takes on a wrinkled appearance. The *endocyst* is a double membrane, the outer is much wrinkled. Myonemes and pellicular striations become indistinguishable and contractile vacuoles pulsate more rapidly. The complete cyst is marked by two scars, one shows the previous point of

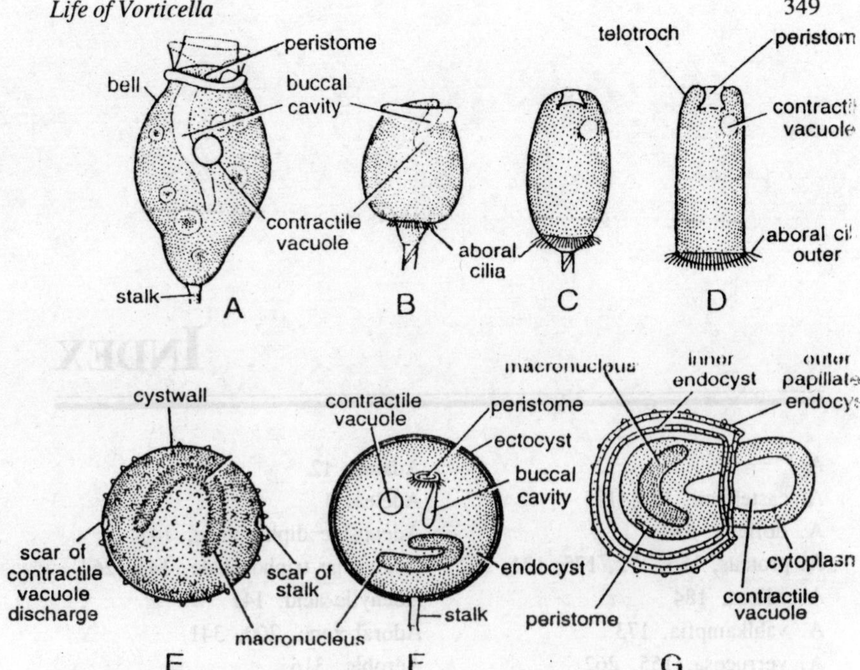

Fig. 17.9. Vorticella. A-D—Formation of telotroch. E—Complete cyst. F—Cyst in section. G—Excystation

attachment to the stalk and opposite one indicates the point of escape for excreted water. During favourable conditions the vacuole enlarge and as a result of pressure, organism comes out of cyst with an explosion, forms a posterior circlet of cilia and swarms off as a telotroch.

INDEX

D

F

R